VOLUME ONE HUNDRED AND SEVENTY

# INTERNATIONAL REVIEW OF NEUROBIOLOGY

Adenosine $A_{2A}$ Receptor Antagonists

# INTERNATIONAL REVIEW OF NEUROBIOLOGY

VOLUME 170

## SERIES EDITOR

### PATRICIA JANAK
*Janak Lab, Department of Neuroscience,*
*Dunning Hall, John Hopkins University,*
*Baltimore, MD, USA*

### PETER JENNER
*Division of Pharmacology and Therapeutics,*
*GKT School of Biomedical Sciences*
*King's College, London, UK*

## EDITORIAL BOARD

RICHARD L. BELL
SHAFIQUR RAHMAN
MARIA STAMELOU
K RAY CHAUDHURI
NATALIYA TITOVA
BAI-YUN ZENG
TODD E. THIELE

HARI SHANKER SHARMA
ARUNA SHARMA
GRAŻYNA SÖDERBOM
NATALIE WITEK
NINA SMYTH
ERIN CALIPARI
LIISA GALEA

VOLUME ONE HUNDRED AND SEVENTY

# INTERNATIONAL REVIEW OF NEUROBIOLOGY

Adenosine $A_{2A}$ Receptor Antagonists

Edited by

**JIANG-FAN CHEN**

*State Key Laboratory of Ophthalmology, Optometry and Visual Science, Eye Hospital; Oujiang Laboratory (Zhejiang Laboratory for Regenerative Medicine, Vision and Brain Health), School of Ophthalmology & Optometry and Eye Hospital, Wenzhou Medical University, Wenzhou, P.R. China*

**AKIHISA MORI**

*SNLD Ltd., Akashi-sho, Chuo-ku, Tokyo, Japan*

Academic Press is an imprint of Elsevier
50 Hampshire Street, 5th Floor, Cambridge, MA 02139, United States
525 B Street, Suite 1650, San Diego, CA 92101, United States
The Boulevard, Langford Lane, Kidlington, Oxford OX5 1GB, United Kingdom
125 London Wall, London, EC2Y 5AS, United Kingdom

First edition 2023

Copyright © 2023 Elsevier Inc. All rights reserved.

No part of this publication may be reproduced or transmitted in any form or by any means, electronic or mechanical, including photocopying, recording, or any information storage and retrieval system, without permission in writing from the publisher. Details on how to seek permission, further information about the Publisher's permissions policies and our arrangements with organizations such as the Copyright Clearance Center and the Copyright Licensing Agency, can be found at our website: www.elsevier.com/permissions.

This book and the individual contributions contained in it are protected under copyright by the Publisher (other than as may be noted herein).

**Notices**
Knowledge and best practice in this field are constantly changing. As new research and experience broaden our understanding, changes in research methods, professional practices, or medical treatment may become necessary.

Practitioners and researchers must always rely on their own experience and knowledge in evaluating and using any information, methods, compounds, or experiments described herein. In using such information or methods they should be mindful of their own safety and the safety of others, including parties for whom they have a professional responsibility.

To the fullest extent of the law, neither the Publisher nor the authors, contributors, or editors, assume any liability for any injury and/or damage to persons or property as a matter of products liability, negligence or otherwise, or from any use or operation of any methods, products, instructions, or ideas contained in the material herein.

ISBN: 978-0-443-18667-7
ISSN: 0074-7742

For information on all Academic Press publications
visit our website at https://www.elsevier.com/books-and-journals

*Publisher:* Zoe Kruze
*Acquisitions Editor:* Mariana Kuhl
*Editorial Project Manager:* Federico Paulo Mendoza
*Production Project Manager:* Abdulla Sait
*Cover Designer:* Matthew Limbert

Typeset by MPS Limited, India

# Contents

| | |
|---|---|
| Contributors | xi |
| Preface | xv |

**1. $A_{2A}$ adenosine receptor agonists, antagonists, inverse agonists and partial agonists** — 1

Kenneth A. Jacobson, R. Rama Suresh, and Paola Oliva

| | | |
|---|---|---|
| 1. | Introduction | 2 |
| 2. | $A_{2A}AR$ agonists | 6 |
| 3. | Positive allosteric enhancers (PAMs) | 13 |
| 4. | $A_{2A}AR$ antagonists and inverse agonists | 14 |
| 5. | Negative allosteric enhancers (NAMs) | 19 |
| 6. | $A_{2A}AR$ partial agonists | 19 |
| 7. | Conclusions | 19 |
| | References | 20 |

**2. Adenosine $A_{2A}$ receptor and glia** — 29

Zhihua Gao

| | | |
|---|---|---|
| 1. | Preface | 29 |
| 2. | $A_{2A}R$ in astrocytes | 30 |
| 3. | Astrocyte $A_{2A}R$ and Alzheimer's disease | 31 |
| 4. | Astrocyte $A_{2A}R$ and PD | 33 |
| 5. | $A_{2A}R$ and microglia | 34 |
| 6. | Microglial $A_{2A}$ receptor and process retraction | 35 |
| 7. | Microglial $A_{2A}$ receptor and neuroinflammation | 36 |
| 8. | Microglial $A_{2A}R$ and AD | 37 |
| 9. | Microglial adenosine $A_{2A}R$ and PD | 38 |
| 10. | Microglial adenosine $A_{2A}$ receptor and stroke | 39 |
| 11. | Microglial $A_{2A}R$ in other diseases | 41 |
| 12. | Role of other microglial adenosine $A_1$ and $A_3$ receptors | 42 |
| 13. | $A_{2A}R$ and oligodendrocytes | 43 |
| 14. | Perspective | 44 |
| | References | 45 |

## 3. The adenosine $A_{2A}$ receptor in the basal ganglia: Expression, heteromerization, functional selectivity and signalling  49
Rafael Franco, Gemma Navarro, and Eva Martínez-Pinilla

1. Introduction  50
2. Expression of $A_{2A}R$ in the basal ganglia  51
3. Functional selectivity and signalling of $A_{2A}R$  53
    3.1 $A_{2A}Rs$ as monomers  53
    3.2 $A_{2A}Rs$ as heteromers  54
    3.3 $A_{2A}R$-containing heteromers with relevance in basal ganglia circuits  55
    3.4 Multiple mechanisms of action of $A_{2A}R$ antagonists  58
    3.5 $A_{2A}R$ mediates the adenosine-dopamine functional interactions in the basal ganglia  59
4. Heteromer expression in neurological diseases affecting the basal ganglia and in alterations of the reward circuits by drugs of abuse  61
5. Concluding remarks  62
Conflict of interests  63
Author contribution  63
Funding  63
References  63

## 4. How and why the adenosine $A_{2A}$ receptor became a target for Parkinson's disease therapy  73
Peter Jenner, Tomoyuki Kanda, and Akihisa Mori

1. Introduction  74
2. The relevance of the adenosinergic system and the $A_{2A}$ adenosine receptor  77
3. Adenosine $A_{2A}$ receptors and motor function  80
4. Adenosine $A_{2A}$ antagonists and dyskinesia  84
5. Adenosine $A_{2A}$ antagonists and motor function in Parkinson's disease  85
6. Other potential mechanisms of action of $A_{2A}$ adenosine antagonists  88
7. Adenosine antagonists and non-motor symptoms of Parkinson's disease  90
8. Exploring the future potential of adenosine $A_{2A}$ antagonists in Parkinson's disease  92
References  95

## 5. Adenosine A$_{2A}$ antagonists and Parkinson's disease     105
Michelle Offit, Brian Nagle, Gonul Ozay, Irma Zhang,
Anastassia Kerasidis, Yasar Torres-Yaghi, and Fernando Pagan

1. Introduction     106
2. Receptor agonists     108
    2.1 Dopamine receptor agonists     108
3. Inverse agonists     112
    3.1 Serotonin receptor 2A inverse agonists     112
4. Receptor antagonists     112
    4.1 Dopamine receptor antagonists     113
    4.2 NMDA receptor antagonists     114
    4.3 Acetylcholine receptor antagonists     115
    4.4 Adenosine receptor antagonists     115
References     117

## 6. Effects of adenosine A$_{2A}$ receptors on cognitive function in health and disease     121
Cinthia P. Garcia, Avital Licht-Murava, and Anna G. Orr

1. Introduction     122
2. Effects of A$_{2A}$ receptor modulation on normal cognitive function     124
    2.1 Pharmacological manipulations     124
    2.2 Global genetic manipulations     126
    2.3 Genetic manipulations targeted to neurons     127
    2.4 Genetic manipulations targeted to astrocytes     128
    2.5 Sex-dependent effects     129
3. Effects of A$_{2A}$ receptor modulation on cognitive impairments in aging and disease     131
    3.1 Aging and dementia     131
    3.2 Multiple sclerosis     134
    3.3 Schizophrenia     135
    3.4 Neurodevelopmental disorders     136
    3.5 Brain trauma and stroke     137
4. Effects of A$_{2A}$ receptor modulation on cognitive impairments in stress     137
5. Effects of A$_{2A}$ receptor modulation on drug use and addiction     138
6. Conclusions     139
Conflicts of interests     141
Acknowledgements     141
References     141

## 7. Adenosine $A_{2A}$ receptors and sleep — 155
Mustafa Korkutata and Michael Lazarus

1. Introduction — 155
2. Adenosine and its receptors — 156
   2.1 Adenosine — 156
   2.2 Purinergic P1 receptors — 157
   2.3 Adenosine $A_{2A}$ receptors — 158
3. The role of adenosine and $A_{2A}$Rs in sleep-wake regulation — 160
   3.1 Adenosine and sleep — 160
   3.2 $A_{2A}$R agonists and sleep — 162
   3.3 Adenosine $A_{2A}$R antagonists and sleep — 163
   3.4 Natural $A_{2A}$R agonists and sleep — 164
   3.5 Allosteric modulation of adenosine $A_{2A}$Rs and sleep — 165
4. Conclusion — 169
References — 169

## 8. Adenosine $A_{2A}$ signals and dystonia — 179
Makio Takahashi

Body — 179
References — 183

## 9. $A_{2A}$R antagonist treatment for multiple sclerosis: Current progress and future prospects — 185
Chenxing Qi, Yijia Feng, Yiwei Jiang, Wangchao Chen, Serhii Vakal, Jiang-Fan Chen, and Wu Zheng

1. Introduction — 187
2. Adenosine metabolism — 188
3. $A_{2A}$R distribution in MS/EAE pathology — 190
4. The confusing effect of $A_{2A}$R signaling in MS/EAE — 191
5. $A_{2A}$R signaling in macrophages and dendritic cells — 193
6. The complex effects of $A_{2A}$R signaling in T cells — 195
7. $A_{2A}$R signaling is essential for maintaining the suppressive capacity of Tregs — 198
8. $A_{2A}$R antagonist attenuates the pro-inflammation phenotype of microglia — 199
9. $A_{2A}$R antagonist inhibits the activation of astrocytes — 201
10. The $A_{2A}$R antagonist protects oligodendrocytes from damage — 203
11. $A_{2A}$R antagonist protects against lymphocyte infiltration into the CNS across the BBB — 205

| 12. $A_{2A}R$ antagonist suppresses T-cell infiltration by decreasing the ChP gateway activity | 207 |
|---|---|
| 13. Safety of the $A_{2A}R$ antagonist in clinical trials | 209 |
| 14. The limitations of current studies on the potential efficacy of $A_{2A}R$ antagonists in MS pathology | 213 |
| 15. Conclusion | 214 |
| Acknowledgments | 214 |
| Declarations | 214 |
| Ethics approval and consent to participate | 214 |
| Consent for publication | 214 |
| Funding | 215 |
| Competing interests | 215 |
| References | 215 |

## 10. $A_{2A}R$ and traumatic brain injury    225

Yan Zhao, Ya-Lei Ning, and Yuan-Guo Zhou

| 1. Preface | 226 |
|---|---|
| 2. $A_{2A}R$ modulation of neuroinflammation at the early stage of TBI | 228 |
|     2.1 Regulation of glial cells by $A_{2A}R$ antagonists | 228 |
|     2.2 Regulation of neuroinflammation by BMDC $A_{2A}R$ | 231 |
|     2.3 Regulation of NLRP3 inflammasome activation by $A_{2A}R$ | 231 |
|     2.4 Bidirectional modulation of neuroinflammation by $A_{2A}R$ | 232 |
| 3. $A_{2A}R$ modulation of glutamate excitotoxicity at the early stage of TBI | 235 |
|     3.1 Potentiation of presynaptic glutamate release by $A_{2A}R$ activation | 236 |
|     3.2 Exacerbation of postsynaptic glutamate excitotoxicity by $A_{2A}R$ activation | 236 |
|     3.3 Control of cerebral glutamate homeostasis by astroglial $A_{2A}R$ | 237 |
| 4. $A_{2A}R$ modulation of aberrant proteins | 240 |
|     4.1 Modulation of tau hyperphosphorylation after TBI | 240 |
|     4.2 Modulation of autophagy after TBI | 241 |
| 5. $A_{2A}R$ modulation of cognitive dysfunction after TBI | 241 |
|     5.1 Alteration of adenosine and its receptors during a chronic period of TBI | 242 |
|     5.2 $A_{2A}R$ regulation in pathological proteins | 244 |
|     5.3 $A_{2A}R$ and delayed cognitive dysfunction of TBI | 245 |
| 6. $A_{2A}R$ modulation of chronic neuroinflammation in TBI | 247 |
|     6.1 Long-lasting neuroinflammation | 247 |
|     6.2 Impairment of synaptic structure and function | 249 |

7. $A_{2A}R$ modulation of other long-lasting consequences of TBI ... 250
　7.1　Sleep disorders ... 250
　7.2　PTSD ... 251
　7.3　Posttraumatic seizures and epilepsy ... 252
References ... 253

## 11. Chemobrain: An accelerated aging process linking adenosine $A_{2A}$ receptor signaling in cancer survivors ... **267**

Alfredo Oliveros, Michael Poleschuk, Peter D. Cole, Detlev Boison, and Mi-Hyeon Jang

1. Introduction ... 268
2. Adenosine $A_{2A}R$ in cognitive function ... 271
　2.1　Physiological role of $A_{2A}R$ and cognitive improvement ... 271
　2.2　$A_{2A}R$ and cognitive improvement in pathological conditions ... 272
　2.3　Convergence of the adenosine $A_{2A}R$, aging and AD ... 272
3. Findings of cognitive dysfunction in chemobrain ... 274
　3.1　Clinical imaging studies ... 274
4. Chemobrain as an accelerated aging phenotype in adult and pediatric cancer ... 275
　4.1　Evidence of accelerated aging in adult cancer patients ... 275
　4.2　Evidence of accelerated aging phenotypes in pediatric cancer patients ... 278
　4.3　Accelerated aging mitochondrial defects associated with cancer and chemobrain ... 279
　4.4　Accelerated aging, telomeres, and chemobrain ... 281
　4.5　Epigenetic related alterations in chemobrain and accelerated aging ... 283
5. The adenosine $A_{2A}$ receptor as a potential mechanism of accelerated aging in chemobrain and cancer ... 285
　5.1　Anticancer effects and immune cell modulation of the adenosine $A_{2A}R$ in malignancies ... 285
　5.2　Interaction between chemotherapy, adenosine $A_{2A}R$ and chemobrain ... 288
　5.3　Adenosine $A_{2A}R$ and chemobrain ... 289
6. Concluding remarks ... 291
Acknowledgments ... 292
Conflict of interest statement ... 292
References ... 292

# Contributors

**Detlev Boison**
Department of Neurosurgery, Robert Wood Johnson Medical School, Rutgers, The State University of New Jersey, Piscataway, NJ, United States

**Jiang-Fan Chen**
State Key Laboratory of Ophthalmology, Optometry and Visual Science, Eye Hospital; Oujiang Laboratory (Zhejiang Laboratory for Regenerative Medicine, Vision and Brain Health), School of Ophthalmology & Optometry and Eye Hospital, Wenzhou Medical University, Wenzhou, P.R. China

**Wangchao Chen**
State Key Laboratory of Ophthalmology, Optometry and Visual Science, Eye Hospital, Wenzhou Medical University, Wenzhou, P.R. China

**Peter D. Cole**
Division of Pediatric Hematology/Oncology, Rutgers Cancer Institute of New Jersey, New Brunswick, NJ, USA

**Yijia Feng**
Key Laboratory of Alzheimer's Disease of Zhejiang Province, Institute of Aging, Wenzhou Medical University, Wenzhou, Zhejiang, P.R. China

**Rafael Franco**
Molecular Neurobiology laboratory, Department of Biochemistry and Molecular Biomedicine, Faculty of Biology, Universitat de Barcelona, Barcelona, Spain; CiberNed, Network Center for Neurodegenerative diseases, National Spanish Health Institute Carlos III, Madrid, Spain; School of Chemistry, Universitat de Barcelona, Barcelona, Spain

**Zhihua Gao**
Department of Neurobiology and Department of Neurology of Second Affiliated Hospital, Zhejiang University School of Medicine; Liangzhu Laboratory, Zhejiang University Medical Center, MOE Frontier Science Center for Brain Science and Brain-machine Integration, State Key Laboratory of Brain-machine Intelligence, Zhejiang University, West Wenyi Road; NHC and CAMS Key Laboratory of Medical Neurobiology, Zhejiang University, Hangzhou, P.R. China

**Cinthia P. Garcia**
Appel Alzheimer's Disease Research Institute; Feil Family Brain and Mind Research Institute; Pharmacology Graduate Program, Weill Cornell Medicine, New York, NY, United States

**Kenneth A. Jacobson**
Molecular Recognition Section, Laboratory of Bioorganic Chemistry, National Institute of Diabetes and Digestive and Kidney Diseases, Bethesda, MD, United States

**Mi-Hyeon Jang**
Department of Neurosurgery, Robert Wood Johnson Medical School, Rutgers, The State University of New Jersey, Piscataway, NJ, United States

**Peter Jenner**
Institute of Pharmaceutical Sciences, King's College London, London, United Kingdom

**Yiwei Jiang**
Alberta Institute, Wenzhou Medical University, Wenzhou, P.R. China

**Tomoyuki Kanda**
Kyowa Kirin Co., Ltd., Otemachi. Chiyoda-ku, Tokyo, Japan

**Anastassia Kerasidis**
MedStar Georgetown University Hospital, Neurology Department, Reservoir Rd, Washington, DC, United States

**Mustafa Korkutata**
Department of Neurology, Division of Sleep Medicine, Beth Israel Deaconess Medical Center and Harvard Medical School, Boston, MA, USA

**Michael Lazarus**
International Institute for Integrative Sleep Medicine (WPI-IIIS) and Institute of Medicine, University of Tsukuba, Tsukuba, Japan

**Avital Licht-Murava**
Appel Alzheimer's Disease Research Institute; Feil Family Brain and Mind Research Institute, Weill Cornell Medicine, New York, NY, United States

**Eva Martínez-Pinilla**
Department of Morphology and Cell Biology, Faculty of Medicine, University of Oviedo; Instituto de Neurociencias del Principado de Asturias (INEUROPA); Instituto de Investigación Sanitaria del Principado de Asturias (ISPA), Asturias, Spain

**Akihisa Mori**
SNLD Ltd., Akashi-sho, Chuo-ku, Tokyo, Japan

**Brian Nagle**
MedStar Georgetown University Hospital, Neurology Department, Reservoir Rd, Washington, DC, United States

**Gemma Navarro**
CiberNed, Network Center for Neurodegenerative diseases, National Spanish Health Institute Carlos III, Madrid, Spain; Department of Biochemistry and Physiology, School of Pharmacy and Food Science Universitat de Barcelona, Barcelona, Spain; Institute of Neurosciences, Universitat de Barcelona, Barcelona, Spain

**Ya-Lei Ning**
Department of Army Occupational Disease, State Key Laboratory of Trauma and Chemical Poisoning, Research Institute of Surgery and Daping Hospital, Army Medical University, P.R. China; Institute of Brain and Intelligence, Army Medical University, Chongqing, P.R. China

**Michelle Offit**
MedStar Georgetown University Hospital, Neurology Department, Reservoir Rd, Washington, DC, United States

**Paola Oliva**
Molecular Recognition Section, Laboratory of Bioorganic Chemistry, National Institute of Diabetes and Digestive and Kidney Diseases, Bethesda, MD, United States

**Alfredo Oliveros**
Department of Neurosurgery, Robert Wood Johnson Medical School, Rutgers, The State University of New Jersey, Piscataway, NJ, United States

**Anna G. Orr**
Appel Alzheimer's Disease Research Institute; Feil Family Brain and Mind Research Institute, Weill Cornell Medicine, New York, NY, United States

**Gonul Ozay**
MedStar Georgetown University Hospital, Neurology Department, Reservoir Rd, Washington, DC, United States

**Fernando Pagan**
MedStar Georgetown University Hospital, Neurology Department, Reservoir Rd, Washington, DC, United States

**Michael Poleschuk**
Department of Neurosurgery, Robert Wood Johnson Medical School, Rutgers, The State University of New Jersey, Piscataway, NJ, United States

**Chenxing Qi**
State Key Laboratory of Ophthalmology, Optometry and Visual Science, Eye Hospital; Oujiang Laboratory (Zhejiang Laboratory for Regenerative Medicine, Vision and Brain Health), School of Ophthalmology & Optometry and Eye Hospital, Wenzhou Medical University, Wenzhou, P.R. China

**R. Rama Suresh**
Molecular Recognition Section, Laboratory of Bioorganic Chemistry, National Institute of Diabetes and Digestive and Kidney Diseases, Bethesda, MD, United States

**Makio Takahashi**
Department of Neurodegenerative disorders, Kansai Medical University, Hirakata, Osaka, Japan

**Yasar Torres-Yaghi**
MedStar Georgetown University Hospital, Neurology Department, Reservoir Rd, Washington, DC, United States

**Serhii Vakal**
Structural Bioinformatics Laboratory, Biochemistry, Faculty of Science and Engineering, Åbo Akademi University, Turku, Finland

**Irma Zhang**
MedStar Georgetown University Hospital, Neurology Department, Reservoir Rd, Washington, DC, United States

**Yan Zhao**
Department of Army Occupational Disease, State Key Laboratory of Trauma and Chemical Poisoning, Research Institute of Surgery and Daping Hospital, Army Medical University, P.R. China; Institute of Brain and Intelligence, Army Medical University, Chongqing, P.R. China

**Wu Zheng**
State Key Laboratory of Ophthalmology, Optometry and Visual Science, Eye Hospital; Oujiang Laboratory (Zhejiang Laboratory for Regenerative Medicine, Vision and Brain Health), School of Ophthalmology & Optometry and Eye Hospital, Wenzhou Medical University, Wenzhou, P.R. China

**Yuan-Guo Zhou**
Department of Army Occupational Disease, State Key Laboratory of Trauma and Chemical Poisoning, Research Institute of Surgery and Daping Hospital, Army Medical University, P.R. China; Institute of Brain and Intelligence, Army Medical University, Chongqing, P.R. China

# Preface

Over the last 60 years, evidence has continued to accumulate that extracellular adenosine is an important intermediary metabolite, acting not only as a building block for nucleic acids and a component of the biological energy currency adenosine triphosphate but also an important modulator of physiological and pathological processes. In the 1990s, four G protein–coupled receptor subtypes of adenosine receptors (namely $A_1$, $A_{2A}$, $A_{2B}$, and $A_3$) were identified and demonstrated to exert a variety of physiological and pathophysiological functions. In the meantime, medicinal chemistry generated agonists and antagonists with high affinity and with selectivity for the human variants of the adenosine $A_{2A}$ receptor. Over the last 25 years, genetic knockout of these genes coupled with molecular and pharmacological characterization has significantly expanded the potential indications for adenosine receptor–based drug action. It has become clear that adenosine itself can be safely used (as Adenoscan) for myocardial perfusion imaging and for the treatment of supraventricular tachycardia, and several FDA-approved drugs (including dipyridamole and methotrexate) may exert their therapeutic effects by altering extracellular adenosine signaling. Over the last two decades, $A_{2A}$ receptor antagonists have emerged as leading nondopaminergic drugs for the treatment of Parkinson disease (PD) for its unique expression pattern and molecular/functional interactions with dopamine receptor in the striatum and for its robust motor benefit in rodent and nonhuman primate models of PD. Since 2000, several $A_{2A}$ receptor antagonists, especially istradefylline (KW-6002), were developed and tested in preclinical and clinical studies for the treatment of PD.

This special volume recognizes the scientific contribution of Akihisa Mori, PhD, who has played an indispensable role in leading a global project at Kyowa Kirin Co., Ltd (Japan), from the early exploratory stages in the 1990s to the successful marketing approval of a first-in class antiparkinsonian product $A_{2A}$ receptor antagonist, istradefylline, in the United States (2019) and Japan (2013). Dr. Mori's training in pharmaceutical sciences (BS) from Kyoto University, and neuroscience (PhD) from the University of Tokyo, uniquely positioned him as the global project leader of the $A_{2A}$ receptor antagonist project aimed at the treatment of PD. Dr. Mori's research specialty focused entirely on $A_{2A}$ receptor neurobiology, basal ganglia physiology, and PD, with the discovery of $A_{2A}$ receptor function in the basal ganglia and a mode of action of $A_{2A}$ receptor antagonist in antiparkinsonian therapy, as evidenced in more than 50 articles published in peer-review journals. Importantly, he handled the clinical development and global project leadership at the Kyowa Kirin Co. Ltd., working on the global portfolio strategy and global medical affairs. His global project leadership has been instrumental in the successful completion of the entire US/EU/Japan clinical phase II-III development program for istradefylline in advanced PD patients. As a result, in August 27, 2019, the US FDA granted approval for Nourianz (istradefylline) to be used as an add-on treatment to L-DOPA/carbidopa in adult patients with PD experiencing "OFF" episodes. This is an important milestone achievement, with a critical contribution from Dr. Mori that has provided PD patients with a novel nondopaminergic treatment option in conjunction with L-DOPA.

This approval also provided some important lessons on future adenosine receptor–based drug development concerning disease-specific adenosine signaling and targeting subpopulations of patients. This approval paves the way to foster entirely novel therapeutic opportunities for $A_{2A}$ receptor antagonists, such as neuroprotection or reversal of mood and cognitive deficits in PD. Recent years have witnessed an acceleration of medicinal chemical activity for $A_{2A}$ receptor ligands through structure-based drug design with successful $A_{2A}$ receptor X-ray structures for PD and cancer immunotherapy. Yet, the great challenge remains in developing adenosine receptor ligands for the wide-spread distribution of adenosine and adenosine receptors. These developments promote us to revisit and discuss, in this special volume, the therapeutic potential of adenosine receptor drugs, focusing on the key biological factors limiting their clinical development for brain disorders and the hurdles that could and should be overcome.

Thanks to the significant efforts by purine medicinal chemists and their success over the last two decades, there are long lists of $A_{2A}$ receptor agonists and antagonists with high affinity and selectivity as described by Dr. Kenneth A. Jacobson and his colleagues (Chapter 1). Ongoing development in their medical chemistry and clinical application of these antagonists offers new opportunity for the treatment of PD and cancer immunotherapy. The chapter by Dr. Gao (Chapter 2) provides a comprehensive review on glial $A_{2A}$ receptors in brain functions and disorders, covering the $A_{2A}$ receptors in astrocytes and microglial cells with its particular focus on their role in Alzheimer disease and PD, stroke and visual pathway, and myelination and dysmyelination. The chapter by Dr. Franco and colleagues (Chapter 3) is a comprehensive overview on the $A_{2A}$ receptor function in the basal ganglia, clarifying diversity of signaling on molecular/functional interaction with other receptors. PD has become to be targeted by adenosine $A_{2A}$ antagonist as actual therapy so that we have three chapters addressing applications to the disease. Dr. Jenner and colleagues (Chapter 4) has described the entire story and rationale for $A_{2A}$ receptor antagonist development as nondopaminergic therapy with future aspects/possibilities. Michelle Offit and colleagues (Chapter 5) has provided an overview of PD therapy including an aspect for nondopaminergic vs. dopaminergic drugs. Dr. Takahashi (Chapter 8) has discussed about therapeutic possibility of $A_{2A}$ receptor antagonists on dystonia with signaling, introducing his fantastic clinical research evidence. The chapter by Cinthia P. Garcia and colleagues (Chapter 6) has provided an exciting view on an involvement of $A_{2A}$ receptor in cognitive function with multiple aspects as well as what happens in dysfunction. Drs. Korkutata and Lazarus (Chapter 7) provided a chapter indicating sleep–wake cycle with physiological and therapeutic involvement of $A_{2A}$ receptor as well as for $A_{2A}$ allosteric modulators in sleep disturbance. Chenxing Qi and his colleagues (Chapter 9) survey the recent developments on $A_{2A}$ receptor antagonists in the treatment of multiple sclerosis with a detailed dissection of the complex actions of the $A_{2A}$ receptors in neuronal and non-neuronal cells in experimental autoimmune encephalomyelitis. The chapter by Yan Zhao and colleagues (Chapter 10) summarizes acute/subacute and chronic changes in adenosine and adenosine $A_{2A}$ receptor–related signals in traumatic brain injury and on recent development in $A_{2A}$ receptor modulation of TBI by control of neuroinflammation, excitotoxicity, and cognition. The chapter by Alfredo Oliveros and colleagues (Chapter 11) introduces on an emerging topic of the "chemobrain" or "chemofog" in relationship with the $A_{2A}$ receptor as potential target for combat it. The review provides thorough discussion on chemobrain as an accelerated aging phenotype in cancer surviving children and adults suffer from aberrant chemotherapy. Furthermore, it is proposed that adenosine $A_{2A}$ receptors may also serve as potential targets for combating the chemobrain by $A_{2A}$ receptor antagonism, not only enhancing cognitions in various learning and memory behaviors under physiological conditions but also reversing the cognitive impairments in aging and Alzheimer disease.

As the guest editors of this volume, we greatly appreciate the work of all authors and reviewers who contributed with enthusiasm. We strongly hope that these reviews enlisted in this volume will be useful to understand better some of the emerging issues, challenges, and opportunities before us and will facilitate the introduction of additional clinical applications of adenosine receptor drugs into the care of patients.

JIANG-FAN CHEN AND AKIHISA MORI

# CHAPTER ONE

# $A_{2A}$ adenosine receptor agonists, antagonists, inverse agonists and partial agonists

### Kenneth A. Jacobson*, R. Rama Suresh, and Paola Oliva

Molecular Recognition Section, Laboratory of Bioorganic Chemistry, National Institute of Diabetes and Digestive and Kidney Diseases, Bethesda, MD, United States
*Corresponding author. e-mail address: kennethj@niddk.nih.gov

## Contents

| | |
|---|---|
| 1. Introduction | 2 |
| 2. $A_{2A}AR$ agonists | 6 |
| 3. Positive allosteric enhancers (PAMs) | 13 |
| 4. $A_{2A}AR$ antagonists and inverse agonists | 14 |
| 5. Negative allosteric enhancers (NAMs) | 19 |
| 6. $A_{2A}AR$ partial agonists | 19 |
| 7. Conclusions | 19 |
| References | 20 |

## Abstract

The Gs-coupled $A_{2A}$ adenosine receptor ($A_{2A}AR$) has been explored extensively as a pharmaceutical target, which has led to numerous clinical trials. However, only one selective $A_{2A}AR$ agonist (regadenoson, Lexiscan) and one selective $A_{2A}AR$ antagonist (istradefylline, Nouriast) have been approved by the FDA, as a pharmacological agent for myocardial perfusion imaging (MPI) and as a cotherapy for Parkinson's disease (PD), respectively. Adenosine is widely used in MPI, as Adenoscan. Despite numerous unsuccessful clinical trials, medicinal chemical activity around $A_{2A}AR$ ligands has accelerated recently, particularly through structure-based drug design. New drug-like $A_{2A}AR$ antagonists for PD and cancer immunotherapy have been identified, and many clinical trials have ensued. For example, imaradenant (AZD4635), a compound that was designed computationally, based on $A_{2A}AR$ X-ray structures and biophysical mapping. Mixed $A_{2A}AR/A_{2B}AR$ antagonists are also hopeful for cancer treatment. $A_{2A}AR$ antagonists may also have potential as neuroprotective agents for treatment of Alzheimer's disease.

## 1. Introduction

Adenosine receptors (ARs) have been explored extensively as pharmaceutical targets since the late 1960s (IJzerman, Jacobson, Müller, Cronstein, & Cunha, 2022; Jacobson & Gao, 2006). Endogenous extracellular adenosine, the native agonist, is produced as a hydrolysis product of ATP by cells, tissues or organs under stress (Borea, Gessi, Merighi, Vincenzi, & Varani, 2018; Cekic & Linden, 2016; Haskó, Antonioli, & Cronstein, 2018; Linden, 2005). ARs have a role in allostasis, the body's stress response to correct imbalances, rather than in homeostasis. Their biological roles have been explored extensively using genetic receptor knockout mouse lines (Lopes, Lourenço, Tomé, Cunha, & Canas, 2021), including one line in which all four of the ARs were deleted, which does not display a strong phenotype compared to WT mice (Xiao, Liu, Jacobson, Gavrilova, & Reitman, 2019). The $A_{2A}AR$ subtype is coupled to $G_s$ guanine nucleotide-regulatory proteins to stimulate intracellular cyclic AMP (cAMP) production. The $A_{2A}AR$ is found, among other locations, in immune cells and platelets, where adenosine suppresses their activation, in vascular smooth muscle, where adenosine induces relaxation and vasodilation to increase circulation and oxygen supply, and in the brain. The combination of $A_{2A}AR$ agonists and other antithrombotic agents, such as $P2Y_{12}R$ antagonists, might be useful clinically (Wolska et al., 2020). In the basal ganglia, adenosine has a role opposite that of dopamine in movement control. Specifically, the $A_{2A}AR$ suppresses postsynaptic dopamine $D_2R$ effects (Ferré, Díaz-Ríos, Salamone, & Prediger, 2018). Locations of high $A_{2A}AR$ gene expression in the human analyzed by RNA-seq are available in online databases. For example, high expression (>0.8 transcript per million, TPM, GeneID: ENSG00000128271) was determined in: EBV-transformed lymphocyte, blood, caudate nucleus, cerebellar hemisphere, cerebellum, liver, nucleus accumbens and testis (https://www.ebi.ac.uk/gxa/home, accessed Dec. 26, 2022). Medium expression (0.5–0.7 TMP) was noted in: lung, ovary, putamen and spleen.

The alkylxanthines caffeine and theophylline are the prototypical antagonists of action of endogenous adenosine through these receptors (Fredholm, 1985; Huang & Daly, 1974; Sattin & Rall, 1970). One of the first medical applications of an adenosine antagonist was the use of the non-subtype-selective theophylline (now as aminophylline, i.e. the salt with ethylenediamine) for treating asthma (Schultze-Werninghaus & Meier-Sydow, 1982), but its AR-subtype related mechanism is still under discussion (Fozard, 2003). Both synthetic, selective agonists (Fig. 1) and antagonists (Fig. 2) of the $A_{2A}AR$ subtype have been sought for over three decades (Baraldi et al., 2018). However, until the mid-1990s, even after the cloning of the canine and human $A_{2A}ARs$ (Le, Townsend-Nicholson, Baker, Sutherland, & Schofield, 1996; Maenhaut et al., 1990), little attention was given to the distinction between $A_{2A}$ and $A_{2B}ARs$ in the medicinal chemistry literature. The earliest pharmaceutical focus on this subtype led to the discovery of $A_{2A}AR$-selective agonists for hypertension control, cardiac stress testing, anti-inflammatory effects and the potential use in treating schizophrenia (Jacobson, Tosh, Jain, & Gao, 2019; Singer and Yee, 2023). Adenosine itself, when infused in the coronary artery causes $A_{2A}AR$-mediated dilatation (Belardinelli, Linden, & Berne, 1989), and adenosine was eventually approved as a pharmacological agent for myocardial perfusion imaging (MPI) as the widely used Adenoscan. Later, a synthetic and moderately $A_{2A}AR$-selective agonist, also an adenine nucleoside derivative, regadenoson, known as Lexiscan, was introduced for the same diagnostic application (Gao et al., 2001; Iskandrian et al., 2007).

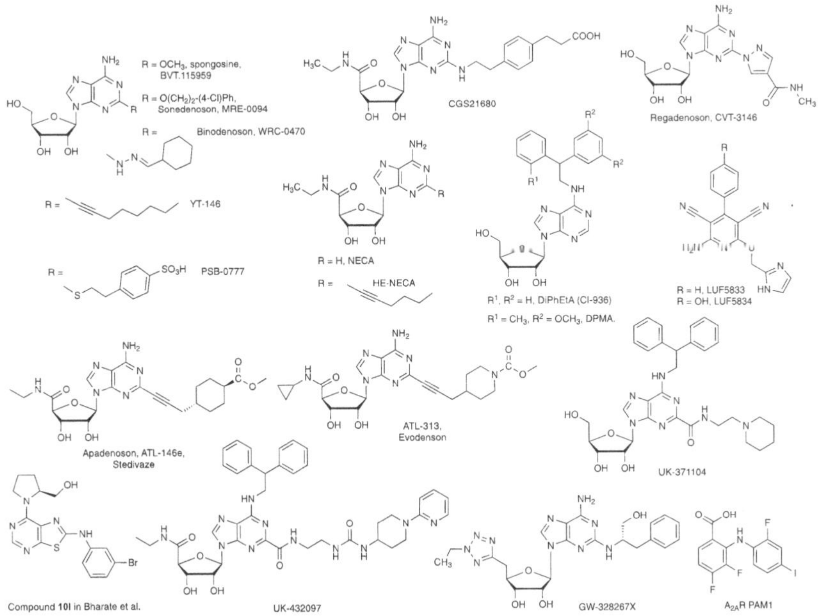

**Fig. 1** Structures of adenosine receptor agonists, partial agonists and allosteric modulators described in the text. Most of the agonists and partial agonists are adenosine derivatives, while all of the allosteric modulators are small heterocyclic molecules. Refer to Table 1 for their binding affinity and selectivity.

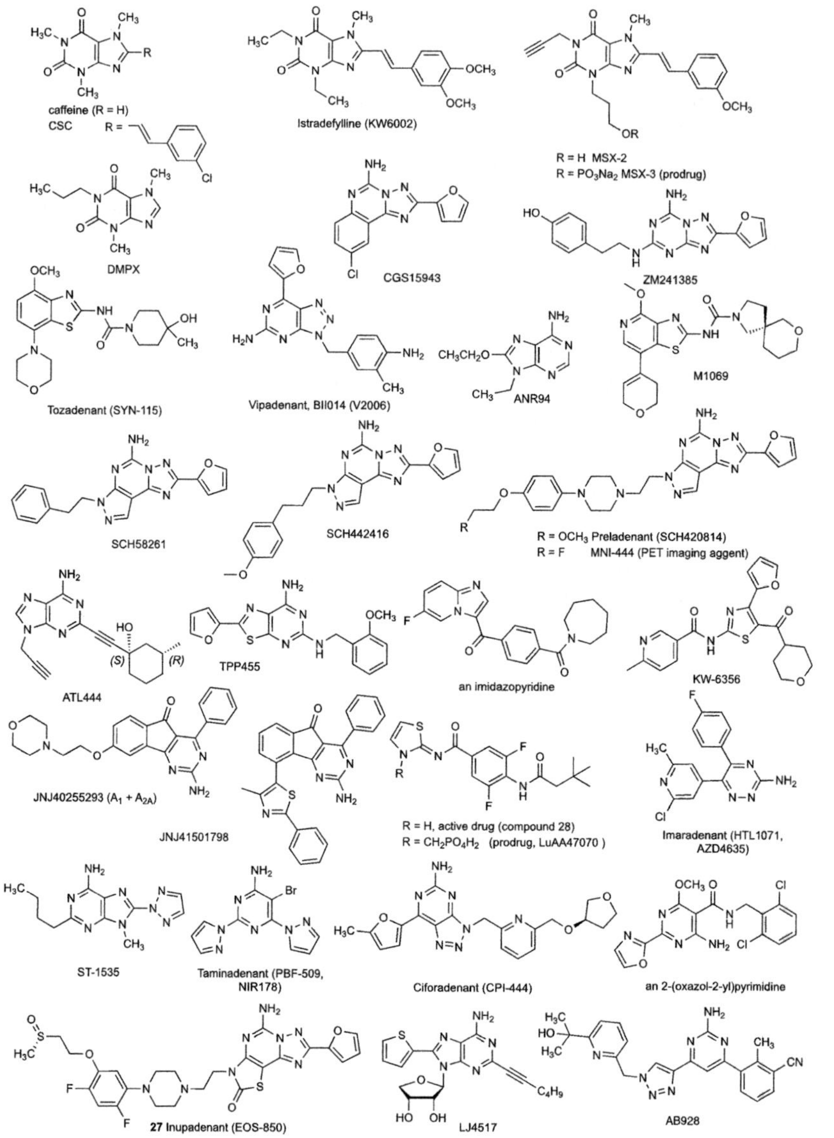

**Fig. 2** Structures of adenosine receptor antagonists described in the text. Refer to Table 1 for their binding affinity and selectivity.

Potent and selective $A_{2A}AR$-selective antagonists were initially developed for CNS applications, particularly in aging and neurodegenerative conditions, including Parkinson's disease (PD), and later for Alzheimer's disease (AD) and attention-deficit/hyperactivity disorder (ADHD)

(Jazayeri, Andrews, & Marshall, 2017; Kanda, Shiozaki, Shimada, Suzuki, & Nakamura, 1994; Merighi et al., 2021; Pinna, 2009; Preti et al., 2015; Schiffmann, Fisone, Moresco, Cunha, & Ferré, 2007). The first synthetic selective $A_{2A}AR$ antagonist was istradefylline, an alkylxanthine, like caffeine, and which was earlier approved in Japan and Korea and since 2019 is now FDA-approved (as Nouriast) as a cotherapy for PD together with dopaminergic therapy (Chen & Cunha, 2020; Ferré et al., 2018). Within recent years, the $A_{2A}AR$ appeared as a drug target for use in cancer immunotherapy (Ohta & Sitkovsky, 2001). There is a pathologically elevated level of adenosine in the tumor microenvironment, and thus blocking $A_{2A}AR$ activation with a selective antagonist, or inhibiting adenosine production locally, removes the adenosine-induced suppression of the immune response in a variety of cell types (Vigano, Alatzoglou, & Irving, 2019). This has led to both the discovery of novel $A_{2A}AR$ antagonists not of the same class (de Lera Ruiz, Lim, & Zheng, 2014; Müller & Jacobson, 2011), as well as repurposing toward cancer of some antagonists that were initially introduced for CNS indications.

Although the earliest medicinal chemistry studies of $A_{2A}AR$ agonists and antagonists were purely based on empirical probing of the structure–activity relationships (SAR), for roughly two decades their discovery has taken into account, been guided by, or even was purely based on structural information about the receptor (Gutiérrez-de-Terán, Sallander, & Sotelo, 2017; Jazayeri et al., 2017). Early $A_{2A}AR$ site-directed mutagenesis identified amino acid residues important for agonist recognition, including those conserved within the AR family, both in the transmembrane region and on the extracellular loops (Jiang et al., 1996; Kim et al., 1996; Kim, Wess, van Rhee, Shöneberg, & Jacobson, 1995). The first antagonist bound $A_{2A}AR$ receptor structure, with ZM241,385, was reported in 2008 by Stevens and colleagues (Jaakola et al., 2008), and the first agonist-bound $A_{2A}AR$ receptor structures in 2011 (Lebon et al., 2011; Xu et al., 2011). A very high-resolution antagonist (ZM241,385)-bound $A_{2A}AR$ structure (1.7 Å) was reported that defines specific water molecules and negative allosteric sodium ion surrounding the binding site (Liu et al., 2012). Notable in this structure-based design of $A_{2A}AR$ antagonists is the work of Heptares (now Sosei-Heptares) that was able to obtain X-ray structures of the receptor in complex with a wide range of antagonists, with affinities ranging from nM to μM (Doré et al., 2011). The key to this ability was the introduction of strategic single amino acid replacements within the transmembrane helical domains of the receptor (TMs) that stabilized either an inactive or active state of the receptor, while maintaining the essential small molecular ligand interactions (Jazayeri et al., 2017). This proprietary approach, termed StaRs (stabilized receptors), was first applied to the $A_{2A}AR$ receptor, and its success was soon extended to many other G protein-coupled receptors (GPCRs). The StaRs approach led to biophysical mapping of the drug binding site, which calculated the energetic contribution of specific molecular interactions and led to a new class of $A_{2A}AR$ antagonists suitable for translation to the clinic. Numerous structures of the $A_{2A}AR$ in complex with antagonists that are in clinical trials are now available (Salmaso, Jain, & Jacobson, 2021). Recently, the electron cryo-microscopy (cryo-EM) structure of the $A_{2A}AR$ in complex with an engineered G protein was solved (García-Nafría, Lee, Bai, Carpenter, & Tate, 2018), and this technique continues to be increasingly popular for AR structures. The $A_{2A}AR$ cryo-EM structures in complex with two agonists are now available (Chen, Zhang, & Weng, 2022). NMR has also been utilized effectively to study the conformational changes associated with $A_{2A}AR$ activation (Eddy et al., 2018). The family of membrane-bound proteins having seven TM helices (7TM receptors), mainly GPCRs, is the largest single class of gene products in the human genome, representing ~4% of the proteins coded. Even before structural approaches to GPCR ligand design were feasible, GPCR modulators were among the most important drug categories introduced, and therefore the feasibility of using a structure-based design approach greatly facilitated the identification of novel drug lead molecules.

## 2. $A_{2A}AR$ agonists

Most known AR agonists are nucleoside derivatives, particularly derivatives of adenosine. When radioligand binding assays for the $A_1AR$ and $A_{2A}AR$ became available (Bruns, Daly, & Snyder, 1980; Jarvis et al., 1989), the pace of medicinal chemical studies of AR agonists accelerated, although most binding was performed at the rat receptors until the mid-1990s. However, there have been several classes of non-nucleosides reported, including 2-thio-3,5-dicyanopyridines and 3-cyano-pyrimidines (Beukers et al., 2004; Catarzi, Varano, & Vigiani, 2022; Rosentreter et, al). Also, a non-nucleoside heterocyclic derivative was reported to activate the $A_{2A}AR$ (Bharate et al., 2016). Some of the major $A_{2A}AR$ agonists, introduced either as receptor probes for pharmacological studies or for translational development are described below.

Table 1 contains affinity determined for $A_{2A}AR$ agonists and antagonists at four AR subtypes. The clinical trial information for agonists that have been tested in humans is referenced in (Jacobson et al., 2019). The agonist CGS21680 was reported as the first highly selective agonist for the $A_{2A}AR$, due to combining substitutions of adenosine at the C2 (aminoalkyl/aryl) and 5′ (uronamide) positions (Hutchison et al., 1989). The terminal carboxylate of CGS21680 served as the basis for functionalized congeners that could be radiolabeled or derivatized for irreversible receptor labeling of fluorescent detection (Jacobson, 2009). Also of note, an early monosubstituted adenosine derivative CV1808 (2-phenylamino-adenosine, structure not shown) had a small degree of $A_{2A}$ selectivity compared to $A_1AR$ ($A_3AR$ had not yet been discovered). SAR studies by Olsson and coworkers identified other (oxy-alkyl/aryl) $A_{2A}AR$-affinity enhancing modifications the C2 positions (Ueeda, Thompson, Arroyo, & Olsson, 1991). CGS21680 was demonstrated to be selective for the $A_{2A}AR$ and totally inactive at the $A_{2B}AR$ (Hide, Padgett, Jacobson, & Daly, 1992), although its rat $A_3AR$ affinity ($K_i$) was 144 nM (Zhou et al., 1992). Later, CGS21680 was found to be relatively potent at the human $A_3AR$ (only 2.7-fold lower affinity than at the human $A_{2A}AR$, Gao, Blaustein, Gross, Melman, & Jacobson, 2003), but it retained its rat $A_{2A}AR$ selectivity. The compound was being evaluated preclinically as an antihypertensive agent, but never progressed to clinical trials. Its binding site in the receptor was later elucidated by Lebon, Edwards, Leslie & Tate (2015). Its use as a chemical precursor functionalized congener for fluorescent probes, radio-iodinated probes, irreversibly binding agonists, nanoparticle conjugates, bivalent drugs, and peptide conjugates was explored by Jacobson and coworkers (Jacobson, 2009). The carboxylate group was found empirically to be in a region of the binding site that was exposed to the extracellular medium and therefore could be coupled to large chemical moieties in these receptor probes without losing $A_{2A}AR$ affinity.

Various mono-substituted adenosine derivatives have been studied as $A_{2A}AR$ agonists, although many are not selective. NECA (5′-(N-ethylcarboxamido)adenosine) was the first highly potent $A_{2A}AR$ agonist reported in the 1970s (Prasad, Bariana, Fung, Savic, & Tietje, 1980). Its original patent claimed it as a rodent poison (Stein, Prasad, & Tietje, 1980). Metrifudil ($N^6$-(2-methylbenzyl)-adenosine), a pan-AR full agonist ($K_i$ 24–60 nM at four AR subtypes, Gao et al., 2003), was in early human trials for glomerulonephritis, but failed to progress (Jacobson et al., 2019).

**Table 1** Affinities of representative $A_{2A}$ agonists and antagonists at the four ARs. The structures are shown in Figs. 1 and 2. cLogP values were predicted using the StarDrop software package (Optibrium Inc., Cambridge, UK) (Segall, 2012). Most values (nM) are from radioligand binding assays at the human ARs, unless indicated.

| Compound | $K_i$ ($A_1AR$) | $K_i$ ($A_{2A}AR$) | $K_i$ ($A_{2B}AR$) | $K_i$ ($A_3AR$) | cLogP |
|---|---|---|---|---|---|
| **Agonists** | | | | | |
| BV.115959 | 150 | 980 | – | 6390 (rat), 28.6 | −0.985 |
| Sonedenoson | >10,000 | 490 | >10,000 | ND | 1.33 |
| Binodeneoson | 48,000 | 270 | 430,000 | 903 | 0.454 |
| CGS21680 | 289 | 27 | 361,000 | 67 | 0.770 |
| Regadenoson | >10,000 | 290 | >10,000 | >10,000 | −1.35 |
| NECA | 14 | 20 | 330 | 6.2 | −0.681 |
| HE-NECA | 60 | 6.4 | 6100 | 2.4 | 0.936 |
| YT-146 | – | 1.2 (EC$_{50}$) | – | – | 1.26 |
| PSB-0777 | 541 | 360 | >10,000 | >10,000 | −0.302 |
| DiPhEtA | 49.9 | 510 | 485 (EC$_{50}$) | 3.9 | 2.71 |
| DPMA (racemic) | 142 (rat), 168 | 4.4 (rat), 153 | – | 106 | 2.66 |
| LUF5834 | 2.6 | 28 | 12 (EC$_{50}$) | 538 | 1.81 |

| | | | | | |
|---|---|---|---|---|---|
| ATL-146e | 77 | 0.5 | >1000 | 45 | 1.15 |
| ATL-313 | 57 | 0.7 | >1000 | 250 | 0.417 |
| GW-328267X | 882 | 2.3 | 51 | 4.2 | 1.03 |
| UK-371104 | 100 | 20 | – | >1000 | 2.67 |
| UK-432097 | – | 4 | – | – | 2.04 |
| Compound 10l | 530 | 153 | – | 1070 | 3.26 |
| **Antagonists** | | | | | |
| Caffeine | 10,700 | 9560 | 10,400 | 133,006 | 0.0231 |
| CSC | >10,000 | 38 | 8200 | >10,000 | 2.24 |
| DMPX | 45,000 (rat) | 16,000 (rat) | 4130 | >10,000 (rat) | 0.629 |
| Istradefylline | 2830 | 36 | 1800 | >3000 | 2.16 |
| MSX-2 | 2500 | 5.4 | >10,000 | >10,000 | 1.12 |
| CGS15943 | 3.5 | 0.15 | 71 | 51 | 3.10 |
| Tozadenant | 1350 | 5.0 | 700 | 1570 | 2.47 |
| Vipadenant (V2006) | 68 | 1.3 | 63 | 1010 | 2.12 |

*(continued)*

**Table 1** Affinities of representative $A_{2A}$ agonists and antagonists at the four ARs. The structures are shown in Figs. 1 and 2. cLogP values were predicted using the StarDrop software package (Optibrium Inc., Cambridge, UK) (Segall, 2012). Most values (nM) are from radioligand binding assays at the human ARs, unless indicated. (cont'd)

| Compound | $K_i$ ($A_1AR$) | $K_i$ ($A_{2A}AR$) | $K_i$ ($A_{2B}AR$) | $K_i$ ($A_3AR$) | cLogP |
|---|---|---|---|---|---|
| ZM241385 | 255 | 0.8 | 50 | >10,000 | 2.13 |
| SCH58241 | 594 | 1.1 | >10,000 | >10,000 | 2.60 |
| SCH442416 | 1110 | 4 | >10,000 | >10,000 | 2.78 |
| Preladenant | 1470 | 1.1 | >1700 | >1000 | 2.41 |
| MNI-444 | – | 2.8 | – | – | 2.73 |
| ATL444 | 7.0 | 2.5 | 61.8 | >1000 | 1.16 |
| ANR94 | 2400 | 46 | >30,000 | 21,000 | 0.958 |
| TPP455 | 3.54 | 0.0058 | 36 | 313 | 3.14 |
| JNJ-40255293 | 48 | 6.5 | 230 | 9200 | 2.72 |
| JNJ-41501798 | 7800 | 11.5 | – | – | 5.14 |
| Active drug of LuAA47070 | 410 | 5.9 | 260 | >10,000 | 2.87 |
| Imaradenant (AZD4635) | 160 | 1.7 | 64 | >10,000 | 3.04 |
| ST1535 | 72 | 6.6 | 352 | >1000 | 1.46 |

| | | | | | |
|---|---|---|---|---|---|
| Taminadenant (PBF-509) | 2500 | 12 | 1000 | 5000 | 1.02 |
| Ciforadenant (CPI-444) | 192 | 3.54 | 1530 | 2460 | 2.21 |
| Inupadenant (EOS-850) | – | – | – | – | 2.66 |
| Etrumadenant (AB928) | 64 | 1.5 | 2.0 | 489 | 2.69 |
| KW-6356 | 100 | 0.12 | 32 | 420 | 2.69 |
| M1069 | 160 ($IC_{50}$) | 0.13 ($IC_{50}$) | 9 ($IC_{50}$) | 1100 ($IC_{50}$) | 2.03 |

Sonedenoson ((2$R$,3$R$,4$S$,5$R$)-2-(6-amino-2-(4-chlorophenethoxy)-9$H$-purin-9-yl)-5-(hydroxymethyl)tetrahydrofuran-3,4-diol) is an A$_{2A}$AR agonist that was tested in MPI (Jacobson et al., 2019). YT-146 (2-(oct-1-yn-1-yl)-adenosine). Following the early A$_{2A}$AR agonist SAR studies of Olsson and colleagues and the introduction of CGS21680 having an extended 2-aminoalkyl/aryl group at the C2 position, several other groups reported 2-alkynyl-adenosine derivatives as A$_{2A}$AR-selective agonists, including YT-146, as potent A$_{2A}$AR agonists (Ono, Matsuoka, Ohkubo, Kimura, & Nakanishi, 1998).

Binodenoson ((2$R$,3$R$,4$S$,5$R$)-2-[6-amino-2-[(2$E$)-2-(cyclohexylmethylidene)hydrazinyl]purin-9-yl]-5-(hydroxymethyl)oxolane-3,4-diol). Other 2-substituted adenosine derivatives were introduced in the early 1990s for A$_{2A}$AR activation, among them binodenoson. This A$_{2A}$AR agonist was administered in patients for MPI, and it did not induce bronchoconstriction (Murray et al., 2009). However, its development was discontinued (Jacobson et al., 2019).

DPMA ($N^6$-(2-(3,5-dimethoxyphenyl)-2-(2-methylphenyl)ethyl)adenosine). One adenosine derivative of adenosine modified only at the $N^6$ position was found to be a moderately rat (but not human) A$_{2A}$AR-selective agonist (Bridges et al., 1988; Gao et al., 2003), i.e. the first reported such compound. At the time, one of the envisioned applications of A$_{2A}$AR agonists was in the treatment of schizophrenia, based on the inverse relationship of CNS dopaminergic and adenosinergic signaling (Boison, Singer, Shen, Feldon, & Yee, 2012). However, this concept was not tested clinically.

Regadenoson (1-[6-amino-9-[(2$R$,3$R$,4$S$,5$R$)-3,4-dihydroxy-5-(hydroxymethyl)oxolan-2-yl]purin-2-yl]-$N$-methylpyrazole-4-carboxamide), an adenosine derivative substituted at the C2 and 5′ positions became a translational success story, with its FDA approval for MPI in 2008. It is short acting, and as such does not display long lasting cardiovascular side effects, and its effectiveness as a diagnostic agent was comparable to adenosine itself (Iskandrian et al., 2007). No other synthetic A$_{2A}$AR agonists have been approved for this indication.

ATL-146e (apadenoson, 4-{3-[6-amino-9-(5-ethylcarbamoyl-3,4-dihydroxy-tetrahydro-furan-2-yl)-9$H$-purin-2-yl]-prop-2-ynyl}-cyclohexanecarboxylic acid methyl ester) and ATL-313 (evodenoson, methyl 4-[3-[6-amino-9-[(2$R$,3$R$,4$S$,5$S$)-5-(cyclopropylcarbamoyl)-3,4-dihydroxyoxolan-2-yl]purin-2-yl]prop-2-ynyl]piperidine-1-carboxylate) from the Linden lab (Lappas, Sullivan, & Joel Linden, 2005) are both adenosine derivatives substituted at the C2 and 5′ positions that have considerable $A_{2A}AR$ agonist selectivity (Jacobson et al., 2019). However, a Phase 2 clinical trial associated with ATL-146e for treatment of sickle cell disease, an ischemic condition, failed to show statistical efficacy (Field et al., 2017). Both ATL146e and ATL313 were shown to inhibit inflammatory damage in transplanted islets, and therefore enhancing islet survival and functional engraftment (Chhabra, Wang, & Zeng, 2010). These results were suggestive of coadministered $A_{2A}AR$ agonists as a novel anti-inflammatory treatment during pancreatic islet transplantation.

UK-432,097 (6-(2,2-diphenylethylamino)-9-((2$R$,3$R$,4$S$,5$S$)-5-(ethyl-carbamoyl)-3,4-dihydroxytetrahydrofuran-2-yl)-$N$-(2-(3-(1-(pyridin-2-yl)piperidin-4-yl)ureido)ethyl)-9$H$-purine-2-carboxamide). This adenosine analog substituted at the N6, C2 and 5′ positions was tested clinically for the indication of COPD, but was found to be non-effective, mainly due to limited bioavailability when administered by inhalation as a dry powder (Jacobson et al., 2019; Mantell, Jones, & Trevethick, 2010).

LUF5834 (2-amino-4-(4-hydroxyphenyl)-6-[(1$H$-imidazol-2-ylmethyl)thio]-3,5-pyridinecarbonitrile) is a non-nucleoside mixed $A_1AR$ and $A_{2A}AR$ agonist (Beukers et al., 2004; Lane et al., 2012). In tritiated form, it was a suitable radioligand for the $A_1AR$, but not the $A_{2A}AR$ or $A_3AR$ (Lane et al., 2010). The X-ray structure of a close analog (LUF5833) in complex with the $A_{2A}AR$ was recently reported (Amelia et al., 2021). The possible trajectory of the ligand approach and mechanism of receptor activation were compared for non-nucleoside agonist LUF5833 and two nucleoside agonists by modeling (Bolcato, Pavan, Bassani, Sturlese, & Moro, 2022).

## 3. Positive allosteric enhancers (PAMs)

The concept of a positive allosteric enhancer (PAM) of the $A_{2A}AR$ is appealing, because it would amplify the effects of endogenous adenosine, in a temporally and spatially selective manner, to avoid the known agonist side effects. There have been several reports of PAMs of the $A_{2A}AR$ (Korkutata, Agrawal, & Lazarus, 2022), including $A_{2A}R$ PAM1 (Fig. 1), but additional pharmacological and structural characterization of their receptor interaction is needed. However, other AR subtypes have more established and chemically explored PAM chemotypes (Vincenzi et al., 2021). Although not acting at the $A_{2A}AR$, the structure activity relationship of prototypical human $A_3AR$ PAM LUF6000 has recently been expanded (Fallot et al., 2022).

## 4. $A_{2A}AR$ antagonists and inverse agonists

Numerous $A_{2A}AR$ antagonists have been reported (Baraldi et al., 2018). Until recently, adenosine derivatives that bound to the $A_{2A}AR$ acted as agonists. However, Shiriaeva et al. (2022) found a 4′-truncated nucleoside derivative LJ4517 that also contained a heteroaryl group (thien-2-yl) at the 8-position and that acted as an $A_{2A}AR$ antagonist. An X-ray structure accompanied the characterization of this atypical antagonist, showing that the truncated ribose moiety was forced into an alternative position by steric crowding in the receptor binding site, thus precluding its accessing the hydrophilic residues needed for receptor activation.

The difference between antagonists and inverse agonists has been studied for various antagonists, but there are contradictions in the literature. Caffeine was revealed to be an inverse agonist (Fernández-Dueñas et al., 2014). However, Bennett et al. (2013) classified theophylline, caffeine, and istradefylline as neutral antagonists because they bind equipotently at active and inactive states. They reported that xanthine amine congener (structure not shown), ZM241385, SCH58261, and preladenant are inverse agonists and bind preferentially to the inactive state.

Caffeine. Caffeine is a prototypical AR antagonist and the most widely consumed drug (Jacobson, Gao, Matricon, Eddy, & Carlsson, 2020). It has significant activity in vivo to block the $A_{2A}AR$ but not selectively. Nevertheless, its $A_{2A}AR$ antagonist effects are being tested clinically for treatment of AD.

DPMX (3,7-dimethyl-1-propargylxanthine). This caffeine derivative from the Daly laboratory at NIH was the first example of a xanthine $A_{2A}AR$ antagonist that showed slight selectivity for that subtype (Daly, 2007). It is predicted to readily enter the CNS and has been used in numerous in vivo studies. However, its only 3-fold $A_{2A}AR$ selectivity has to be emphasized in interpreting experimental results, and its use as a pharmacological tool has been supplanted by more potent and selective antagonists.

CGS15943 (N-9-chloro-2-(2-furanyl)[1,2,4]triazolo[1,5-c]quinazolin-5-amine). This compound was the first example in 1988 of a non-xanthine $A_{2A}AR$ antagonist that showed slight selectivity for that subtype (Francis et al., 1988). The affinity in radioligand binding at the rat $A_1AR$ was 6.4-fold weaker than at the rat $A_{2A}AR$ ($K_i$ 3.3 nM), which was ~500-fold more potent that theophylline in blocking adenosine-stimulated cAMP in guinea pig synaptoneurosomes. It was identified in a broad screen of different heterocyclic derivatives for this inhibitory activity and was considered for potential anxiomodulator activity. The heterocyclic scaffold was distantly related to the xanthine series, except that a N atom of the pyrimidine moiety was moved, and the remaining 6-membered ring was fused to a chloro-substituted phenyl ring. It was the first example of substitution of the heterocyclic antagonist with a fur-2-yl group, which was essential for high $A_{2A}AR$ affinity and later appeared in other $A_{2A}AR$ antagonists (e.g. ZM241385, vipadenant, SCH58261, SCH44226, TPP455, preladenant, inupadenant and ciforadenant).

ZM241385 (4-[2-[7-amino-2-(2-furyl)-1,2,4-triazolo[1,5-*a*][1,3,5]triazin-5-yl-amino]ethyl]phenol) was introduced as a $A_{2A}AR$ antagonist with an alternative heterocyclic core in 1995 (de Lera Ruiz et al., 2014; Poucher et al., 1995). It has become a widely used $A_{2A}AR$ antagonist for both pharmacological and structural studies. However, it should be noted that it has considerable affinity at the $A_{2A}AR$. When receptor bound (Liu et al., 2012), its furyl group points downward toward the hydrophilic region of the $A_{2A}AR$ that is the binding site of the ribose moiety of $A_{2A}AR$ agonists. Its phenol side chain points toward the extracellular region of the receptor (Salmaso et al., 2021). In fact, in different X-ray structure in complex with $A_{2A}AR$, this phenolic chain attains different conformations. The phenolic side chain can be radioiodinated to generate a high-affinity $A_{2A}AR$ radioligand (Palmer, Poucher, Jacobson, & Stiles, 1996).

Istradefylline (8-[(*E*)-2-(3,4-dimethoxyphenyl)ethenyl]-1,3-diethyl-7-methylpurine-2,6-dione, KW-6002). The class of 8-styrylxanthines was first reported by Suzuki and colleagues as $A_{2A}AR$ antagonists (Kanda et al., 1994). Over 52 weeks of once-daily oral administration in levodopa-treated PD patients it was well tolerated and remained effective in reducing off time in PD (Kondo & Mizuno, 2015). Its recent approval by the FDA followed clinical studies in > 4000 PD patients (Chen & Cunha, 2020). It will be interesting to see future data indicating if there is a slower depression of disease progression in PD patients or improvement of mood and memory in aging patients with istradefylline or other $A_{2A}AR$ antagonists (Merighi et al. 2022; Merighi et al., 2021). Istradefylline will begin a clinical trial in combination with low oxygen therapy, aiming to improve functional mobility after spinal cord injury. Another $A_{2A}AR$-selective antagonist from the same source (Kyowa Hakko Kirin), KW-6356 (Ohno et al., 2023a,b; Chen & Cunha, 2020), was in a Phase 2 trial for PD, and was well tolerated. However, it is no longer in development for PD but might yet prove useful for cancer.

CSC (8-(3-chlorostyryl)caffeine). CSC is an early reported 8-styrylxanthine that displays relatively high affinity and selectivity as an $A_{2A}AR$ antagonist (Jacobson et al., 1993). However, it was later shown to inhibit MAO-B ($K_i$ 80.6 nM) as well (Vlok, Malan, Castagnoli Jr, Bergh, & Petzer, 2006). It acts in the CNS following peripheral administration (Aguiar et al., 2008) to ameliorate symptoms in 6-OHDA-lesioned rats, leading to the conclusion that CSC acts in the brain as both $A_{2A}AR$ antagonist and MAO-B inhibitor, both being beneficial activities in this neurodegenerative model.

MSX-3 (disodium 3-[8-[(E)-2-(3-methoxyphenyl)ethenyl]-7-methyl-2,6-dioxo-1-prop-2-ynylpurin-3-yl]propyl phosphate) was introduced by the Müller lab as water-soluble prodrugs in the 8-styrylxanthine series. The enzymatic hydrolysis of a phosphoester group in vivo generates the active drug MSX-2 (Mott et al., 2009).

ATL-444 ((1S,3R)-1-((6-amino-9-(prop-2-yn-1-yl)-4,5-dihydro-9H-purin-2-yl)ethynyl)-3-methylcyclohexan-1-ol), a mixed $A_1AR$ and $A_{2A}AR$ antagonist, was studied in relation to dopaminergic and glutamatergic psychostimulant action in rats and increased cocaine self-administration (Doyle, Breslin, Rieger, Beauglehole, & Lynch, 2012).

A series of SCH antagonists of the 7H-pyrazolo[4,3-e][1,2,4]triazolo[1,5-c] pyrimidine-5-amine family have been reported as pharmacological, imaging and therapeutic agents. SCH-58261 was an early $A_{2A}AR$ antagonist that proved useful in numerous in vitro and in vivo studies, including as a tritiated radioligand (Pinna, Fenu, & Morelli, 2001). SCH-412,348 showed anti-PD activity in rodent and primate models (Hodgson et al., 2009; Smith, Browne, & Jayaraman, 2014), but did not progress to clinical trials. SCH-442,416 (2-(2-furyl)-7-[3-(4-methoxyphenyl)propyl]-7H-pyrazolo[4,3-e][1,2,4]triazolo[1,5-c] pyrimidin-5-amine) is a selective antagonist that has been labeled with $^{11}C$ for PET imaging (Todde et al., 2000). Preladenant (SCH-420,814, 2-(2-furanyl)-7-[2-[4-[4-(2-methoxyethoxy)phenyl]-1-piperazinyl]ethyl]7H-pyrazolo[4,3-e] [1,2,4]triazolo[1,5-c]pyrimidine-5-amine) (Hodgson et al., 2009) was in multiple clinical trials for PD, but its clinical efficacy was variable and development discontinued. However, it is now being repurposed for cancer treatment as (MK-3814A) alone and in combination with pembrolizumab. MNI-444, is a derivative of SCH-420,814 that when labeled with $^{18}F$, is a useful as in vivo imaging agent showing high labeling of the $A_{2A}AR$ in the monkey putamen (Vala et al., 2016).

ST1535 (2-butyl-9-methyl-8-(2H-1,2,3-triazol 2-yl)-9Hpurin-6-xylamine) is a selective $A_{2A}AR$ antagonist that binds with higher affinity in the striatum than in the hippocampus, suggestive of the formation of AR-heterodimers in the latter brain region (Riccioni, Leonardi, & Borsini, 2010).

Vipadenant (3-(4-amino-3-methylbenzyl)-7-(furan-2-yl)-3$H$-[1,2,3]triazolo[4,5-$d$]pyrimidin-5-amine, BIIB-014) is a moderately $A_{2A}$AR-selective antagonist (52-fold) that was shown to be active in vivo in a PD model without inducing dyskinesias in either naïve or L-DOPA pre-sensitized rats, but did not block dyskinesias resulting from L-DOPA (Jones, Bleickardt, Mullins, Parker, & Hodgson, 2013).

Tozadenant (4-hydroxy-$N$-(4-methoxy-7-morpholinobenzo[$d$]thiazol-2-yl)-4-methylpiperidine-1-carboxamide, SYN115) was in a Phase 2 clinical trial that was abruptly ceased due to patient deaths. Out of ~890 patients receiving tozadenant, there were seven sepsis cases (including five fatalities due to agranulocytosis) (LeWitt, Aradi, Hauser, & Rascol, 2020).

ANR94 (Spinaci et al., 2022) is an $A_{2A}$AR-selective 8-ethoxy-adenine derivative that crosses the blood-brain barrier and reduces the motor deficits and tremors in a PD model without inducing long-term plastic changes in the striatum.

TPP455 (2-(2-furanyl)-$N^5$-(2-methoxybenzyl)[1,3]thiazolo[5,4-$d$]pyrimidine-5,7-diamine) demonstrated an unusually high $A_{2A}$AR affinity in the pM range (Gessi et al., 2017). It reduced acute pain in the mouse and inhibited proliferation of cancer cells.

Lu AA47070 (phosphoric acid mono-{2-[($E/Z$)-4-(3,3-dimethyl-butyrylamino)-3,5-difluoro-benzoylimino]-thiazol-3-ylmethyl} ester) is a water-soluble (8.6 mg/mL) prodrug of a potent and selective $A_{2A}$AR antagonist (compound 28) (Sams et al., 2011). It was dosed orally in solution to show efficacy in reversing haloperidol-induced hypolocomotion in the mouse. [$^3$H]SCH-58261 ex-vivo binding in the brain was used to demonstrate target $A_{2A}$AR engagement at the efficacious dose.

Imaradenant (6-(2-chloro-6-methylpyridin-4-yl)-5-(4-fluorophenyl)-1,2,4-triazin-3-amine, AZD4635) was designed through a combination of X-ray crystallography and biophysical mapping (Jazayeri et al., 2017). It was in a Phase 2 clinical trial for progressive metastatic castrate-resistant prostate cancer (in combination with durvalumab and in combination with cabazitaxel and durvaluma) and Phase 1 trial in Japan for advanced solid malignancies. Its target is to inhibit the immunosuppressive $A_{2A}$AR on T-lymphocytes in the tumor microenvironment.

Inupadenant (7-amino-10-[2-[4-[2,4-difluoro-5-[2-[($S$)-methylsulfinyl]ethoxy]phenyl]piperazin-1-yl]ethyl]-4-(furan-2-yl)-12-thia-3,5,6,8,10-pentazatricyclo[7.3.0.02,6]dodeca-1(9),2,4,7-tetraen-11-one, EOS-850) is a highly selective, non-brain-penetrant $A_{2A}AR$ antagonist that is being developed for the treatment of advanced lung cancer, head and neck cancer and melanoma (Buisseret et al., 2021).

Ciforadenant (($R$)-7-(5-methylfuran-2-yl)-3-((6-(((tetrahydrofuran-3-yl)oxy)methyl)pyridin-2-yl)methyl)-3$H$-[1,2,3]triazolo[4,5-$d$]pyrimidin-5-amine, CPI-444) (Zhang, Yan, Duan, Wüthrich, & Cheng, 2020) completed Phase 1B clinical trials for renal cell cancer, as a monotherapy or in combination with anti-CTLA-4 and anti-PD1 treatment.

Taminadenant (5-bromo-2,6-di(1$H$-pyrazol-1-yl)pyrimidin-4-amine, PBF-509) was well tolerated in a Phase 1 clinical trial for advanced non-small cell lung cancer, either as a monotherapy or in combination with spartalizumab, with only with 20% and 12% discontinuation, respectively, due to side events (Chiappori et al., 2022).

Several AR antagonists of the 2-amino-9-(4-methyl-2-phenylthiazol-5-yl)-4-phenyl-5$H$-indeno[1,2-$d$]pyrimidin-5-one family have been developed. JNJ-40255293, a mixed $A_1AR$ and $A_{2A}AR$ antagonist that shows efficacy in PD animal models (Atack et al., 2014). In 6-hydroxydopamine-lesioned rat, there was agreement of receptor occupancy (60–90% required) and potentiation of L-DOPA effects. JNJ-41501798 (695-fold selective for binding to the $A_{2A}AR$ compared to $A_1AR$), as well as JNJ-40255293, reduced meningeal arterial dilation induced by an $A_{2A}AR$ agonist (CGS21680) or by electrical stimulation, indicative of use in treating migraine, without affecting neurogenic vasodilation (Haanes et al., 2018).

Among the newest AR antagonists to enter human testing, ILB2109 (Innolake Biopharma), an $A_{2A}AR$ antagonist, is in a Phase 1 clinical trial in China in patients with locally advanced or metastatic solid malignancies. Etrumadenant (AB928, Arcus Biosciences) is a mixed $A_{2A}AR/A_{2B}AR$ antagonist in Phase 2 clinical trials for colorectal cancer that increased tumor cell cytolysis through CAR T cell secretion and proliferation (Seifert, Benmebarek, & Briukhovetska, 2022). The structure of its $A_{2A}AR$-bound complex was reported (Claff et al., 2023). However, its further clinical development was recently terminated. A Phase 2B PD clinical trial of a highly potent and selective $A_{2A}AR$ inverse agonist KW-6356 (sipagladenant) was also halted. Nevertheless, a recent X-ray structure of its receptor complex highlighted the stabilizing interactions that contribute to its long receptor residence time (Ohno et al., 2023b). Its anti-parkinsonian activity in combination with L-DOPA has been reported (Ohno et al., 2023a).

$A_{2A}AR/A_{2B}AR$ antagonist M1069 (Domain Therapeutics licensed to Merck-EMD Serono, administered as the fumarate salt) is in a Phase 1 clinical trial as an oral drug for metastatic or locally advanced unresectable solid tumors. Other adenosinergic signaling blockers, including CD73 inhibitors, that are in clinical trials for various cancers were recently reviewed (Wang, Du, & Chen, 2022).

## 5. Negative allosteric enhancers (NAMs)

Sodium ion act as a negative allosteric enhancer (NAM) of the $A_{2A}AR$ (Liu et al., 2012). A mass spectroscopy approach was used to identify an apparent NAM that bound to the $A_{2A}AR$ sodium site (Lu, Liu, & Yang, 2021). Cholesterol has weak negative effect on $A_{2A}AR$ activation, but this is now ascribed to indirect membrane effects (Huang et al., 2022). Other potential $A_{2A}AR$ NAMs, such as amiloride analogs, are described in Korkutata et al. (2022).

## 6. $A_{2A}AR$ partial agonists

8-Aminoalkyl-substituted adenosine derivatives that act as $A_{2A}AR$ partial agonists were reported (van Tilburg, J.von Frijtag Drabbe Künzel, de Groote, & IJzerman, 2002). More recently, a series of 7-(prolinol-N-yl)-2-phenylaminothiazolo[5,4-d]pyrimidines, including compound 10l as were reported to act as non-nucleoside $A_{2A}AR$ partial agonists (Bharate et al., 2016).

## 7. Conclusions

Despite the setbacks of many unsuccessful clinical trials, $A_{2A}AR$ antagonists, agonists and partial agonists have promise as future therapeutic agents. Some $A_{2A}AR$ agonists have a liability of cardiovascular side effects, but there are encouraging results for their use and use of PAMs in treating inflammatory and autoimmune conditions. $A_{2A}AR$ antagonists have been avidly pursued by the pharmaceutical industry for treatment of neurodegenerative conditions and for cancer immunotherapy. The use of receptor structures in the rational design of new ligands is increasingly popular and productive.

# References

Aguiar, L. M. V., Macêdo, D. S., Vasconcelos, S. M. M., Oliveira, A. A., de Sousa, F. C. F., & Viana, G. S. B. (2008). CSC, an adenosine $A_{2A}$ receptor antagonist and MAO B inhibitor, reverses behavior, monoamine neurotransmission, and amino acid alterations in the 6-OHDA-lesioned rats. *Brain Research, 1191*, 192–199. https://doi.org/10.1016/j.brainres.2007.11.051.

Amelia, T., van Veldhoven, J. P. D., Falsini, M., Liu, R., Heitman, L. H., van Westen, G. J. P., ... IJzerman, A. P. (2021). Crystal structure and subsequent ligand design of a nonriboside partial agonist bound to the adenosine $A_{2A}$ receptor. *Journal of Medicinal Chemistry, 64*(7), 3827–3842. https://doi.org/10.1021/acs.jmedchem.0c01856.

Atack, J. R., Shook, B. C., Rassnick, S., Jackson, P. F., Rhodes, K., Drinkenburg, W. H., ... Megens, A. A. (2014). JNJ-40255293, a novel adenosine A2A/A1 antagonist with efficacy in preclinical models of Parkinson's disease. *ACS Chemical Neuroscience, 5*(10), 1005–1019. https://doi.org/10.1021/cn5001606.

Baraldi, S., Baraldi, P. G., Oliva, P., Toti, K. S., Ciancetta, A., Jacobson K. A. (2018). Chapter 5. A2A adenosine receptor: Structures, modeling and medicinal chemistry. In K. Varani (Vol. Ed.), *The Adenosine Receptors, The Receptors* (pp. 91–136). Springer. https://doi.org/10.1007/978-3-319-90808-3_5.

Bennett, K. A., Tehan, B., Lebon, G., Tate, C. G., Weir, M., Marshall, F. H., & Langmead, C. J. (2013). Isolated $A_{2A}$-R conformation pharmacology defines efficacy. *Molecular Pharmacology, 83*(5), 949–958. https://doi.org/10.1124/mol.112.084509.

Belardinelli, L., Linden, J., & Berne, R. M. (1989). The cardiac effects of adenosine. *Progress in Cardiovascular Diseases, 32*(1), 73–97. https://doi.org/10.1016/0033-0620(89)90015-7.

Beukers, M. W., Chang, L. C., von Frijtag Drabbe Künzel, J. K., Mulder-Krieger, T., Spanjersberg, R. F., Brussee, J., & IJzerman, A. P. (2004). New, non-adenosine, high-potency agonists for the human adenosine $A_{2B}$ receptor with an improved selectivity profile compared to the reference agonist N-ethylcarboxamidoadenosine. *Journal of Medicinal Chemistry, 47*(15), 3707–3709. https://doi.org/10.1021/jm049947s.

Bharate, S. B., Singh, B., Kachler, S., Oliveira, A., Kumar, V., Bharate, S. S., ... Gutiérrez de Terán, H. (2016). Discovery of 7-(prolinol-N-yl)-2-phenylamino-thiazolo[5,4-d] pyrimidines as novel non-nucleoside partial agonists for the A2A adenosine receptor: Prediction from molecular modeling. *Journal of Medicinal Chemistry, 59*(12), 5922–5928. https://doi.org/10.1021/acs.jmedchem.6b00552.

Boison, D., Singer, P., Shen, H. Y., Feldon, J., & Yee, B. K. (2012). Adenosine hypothesis of schizophrenia—Opportunities for pharmacotherapy. *Neuropharmacology, 62*(3), 1527–1543. https://doi.org/10.1016/j.neuropharm.2011.01.048.

Bolcato, G., Pavan, M., Bassani, D., Sturlese, M., & Moro, S. (2022). Ribose and non-ribose A2A adenosine receptor agonists: Do they share the same receptor recognition mechanism? *Biomedicines, 10*(2), 515. https://doi.org/10.3390/biomedicines10020515.

Borea, P. A., Gessi, S., Merighi, S., Vincenzi, F., & Varani, K. (2018). Pharmacology of adenosine receptors: The state of the art. *Physiological Reviews, 98*, 1591–1625.

Bridges, A. J., Bruns, R. F., Ortwine, D. F., Priebe, S. R., Szotek, D. L., & Trivedi, B. K. (1988). $N^6$-[2-(3,5-dimethoxyphenyl)-2-(2-methylphenyl)ethyl]adenosine and its uronamide derivatives. Novel adenosine agonists with both high affinity and high selectivity for the adenosine $A_2$ receptor. *Journal of Medicinal Chemistry, 31*(7), 1282–1285.

Bruns, R. F., Daly, J. W., & Snyder, S. H. (1980). Adenosine receptors in brain membranes: Binding of $N^6$-cyclohexyl[$^3$H]adenosine and 1,3-diethyl-8-[$^3$H]phenylxanthine. *Proceedings of the National Academy of Sciences of the United States of America, 77*(9), 5547–5551. https://doi.org/10.1073/pnas.77.9.5547.

Buisseret, L., Rottey, S., De Bono, J. S., Migeotte, A., Delafontaine, B., Manickavasagar, T., ... Gangolli, E. A. (2021). Phase 1 trial of the adenosine A2A receptor antagonist inupadenant (EOS-850): Update on tolerability, and antitumor activity potentially associated with the expression of the $A_{2A}$ receptor within the tumor. *Journal of Clinical Oncology: Official Journal of the American Society of Clinical Oncology, 39*, 2562–2562.

Catarzi, D., Varano, F., Vigiani, E., et al. (2022). 4-Heteroaryl substituted amino-3,5-dicyanopyridines as new adenosine receptor ligands: Novel insights on structure-activity relationships and perspectives. *Pharmaceuticals (Basel), 15*(4), 478. https://doi.org/10.3390/ph15040478.

Cekic, C., & Linden, J. (2016). Purinergic regulation of the immune system. *Nature Reviews. Immunology, 16*(3), 177–192. https://doi.org/10.1038/nri.2016.4.

Chen, J. F., & Cunha, R. A. (2020). The belated US FDA approval of the adenosine $A_{2A}$ receptor antagonist istradefylline for treatment of Parkinson's disease. *Purinergic Signalling, 16*, 167–174.

Chen, Y., Zhang, J., Weng, Y., et al. (2022). Cryo-EM structure of the human adenosine $A_{2B}$ receptor–$G_s$ signaling complex. *Science Advances, 8*(51), eadd3709. https://doi.org/10.1126/sciadv.add3709.

Chhabra, P., Wang, K., Zeng, Q., et al. (2010). Adenosine $A_{2A}$ agonist administration improves islet transplant outcome: Evidence for the role of innate immunity in islet graft rejection. *Cell Transplantation, 19*(5), 597–612. https://doi.org/10.3727/096368910X491806.

Chiappori, A. A., Creelan, B., Tanvetyanon, T., Gray, J. E., Haura, E. B., Thapa, R., ... Antonia, S. (2022). Phase I study of taminadenant (PBF509/NIR178), an adenosine 2A receptor antagonist, with or without spartalizumab (PDR001), in patients with advanced non-small cell lung cancer. *Clinical Cancer Research: An Official Journal of the American Association for Cancer Research, 28*(11), 2313–2320. https://doi.org/10.1158/1078-0432.CCR-21-2742.

Claff, T., Schlegel, J. G., Voss, J. H., Vaaßen, V. J., Weiße, R. H., Cheng, R. K. Y., ... Müller, C. E. (2023). Crystal structure of adenosine $A_{2A}$ receptor in complex with clinical candidate Etrumadenant reveals unprecedented antagonist interaction. *Commun Chem, 6*, 106. https://doi.org/10.1038/s42004-023-00894-6.

Daly, J. W. (2007). Caffeine analogs: Biomedical impact. *Cellular and Molecular Life Sciences: CMLS, 64*(16), 2153–2169. https://doi.org/10.1007/s00018-007-7051-9.

de Lera Ruiz, M., Lim, Y. H., & Zheng, J. (2014). Adenosine $A_{2A}$ receptor as a drug discovery target. *Journal of Medicinal Chemistry, 57*, 3623–3650. https://doi.org/10.1021/jm4011669.

Doré, A. S., Robertson, N., Errey, J. C., Ng, I., Hollenstein, K., Tehan, B., ... Tate, C. G. (2011). Structure of the adenosine $A_{2A}$ receptor in complex with ZM241385 and the xanthines XAC and caffeine. *Structure (London, England: 1993), 19*(9), 1283–1293.

Doyle, S. E., Breslin, F. J., Rieger, J. M., Beauglehole, A., & Lynch, W. J. (2012). Time and sex-dependent effects of an adenosine $A_{2A}/A_1$ receptor antagonist on motivation to self-administer cocaine in rats. *Pharmacology, Biochemistry, and Behavior, 102*(2), 257–263. https://doi.org/10.1016/j.pbb.2012.05.001.

Eddy, M. T., Gao, Z. G., Mannes, P., Patel, N., Jacobson, K. A., Katritch, V., ... Wüthrich, K. (2018). Extrinsic tryptophans as NMR probes of allosteric coupling in membrane proteins: Application to the $A_{2A}$ adenosine receptor. *Journal of the American Chemical Society, 140*, 8228–8235. https://doi.org/10.1021/jacs.8b03805.

Fallot, L. B., Suresh, R. R., Fisher, C. L., Salmaso, V., O'Connor, R. D., Kaufman, N., ... Jacobson, K. A. (2022). Structure activity studies of 1*H*-imidazo[4,5-c]quinolin-4-amine derivatives as $A_3$ adenosine receptor positive allosteric modulators. *J. Med. Chem. 65*(22), 15238–15262. https://doi.org/10.1021/acs.jmedchem.2c01170.

Fernández-Dueñas, V., Gómez-Soler, M., López-Cano, M., Taura, J., Ledent, C., Watanabe, M., ... Ciruela, F. (2014). Uncovering caffeine's adenosine $A_{2A}$ receptor inverse agonism in experimental parkinsonism. *ACS Chemical Biology, 9*, 2496–2501. https://doi.org/10.1021/cb5005383.

Ferré, S., Díaz-Ríos, M., Salamone, J. D., & Prediger, R. D. (2018). New developments on the adenosine mechanisms of the central effects of caffeine and their implications for neuropsychiatric disorders. *Journal of Caffeine and Adenosine Research, 8*(4), 121–131. https://doi.org/10.1089/caff.2018.0017.

Field, J. J., Majerus, E., Gordeuk, V. R., Gowhari, M., Hoppe, C., Heeney, M. M., et al. (2017). Randomized phase 2 trial of regadenoson for treatment of acute vaso-occlusive crises in sickle cell disease. *Blood Advances, 1*, 1645–1649. https://doi.org/10.1182/bloodadvances.2017009613.

Fozard, J. R. (2003). The case for a role for adenosine in asthma: Almost convincing? *Current Opinion in Pharmacology, 3*(3), 264–269. https://doi.org/10.1016/S1471-4892(03)00039-0.

Francis, J. E., Cash, W. D., Psychoyos, S., Ghai, G., Wenk, P., Friedmann, R. C., ... Furness, P. (1988). Structure-activity profile of a series of novel triazoloquinazoline adenosine antagonists. *Journal of Medicinal Chemistry, 31*, 1014–1020. https://doi.org/10.1021/jm00400a022.

Fredholm, B. B. (1985). On the mechanism of action of theophylline and caffeine. *Acta Medica Scandinavica, 217*(2), 149–153. https://doi.org/10.1111/j.0954-6820.1985.tb01650.x.

Gao, Z., Li, Z., Baker, S. P., Lasley, R. D., Meyer, S., Elzein, E., ... Belardinelli, L. (2001). Novel short-acting $A_{2A}$ adenosine receptor agonists for coronary vasodilation: Inverse relationship between affinity and duration of action of $A_{2A}$ agonists. *The Journal of Pharmacology and Experimental Therapeutics, 298*(1), 209–218.

Gao, Z. G., Blaustein, J., Gross, A. S., Melman, N., & Jacobson, K. A. (2003). $N^6$-Substituted adenosine derivatives: Selectivity, efficacy, and species differences at $A_3$ adenosine receptors. *Biochemical Pharmacology, 65*, 1675–1684.

García-Nafría, J., Lee, Y., Bai, X., Carpenter, B., & Tate, C. G. (2018). Cryo-EM structure of the adenosine A2A receptor coupled to an engineered heterotrimeric G protein. *eLife, 7*, e35946. https://doi.org/10.7554/eLife.35946.

Gessi, S., Bencivenni, S., Battistello, E., Vincenzi, F., Colotta, V., Catarzi, D., ... Varani, K. (2017). Inhibition of $A_{2A}$ adenosine receptor signaling in cancer cells proliferation by the novel antagonist TP455. *Frontiers in Pharmacology, 8*, 888.

Gutiérrez-de-Terán, H., Sallander, J., & Sotelo, E. (2017). Structure-based rational design of adenosine receptor ligands. *Current Topics in Medicinal Chemistry, 17*(1), 40–58. https://doi.org/10.2174/1568026616666160719164207.

Haanes, K. A., Labastida-Ramírez, A., Chan, K. Y., de Vries, R., Shook, B., Jackson, P., ... MaassenVanDenBrink, A. (2018). Characterization of the trigeminovascular actions of several adenosine $A_{2A}$ receptor antagonists in an in vivo rat model of migraine. *The Journal of Headache and Pain, 19*(1), 41. https://doi.org/10.1186/s10194-018-0867-x.

Haskó, G., Antonioli, L., & Cronstein, B. N. (2018). Adenosine metabolism, immunity and joint health. *Biochemical Pharmacology, 151*, 307–313. https://doi.org/10.1016/j.bcp.2018.02.002.

Hide, I., Padgett, W. L., Jacobson, K. A., & Daly, J. W. (1992). $A_{2a}$-Adenosine receptors from rat striatum and rat pheochromocytoma PC12 cells: Characterization with radioligand binding and by activation of adenylate cyclase. *Molecular Pharmacology, 41*, 352–359.

Hodgson, R. A., Bertorelli, R., Varty, G. B., Lachowicz, J. E., Forlani, A., Fredduzzi, S., et al. (2009). Characterization of the potent and highly selective $A_{2A}$ receptor antagonists preladenant and SCH 412348 [7-[2-[4-2,4-difluorophenyl]-1-piperazinyl]ethyl]-2-(2-furanyl)-7H-pyrazolo[4,3-e][1,2,4]triazolo[1,5-c]pyrimidin-5-amine] in rodent models of movement disorders and depression. *The Journal of Pharmacology and Experimental Therapeutics, 330*(1), 294–303. https://doi.org/10.1124/jpet.108.149617.

Huang, M., & Daly, J. W. (1974). Adenosine-elicited accumulation of cyclic AMP in brain slices: Potentiation by agents which inhibit uptake of adenosine. *Life Sciences, 14*(3), 489–503. https://doi.org/10.1016/0024-3205(74)90364-6.

Huang, S. K., Almurad, O., Pejana, R. J., Morrison, Z. A., Pandey, A., Picard, L. P., ... Prosser, R. S. (2022). Allosteric modulation of the adenosine $A_{2A}$ receptor by cholesterol. *eLife, 11*, e73901. https://doi.org/10.7554/eLife.73901.

Hutchison, A. J., Webb, R. L., Oei, H. H., Ghai, G. R., Zimmerman, M. B., & Williams, M. (1989). CGS 21680C, an $A_2$ selective adenosine receptor agonist with preferential hypotensive activity. *The Journal of Pharmacology and Experimental Therapeutics, 251*, 47–55.

IJzerman, A. P., Jacobson, K. A., Müller, C. E., Cronstein, B. N., & Cunha, R. A. (2022). International Union of Basic and Clinical Pharmacology. CXII: Adenosine receptors – A further update. *Pharmacological Reviews, 74*, 340–372 PMC8973513.

Iskandrian, A. E., Bateman, T. M., Belardinelli, L., Blackburn, B., Cerqueira, M. D., Hendel, R. C., ... Wang, W. ADVANCE MPI Investigators. (2007). Adenosine versus regadenoson comparative evaluation in myocardial perfusion imaging: Results of the ADVANCE phase 3 multicenter international trial. *Journal of Nuclear Cardiology: Official Publication of the American Society of Nuclear Cardiology, 14*(5), 645–658. https://doi.org/10.1016/j.nuclcard.2007.06.114.

Jaakola, V. P., Griffith, M. T., Hanson, M. A., Cherezov, V., Chien, E. Y., Lane, J. R., ... Stevens, R. C. (2008). The 2.6 angstrom crystal structure of a human $A_{2A}$ adenosine receptor bound to an antagonist. *Science (New York, N. Y.), 322*(5905), 1211–1217. https://doi.org/10.1126/science.1164772.

Jacobson, K. A., Gallo-Rodriguez, C., Melman, N., Fischer, B., Maillard, M., van Bergen, A., ... Karton, Y. (1993). Structure-activity relationships of 8-styrylxanthines as $A_2$-selective adenosine antagonists. *Journal of Medicinal Chemistry, 36*, 1333–1342. https://doi.org/10.1021/jm00062a005.

Jacobson, K. A., & Gao, Z. G. (2006). Adenosine receptors as therapeutic targets. *Nature Reviews. Drug Discovery, 5*, 247–264. https://doi.org/10.1038/nrd1983.

Jacobson, K. A. (2009). Functionalized congener approach to the design of ligands for G protein–coupled receptors (GPCRs). *Bioconjugate Chemistry, 20*, 1816–1835.

Jacobson, K. A., Tosh, D. K., Jain, S., & Gao, Z. G. (2019). Historical and current adenosine receptor agonists in preclinical and clinical development. *Frontiers in Cellular Neuroscience, 13*, 124. https://doi.org/10.3389/fncel.2019.00124 PMC6447611.

Jacobson, K. A., Gao, Z. G., Matricon, P., Eddy, M. T., & Carlsson, J. (2020). Adenosine $A_{2A}$ receptor antagonists: From caffeine to selective non-xanthines. *British Journal of Pharmacology*, 1–16. https://doi.org/10.1111/bph.15103.

Jarvis, M. F., Schulz, R., Hutchison, A. J., Do, U. H., Sills, M. A., & Williams, M. (1989). [$^3$H]CGS 21680, a selective A2 adenosine receptor agonist directly labels $A_2$ receptors in rat brain. *The Journal of Pharmacology and Experimental Therapeutics, 251*(3), 888–893.

Jazayeri, A., Andrews, S. P., & Marshall, F. H. (2017). Structurally enabled discovery of adenosine $A_{2A}$ receptor antagonists. *Chemical Reviews, 117*(1), 21–37. https://doi.org/10.1021/acs.chemrev.6b00119.

Jiang, Q., van Rhee, A. M., Kim, J., Yehle, S., Wess, J., & Jacobson, K. A. (1996). Hydrophilic side chains in the third and seventh transmembrane helical domains of human $A_{2a}$ adenosine receptors are required for ligand recognition. *Molecular Pharmacology, 50*, 512–521.

Jones, N., Bleickardt, C., Mullins, D., Parker, E., & Hodgson, R. (2013). $A_{2A}$ receptor antagonists do not induce dyskinesias in drug-naive or L-dopa sensitized rats. *Brain Research Bulletin, 98*, 163–169. https://doi.org/10.1016/j.brainresbull.2013.07.001.

Kanda, T., Shiozaki, S., Shimada, J., Suzuki, F., & Nakamura, J. (1994). A novel selective adenosine $A_{2A}$ antagonist with anticataleptic activity. *European Journal of Pharmacology, 256*, 263–268. https://doi.org/10.1002/ana.410430415.

Kim, J., Wess, J., van Rhee, A. M., Shöneberg, T., & Jacobson, K. A. (1995). Site-directed mutagenesis identifies residues involved in ligand recognition in the human $A_{2a}$ adenosine receptor. *The Journal of Biological Chemistry, 270*, 13987–13997. https://doi.org/10.1074/jbc.270.23.13987.

Kim, J., Jiang, Q., Glashofer, M., Yehle, S., Wess, J., & Jacobson, K. A. (1996). Glutamate residues in the second extracellular loop of the human $A_{2a}$ adenosine receptors are required for ligand recognition. *Molecular Pharmacology, 49*, 683–691.

Kondo, T., & Mizuno, Y. (2015). Japanese Istradefylline Study Group A long-term study of istradefylline safety and efficacy in patients with Parkinson disease. *Clinical Neuropharmacology, 38*, 41–46.

Korkutata, M., Agrawal, L., & Lazarus, M. (2022). Allosteric modulation of adenosine $A_{2A}$ receptors as a new therapeutic avenue. *International Journal of Molecular Sciences, 23*(4), 2101. https://doi.org/10.3390/ijms23042101.

Lane, J. R., Klaasse, E., Lin, J., van Bruchem, J., Beukers, M. W., & IJzerman, A. P. (2010). Characterization of [$^3$H]LUF5834: A novel non-ribose high-affinity agonist radioligand for the adenosine A1 receptor. *Biochemical Pharmacology, 80*(8), 1180–1189. https://doi.org/10.1016/j.bcp.2010.06.041.

Lane, J. R., Klein Herenbrink, C., van Westen, G. J., Spoorendonk, J. A., Hoffmann, C., & IJzerman, A. P. (2012). A novel nonribose agonist, LUF5834, engages residues that are distinct from those of adenosine-like ligands to activate the adenosine $A_{2a}$ receptor. *Molecular Pharmacology, 81*(3), 475–487.

Lappas, C. M., Sullivan, G. W., & Joel Linden, J. (2005). Adenosine $A_{2A}$ agonists in development for the treatment of inflammation. *Expert Opinion on Investigational Drugs, 14*(7), 797–806. https://doi.org/10.1517/13543784.14.7.797.

Le, F., Townsend-Nicholson, A., Baker, E., Sutherland, G. R., & Schofield, P. R. (1996). Characterization and chromosomal localization of the human A2a adenosine receptor gene: ADORA2A. *Biochemical and Biophysical Research Communications, 223*(2), 461–467. https://doi.org/10.1006/bbrc.1996.0916.

Lebon, G., Warne, T., Edwards, P. C., Bennett, K., Langmead, C. J., Leslie, A. G. W., & Tate, C. G. (2011). Agonist-bound adenosine $A_{2A}$ receptor structures reveal common features of GPCR activation. *Nature, 474*(7352), 521–525. https://doi.org/10.1038/nature10136.

Lebon, G., Edwards, P. C., Leslie, A. G. W., & Tate, C. G. (2015). X-ray structure of CGS2680-bound human $A_{2A}R$. *Molecular Pharmacology, 87*(6), 907–915. https://doi.org/10.1124/mol.114.097360.

LeWitt, P. A., Aradi, S. D., Hauser, R. A., & Rascol, O. (2020). The challenge of developing adenosine A2A antagonists for Parkinson disease: Istradefylline, preladenant, and tozadenant. *Parkinsonism & Related Disorders, 80*, S54–S63.

Linden, J. (2005). Adenosine in tissue protection and tissue regeneration. *Molecular Pharmacology, 67*, 1385–1387.

Liu, W., Chun, E., Thompson, A. A., Chubukov, P., Xu, F., Katritch, V., ... Cherezov, V. (2012). Structural basis for allosteric regulation of GPCRs by sodium ions. *Science (New York, N. Y.), 337*(6091), 232–236.

Lopes, C. R., Lourenço, V. S., Tomé, Â. R., Cunha, R. A., & Canas, P. M. (2021). Use of knockout mice to explore CNS effects of adenosine. *Biochemical Pharmacology, 187*, 114367. https://doi.org/10.1016/j.bcp.2020.114367.

Lu, Y., Liu, H., Yang, D., et al. (2021). Affinity mass spectrometry-based fragment screening identified a new negative allosteric modulator of the adenosine $A_{2A}$ receptor targeting the sodium ion pocket. *ACS Chemical Biology, 16*(6), 991–1002. https://doi.org/10.1021/acschembio.0c00899.

Maenhaut, C., Van Sande, J., Libert, F., Abramowicz, M., Parmentier, M., Vanderhaegen, J. J., ... Schiffmann, S. (1990). RDC8 encodes for an adenosine $A_2$ receptor with physiological constitutive activity. *Biochemical and Biophysical Research Communications, 173*, 1169–1178.

Mantell, S., Jones, R., & Trevethick, M. (2010). Design and application of locally delivered agonists of the adenosine A2A receptor. *Expert Review of Clinical Pharmacology, 3*, 55–72. https://doi.org/10.1586/ecp.09.57.

Merighi, S., Poloni, T. E., Pelloni, L., Pasquini, S., Varani, K., Vincenzi, F., ... Gessi, S. (2021). An open question: Is the $A_{2A}$ adenosine receptor a novel target for Alzheimer's disease treatment? *Frontiers in Pharmacology, 12*, 652455.

Merighi, S., Borea, P. A., Varani, K., Vincenzi, F., Travagli, A., Nigro, M., ... Gessi, S. (2022). Pathophysiological role and medicinal chemistry of $A_{2A}$ adenosine receptor antagonists in Alzheimer's disease. *Molecules (Basel, Switzerland), 27*(9), 2680. https://doi.org/10.3390/molecules27092680.

Mott, A. M., Nunes, E. J., Collins, L. E., Port, R. G., Sink, K. S., Hockemeyer, J., et al. (2009). The adenosine $A_{2A}$ antagonist MSX-3 reverses the effects of the dopamine antagonist haloperidol on effort-related decision making in a T-maze cost/benefit procedure. *Psychopharmacology, 204*(1), 103–112. https://doi.org/10.1007/s00213-008-1441-z.

Müller, C. E., & Jacobson, K. A. (2011). Recent developments in adenosine receptor ligands and their potential as novel drugs. *Biochimica et Biophysica Acta (BBA) - Biomembranes, 1808*, 1290–1308.

Murray, J. J., Weiler, J. M., Schwartz, L. B., Busse, W. W., Katial, R. K., Lockey, R. F., et al. (2009). Safety of binodenoson, a selective adenosine A2A receptor agonist vasodilator pharmacological stress agent, in healthy subjects with mild intermittent asthma. *Circulation: Cardiovascular Imaging, 2*, 492–498. https://doi.org/10.1161/CIRCIMAGING.108.817932.

Ohno, Y., Okita, E., Kawai-Uchida, M., Shoukei, Y., Soshiroda, K., Kanda, T., & Uchida, S. (2023a). The adenosine $A_{2A}$ receptor antagonist/inverse agonist, KW-6356 enhances the anti-parkinsonian activity of L-DOPA with a low risk of dyskinesia in MPTP-treated common marmosets. *J Pharmacol Sci, 152*(3), 193–199. https://doi.org/10.1016/j.jphs.2023.05.001.

Ohno, Y., Suzuki, M., Asada, H., Kanda, T., Saki, M., Miyagi, H., ... Uchida, S. (2023b). In vitro pharmacological profile of KW-6356, a novel adenosine $A_{2A}$ receptor antagonist/inverse agonist. *Molecular Pharmacology*. https://doi.org/10.1124/molpharm.122.000633.

Ohta, A., & Sitkovsky, M. (2001). Role of G-protein-coupled adenosine receptors in downregulation of inflammation and protection from tissue damage. *Nature, 414*(6866), 916–920. https://doi.org/10.1038/414916a.

Ono, T., Matsuoka, I., Ohkubo, S., Kimura, J., & Nakanishi, H. (1998). Effects of YT-146 [2-(1-octynyl) adenosine], an adenosine $A_{2A}$ receptot agonist, on cAMP production and noradrenaline release in PC12 cells. *Japanese Journal of Pharmacology, 78*(3), 269–278. https://doi.org/10.1254/jjp.78.269.

Palmer, T. M., Poucher, S. M., Jacobson, K. A., & Stiles, G. L. (1996). $^{125}$I-4-(2-[7-Amino-2-{furyl}{1,2,4}triazolo{2,3-a}{1,3,5}triazin-5-ylaminoethyl)phenol ($^{125}$I-ZM241385), a high affinity antagonist radioligand selective for the $A_{2a}$ adenosine receptor. *Molecular Pharmacology, 48*, 970–974.

Pinna, A., Fenu, S., & Morelli, M. (2001). Motor stimulant effects of the adenosine A2A receptor antagonist SCH 58261 do not develop tolerance after repeated treatments in 6-hydroxydopamine-lesioned rats. *Synapse (New York, N. Y.), 39*(3), 233–238. https://doi.org/10.1002/1098-2396(20010301)39:3<233::AID-SYN1004>3.0.CO;2-K

Pinna, A. (2009). Novel investigational adenosine $A_{2A}$ receptor antagonists for Parkinson's disease. *Expert Opinion on Investigational Drugs, 18*(11), 1619–1631. https://doi.org/10.1517/13543780903241615.

Poucher, S. M., Keddie, J. R., Singh, P., Stoggall, S. M., Caulkett, P. W. R., Jones, G., ... Collis, M. G. (1995). The in vitro pharmacology of ZM241385, a potent, non-xanthine, $A_{2a}$ selective adenosine receptor antagonist. *British Journal of Pharmacology, 115*, 1096–1102.

Prasad, R. N., Bariana, D. S., Fung, A., Savic, M., & Tietje, K. (1980). Modification of the 5′ position of purine nucleosides. 2. Synthesis and some cardiovascular properties of adenosine-5′-(N-substituted)carboxamides. *Journal of Medicinal Chemistry, 23*, 313–319.

Preti, et al. (2015). History and perspectives of $A_{2A}$ adenosine receptor antagonists as potential therapeutic agents. *Medicinal Research Reviews, 35*(4), 790–848.

Riccioni, T., Leonardi, F., & Borsini, F. (2010). Adenosine $A_{2A}$ receptor binding profile of two antagonists, ST1535 and KW6002: Consideration on the presence of atypical adenosine A2A binding sites. *Frontiers in Psychiatry, 1*, 22. https://doi.org/10.3389/fpsyt.2010.00022.

Rosentreter, U., Henning, R., Bauser, M., Krämer, T., Vaupel, A., Hübsch, W., ... Krahn T. Substituted 2-Thio-3,5-Dicyano-4-Aryl-6-Aminopyridines and the Use Thereof as Adenosine Receptor Ligands. WO2001025210.

Salmaso, V., Jain, S., Jacobson, K.A. (2021). Purinergic GPCR transmembrane residues involved in ligand recognition and dimerization. In A. Shukla, Ed., *Meth. Cell Biol, 166*, (pp. 133–159). Biomolecular Interactions. https://doi.org/10.1016/bs.mcb.2021.06.001.

Sams, A. G., Mikkelsen, G. K., Larsen, M., Langgård, M., Howells, M. E., Schrøder, T. J., ... Bang-Andersen, B. (2011). Discovery of phosphoric acid mono-{2-[(E/Z)-4-(3,3-dimethyl-butyrylamino)-3,5-difluoro-benzoylimino]-thiazol-3-ylmethyl} ester (Lu AA47070): a phosphonooxymethylene prodrug of a potent and selective hA(2A) receptor antagonist. *Journal of Medicinal Chemistry, 54*(3), 751–764. https://doi.org/10.1021/jm1008659.

Sattin, A., & Rall, T. W. (1970). The effect of adenosine and adenine nucleotides on the cyclic adenosine 3′, 5′-phosphate content of guinea pig cerebral cortex slices. *Molecular Pharmacology, 6*(1), 13–23.

Schiffmann, S. N., Fisone, G., Moresco, R., Cunha, R. A., & Ferré, S. (2007). Adenosine A2A receptors and basal ganglia physiology. *Progress in Neurobiology, 83*(5), 277–292. https://doi.org/10.1016/j.pneurobio.2007.05.001.

Schultze-Werninghaus, G., & Meier-Sydow, J. (1982). The clinical and pharmacological history of theophylline: first report on the bronchospasmolytic action in man by S. R. Hirsch in Frankfurt (Main) 1922. *Clinical Allergy, 12*(2), 211–215. https://doi.org/10.1111/j.1365-2222.1982.tb01641.x.

Segall, M. D. (2012). Multi-parameter optimization: Identifying high quality compounds with a balance of properties. *Current Pharmaceutical Design, 18*, 1292–1310.

Seifert, M., Benmebarek, M. R., Briukhovetska, D., et al. (2022). Impact of the selective $A2_AR$ and $A2_BR$ dual antagonist AB928/etrumadenant on CAR T cell function. *British Journal of Cancer, 127*, 2175–2185. https://doi.org/10.1038/s41416-022-02013-z.

Shiriaeva, A., Park, D. J., Kim, G., Lee, Y., Hou, X., Jarhad, D. B., ... Cherezov, V. (2022). GPCR agonist-to-antagonist conversion: Enabling the design of nucleoside functional switches for the $A_{2A}$ adenosine receptor. *Journal of Medicinal Chemistry, 65*(8), 6325–6337. https://doi.org/10.1021/acs.jmedchem.2c00462.

Singer, P., & Yee, B. K. (2023). The adenosine hypothesis of schizophrenia into its thirddecade: From neurochemical imbalance to early life etiological risks. *Front. Cell. Neurosci. 17*, 1120532. https://doi.org/10.3389/fncel.2023.1120532.

Smith, K. M., Browne, S. E., Jayaraman, S., et al. (2014). Effects of the selective adenosine A2A receptor antagonist, SCH 412348, on the parkinsonian phenotype of MitoPark mice. *European Journal of Pharmacology, 728*, 31–38. https://doi.org/10.1016/j.ejphar.2014.01.052.

Spinaci, A., Lambertucci, C., Buccioni, M., Dal Ben, D., Graiff, C., Barbalace, M. C., ... Marucci, G. (2022). $A_{2A}$ adenosine receptor antagonists: Are triazolotriazine and purine scaffolds interchangeable? *Molecules (Basel, Switzerland), 27*(8), 2386. https://doi.org/10.3390/molecules27082386.

Stein, H. H., Prasad, R. N., & Tietje, K. R. (1979). Adenosine-5′-carboxamides for controlling undesired animals. US 4,167, 565; (1980). *Chemical Abstracts, 92*, P76890.

Todde, S., Moresco, R. M., Simonelli, P., Baraldi, P. G., Cacciari, B., Spalluto, G., ... Fazio, F. (2000). Design, radiosynthesis, and biodistribution of a new potent and selective ligand for in vivo imaging of the adenosine $A_{2A}$ receptor system using positron emission tomography. *Journal of Medicinal Chemistry, 43*, 4359–4362. https://doi.org/10.1021/jm0009843.

Ueeda, M., Thompson, R. D., Arroyo, L. H., & Olsson, R. A. (1991). 2-Alkoxyadenosines: Potent and selective agonists at the coronary artery $A_2$ adenosine receptor. *Journal of Medicinal Chemistry, 34*, 1334–1339.

Vala, C., Morley, T. J., Zhang, X., Papin, C., Tavares, A. A., Lee, H. S., ... Alagille, D. (2016). Synthesis and in vivo evaluation of Fluorine-18 and Iodine-123 Pyrazolo[4,3-e]-1,2,4-triazolo[1,5-c]pyrimidine derivatives as PET and SPECT radiotracers for mapping A2A receptors. *ChemMedChem, 11*, 1936–1943.

van Tilburg, E. W., J.von Frijtag Drabbe Künzel, J., M., de Groote, M., & IJzerman, A. P. (2002). 2,5′-Disubstituted adenosine derivatives: Evaluation of selectivity and efficacy for the adenosine $A_1$, $A_{2A}$, and $A_3$ receptor. *Journal of Medicinal Chemistry, 45*, 420–429.

Vigano, S., Alatzoglou, D., Irving, M., et al. (2019). Targeting adenosine in cancer immunotherapy to enhance T-cell function. *Frontiers in Immunology, 10*, 925. https://doi.org/10.3389/fimmu.2019.00925.

Vincenzi, F., Pasquini, S., Battistello, E., Merighi, S., Gessi, S., Borea, P. A., & Varani, K. (2021). $A_1$ Adenosine receptor partial agonists and allosteric modulators: Advancing toward the clinic? *Front. Pharmacol. 11*, 625134. https://doi.org/10.3389/fphar.2020.625134.

Vlok, N., Malan, S. F., Castagnoli Jr, N., Bergh, J. J., & Petzer, J. P. (2006). Inhibition of monoamine oxidase B by analogues of the adenosine $A_{2A}$ receptor antagonist (*E*)-8-(3-chlorostyryl) caffeine (CSC). *Bioorganic & Medicinal Chemistry, 14*(10), 3512–3521.

Wang, J., Du, L., & Chen, X. (2022). Adenosine signaling: Optimal target for gastric cancer immunotherapy. *Frontiers in Immunology, 13*, 1027838. https://doi.org/10.3389/fimmu.2022.1027838.

Wolska, N., Boncler, M., Polak, D., Wzorek, J., Przygodzki, T., Gapinska, M., ... Rozalski, M. (2020). Adenosine receptor agonists exhibit anti-platelet effects and the potential to overcome resistance to $P2Y_{12}$ receptor antagonists. *Molecules (Basel, Switzerland), 25*(1), 130. https://doi.org/10.3390/molecules25010130.

Xiao, C., Liu, N., Jacobson, K. A., Gavrilova, O., & Reitman, M. L. (2019). Physiology and effects of nucleosides in mice lacking all four adenosine receptors. *PLoS Biology, 17*(3), e3000161. https://doi.org/10.1371/journal.pbio.3000161.

Zhang, J., Yan, W., Duan, W., Wüthrich, K., & Cheng, J. (2020). Tumor immunotherapy using $A_{2A}$ adenosine receptor antagonists. *Pharmaceuticals (Basel), 13*(9), 237. https://doi.org/10.3390/ph13090237.

Zhou, Q. Y., Li, C., Olah, M. E., Johnson, R. A., Stiles, G. L., & Civelli, O. (1992). Molecular cloning and characterization of an adenosine receptor: The $A_3$ adenosine receptor. *Proceedings of the National Academy of Sciences of the United States of America, 89*(16), 7432–7436. https://doi.org/10.1073/pnas.89.16.7432.

CHAPTER TWO

# Adenosine $A_{2A}$ receptor and glia

Zhihua Gao[a,b,c,*]

[a]Department of Neurobiology and Department of Neurology of Second Affiliated Hospital, Zhejiang University School of Medicine, Hangzhou, P.R. China
[b]Liangzhu Laboratory, Zhejiang University Medical Center, MOE Frontier Science Center for Brain Science and Brain-machine Integration, State Key Laboratory of Brain-machine Intelligence, Zhejiang University, West Wenyi Road, Hangzhou, P.R. China
[c]NHC and CAMS Key Laboratory of Medical Neurobiology, Zhejiang University, Hangzhou, P.R. China
*Corresponding author. e-mail address: zhihuagao@zju.edu.cn

## Contents

1. Preface  29
2. $A_{2A}R$ in astrocytes  30
3. Astrocyte $A_{2A}R$ and Alzheimer's disease  31
4. Astrocyte $A_{2A}R$ and PD  33
5. $A_{2A}R$ and microglia  34
6. Microglial $A_{2A}$ receptor and process retraction  35
7. Microglial $A_{2A}$ receptor and neuroinflammation  36
8. Microglial $A_{2A}R$ and AD  37
9. Microglial adenosine $A_{2A}R$ and PD  38
10. Microglial adenosine $A_{2A}$ receptor and stroke  39
11. Microglial $A_{2A}R$ in other diseases  41
12. Role of other microglial adenosine $A_1$ and $A_3$ receptors  42
13. $A_{2A}R$ and oligodendrocytes  43
14. Perspective  44
References  45

## Abstract

The adenosine $A_{2A}$ receptor ($A_{2A}R$) is abundantly expressed in the brain, including both neurons and glial cells. While the expression of $A_{2A}R$ is relative low in glia, its levels elevate robustly in astrocytes and microglia under pathological conditions. Elevated $A_{2A}R$ appears to play a detrimental role in a number of disease states, by promoting neuroinflammation and astrocytic reaction to contribute to the progression of neurodegenerative and psychiatric diseases.

## 1. Preface

The adenosine $A_{2A}$ receptor ($A_{2A}R$) is a G-protein-coupled receptor that binds to adenosine, an important intermediate in the pathway of energy metabolism. $A_{2A}R$ is widely expressed in various tissues, especially in the brain, where it plays a diverse role in regulating neurotransmission, synaptic plasticity, and neuroinflammation. The expression of $A_{2A}R$ in glial cells is relatively low under physiological conditions, compared to its expression in neurons. However, the levels of $A_{2A}R$ can be significantly elevated in astrocytes and microglia during brain injury and diseases. Elevated $A_{2A}R$ are actively involved in the regulation of astrocytic and microglial reactions, contributing to aggravated neuroinflammation and disease progression in various neurological and psychiatric disorders. Therefore, glial $A_{2A}R$ is a potential therapeutic target for ameliorating neuroinflammation and reducing disease pathology in the brain. In fact, the adenosine $A_{2A}R$ antagonist istradefylline has been approved for the treatment of Parkinson's disease (PD) by the US Food and Drug Administration (FDA) in 2019, marking an important step to foster therapeutic application of adenosine $A_{2A}R$ antagonists in the clinical treatment of PD and other brain diseases.

## 2. $A_{2A}R$ in astrocytes

Astrocytes play an important role in maintaining brain homeostasis and modulating synaptic transmission (Trujillo-Estrada et al., 2019). Astrocytes communicate with neurons and other glial cells through various signaling molecules, including ATP and adenosine. ATP is released by astrocytes in response to different stimuli, such as neuronal activity, inflammation or hypoxia, and degraded to adenosine by ectonucleotidases. Adenosine acts on four different subtypes of adenosine receptors ($A_1$, $A_{2A}$, $A_{2B}$ and $A_3$), which are widely expressed in the central nervous system (CNS). Among these receptors, adenosine $A_{2A}$ receptor ($A_{2A}R$) has attracted much attention for its involvement in several neurological disorders, such as Alzheimer's disease (AD), Parkinson's disease (PD), epilepsy and stroke. $A_{2A}R$ is a Gs-coupled receptor that activates adenylate cyclase and increases intracellular cAMP levels. $A_{2A}R$ is predominantly expressed in the striatum, where it regulates motor function and dopamine signaling. $A_{2A}R$ is also present in astrocytes to modulate the morphology and function of astrocyte including glutamate uptake, calcium signaling, cytokine production, and morphological transformation under different conditions.

## 3. Astrocyte $A_{2A}R$ and Alzheimer's disease

Alzheimer's disease (AD) is a progressive neurodegenerative disorder that affects millions of people worldwide. The main pathological hallmarks of AD are the accumulation of amyloid-beta (Aβ) plaques and neurofibrillary tangles (NFTs) in the brain, as well as synaptic loss and dysfunction. Although precise mechanisms underlying AD remain incompletely understood, growing evidence suggests that astrocytes are not only passive bystanders, but also active players in the pathogenesis of AD (Phatnani & Maniatis, 2015).

Astrocytes undergo morphological and functional changes in AD, which can be further classified into atrophic and reactive phases (Phatnani & Maniatis, 2015). Atrophy occurs at early stages of AD and involves a reduction of astrocytic processes and contacts with synapses, whereas reactivity occurs at later stages of AD and involves an increased astrocytic size and expression of glial fibrillary acidic protein (GFAP), a marker of astrocytic reaction (Preeti, Sood, & Fernandes, 2022; Qian et al., 2023). These changes affect the ability of astrocytes to support neuronal survival and function, as well as to regulate synaptic plasticity and transmission. Synaptic plasticity is essential for learning and memory formation, as well as for brain adaptation and repair under stress and disease conditions. Astrocytes can modulate synaptic plasticity by releasing various gliotransmitters. One of these gliotransmitters is ATP, which can be further degraded to adenosine. Adenosine can act on $A_{2A}R$ to regulate long-term potentiation (LTP) and long-term depression (LTD), associated with an increase or a decrease of synaptic efficacy. $A_{2A}R$ activation can enhance LTP by increasing cAMP levels and activating protein kinase A (PKA), which can phosphorylate AMPAR and increase their trafficking to the synaptic membrane. Conversely, $A_{2A}R$ activation can inhibit LTD by preventing NMDAR activation and calcium influx (Cieslak & Wojtczak, 2018; Rahman, 2009).

Studies have shown that $A_{2A}R$ levels are increased in the astrocytes of the AD patients, especially in brain regions with high Aβ deposition, such as the hippocampus and cortex (Orr et al., 2015; Paiva et al., 2019). In aging mice expressing the human amyloid precursor proteins (APP), elevated $A_{2A}R$ expression is also observed in astrocytes (Orr et al., 2015). Genetic ablation of astrocytic $A_{2A}R$ enhanced LTP and memory in these aging mice, whereas chemogenetic activation of astrocytic Gs-coupled signaling reduced long-term memory, suggesting that astrocytic $A_{2A}R$-coupled Gs signaling in the regulation of memory. Upregulation of $A_{2A}R$ in AD may reflect a compensatory mechanism to counteract the deleterious effects of Aβ on synaptic transmission and neuronal survival. However, excessive activation of $A_{2A}R$ appears to have detrimental consequences for astrocytic function and lead to an imbalance between LTP and LTD, resulting in impaired synaptic plasticity and memory formation. and contribute to AD progression. For instance, $A_{2A}R$ stimulation can impair astrocytic glutamate uptake and increase extracellular glutamate levels, leading to excitotoxicity and neuronal death (Matos et al., 2015). Moreover, $A_{2A}R$ activation can enhance astrocytic production of proinflammatory cytokines and reactive oxygen species (ROS), which can exacerbate neuroinflammation, angiogenesis and oxidative stress (Paiva et al., 2019). Furthermore, $A_{2A}R$ stimulation can induce astrocyte atrophy and reactivity, which can alter their morphology and connectivity with neurons. Therefore, targeting astrocytic $A_{2A}R$ may represents a potential therapeutic strategy for AD.

Consumption of caffeine, a nonselective adenosine $A_{2A}$ receptor ($A_{2A}R$) antagonist, reduces the risk of AD onset in humans and ameliorate both amyloid and Tau burden in transgenic mouse models (Espinosa et al., 2013). Several pharmacological agents that can modulate $A_{2A}R$ activity have been developed and tested in animal models of AD. For example, SCH58261, a selective $A_{2A}R$ antagonist, has been shown to improve memory performance and reduce Aβ levels and neuroinflammation in transgenic mice expressing human amyloid precursor protein (APP) (Merighi et al., 2022; Silva et al., 2018). Similarly, KW6002, another selective $A_{2A}R$ antagonist, has been shown to attenuate cognitive impairment and synaptic dysfunction in APP/PS1 mice. On the other hand, CGS21680, a selective $A_{2A}R$ agonist, has been shown to worsen memory deficits and increase astrocyte activation and oxidative stress in APP/PS1 mice (Merighi et al., 2022). In a Tau transgenic AD mouse model, MSX-3, a specific $A_{2A}R$ antagonist also significantly reduced memory loss and Tau hyperphosphorylation (Espinosa et al., 2013), highlighting $A_{2A}$ receptors as important molecular targets in both Aβ and tau-associated pathology in AD. These findings suggest that blocking $A_{2A}R$ signaling may have beneficial effects on AD pathology and cognition. However, more studies are needed to elucidate the molecular mechanisms underlying the modulation of astrocyte function by $A_{2A}R$, as well as to evaluate the safety and efficacy of $A_{2A}R$ antagonists in clinical trials. Moreover, it is important to consider the potential side effects of interfering with $A_{2A}R$ signaling in other cell types and brain regions, as well as the possible interactions with other drugs or treatments.

## 4. Astrocyte $A_{2A}R$ and PD

Epidemiological surveys have suggested that consumption of caffeine, a nonselective adenosine $A_{2A}$ receptor antagonist, reduces the risk of developing Parkinson's disease (PD) by five times. Further studies have shown that caffeine confers neuronal protection in different animal models of PD by inhibiting the $A_{2A}R$, as genetic ablation of $A_{2A}R$ abolished the neuroprotective role of caffeine in 1-methyl-4-phenyl-1,2,3,6 tetra-hydropyridine (MPTP)–induced neurotoxicity models of PD. Interestingly, conditional knockout of $A_{2A}R$ in either postnatal forebrain neurons or in astrocytes appear to have different impacts on caffeine-induced effects. In mice with the deletion of $A_{2A}R$ in forebrain neurons, the neuroprotective effects of caffeine were preserved, but its locomotor stimulant effects were lost. By contrast, both conditional removal of $A_{2A}R$ in postnatal forebrain astrocytes seems to unaffect both the locomotor stimulating and neuroprotective effects (Xu et al., 2016), suggesting that caffeine may act through different cellular substrates to confer its effects.

The expression of $A_{2A}R$ peaks in the striatum, where it overlaps with dopamine D2 receptors (D2R). D2R is Gi-coupled receptor, and its interactions with $A_{2A}R$, a Gs-coupled receptor has been well-recognized in striatal neurons. The lateral inhibition between adenosine $A_{2A}R$ and D2R in striatal medium spiny neurons has presented new perspectives on the molecular mechanisms involved in Parkinson's disease. Recent studies also observed that both $A_{2A}R$ and D2R are expressed in adult rat striatal astrocytes and co-distributed on the same astrocyte processes (Cervetto et al., 2017). $A_{2A}R$ and D2R interactions appear to control glutamate release from astrocytes, with D2R inhibiting the glutamate release, whereas $A_{2A}R$ activation abolishes the effects of D2R, by interfering with the homocysteine allosteric action of D2R (Cervetto et al., 2018). The study suggesting that molecular interactions of $A_{2A}R$–D2R also affect molecular circuits at the plasma membrane of striatal astrocyte processes. The fact that homocysteine reduced D2-mediated inhibition of glutamate release could provide new insights into striatal astrocyte–neuron intercellular communications. Further studies are required to characterize how astrocytic $A_{2A}R$, and $A_{2A}R$–D2R interactions are involved in Parkinson's pathophysiology.

## 5. $A_{2A}R$ and microglia

Microglia are the primary immune resident cells in the brain. They continuously survey the brain parenchyma with their ramified and highly motile processes to maintain the homeostasis. Upon injury, microglia rapidly extend their processes towards the injury site to isolate and scavenge the injured area and to release neurotrophic factors that promote neuronal survival and tissue repair.

Purinergic signaling plays an important role in regulating microglial dynamics and immune responses through ATP and adenosine-mediated purinergic type 2 (P2) and type 1 (P1) receptors. $P2Y_{12}$ receptor, an inhibitory G protein ($G_i$) coupled receptor (Sasaki et al., 2003), is an important homeostatic gene of microglia under physiologic conditions (Butovsky et al., 2014; Hickman et al., 2013). Activation of $P2Y_{12}$ receptor by ATP/ADP induces microglial process extension and morphological changes (Haynes et al., 2006). When there is an injury in the brain, ATP released from the injured cells, triggers rapid chemotaxis of microglia towards the ATP-releasing source in a $P2Y_{12}$-dependent manner (Davalos et al., 2005). ATP-$P2Y_{12}$ receptor signaling also mediates microglial process convergence towards the hyper-activated neurons (Dissing-Olesen et al., 2014; Eyo et al., 2014). $P2Y_{12}$ signaling has also been shown to potentiate the activity of receptor-coupled two-pore potassium channel, THIK-1; however, $P2Y_{12}$ activation evokes process outgrowth toward elevated nucleotides, independent of THIK-1 activation (Madry et al., 2018; Swiatkowski et al., 2016). Moreover, $P2Y_{12}$ receptor mediates interactions between microglial process and neuronal cell bodies under physiological conditions. Interestingly, after stroke, microglial processes contact more neuronal cell bodies originating from somatic microglia–neuron junctions in both mice and human, and central $P2Y_{12}$ receptor inhibition prevented increases in microglial process coverage of neuronal cell bodies and altered neuronal activity and worse neurological outcome (Cserep et al., 2020).

## 6. Microglial A$_{2A}$ receptor and process retraction

A$_{2A}$ receptor is usually low or undetectable in homeostatic microglia. However, when challenged with immune insults, for example, LPS exposure, ischemia, microglia quickly upregulate A$_{2A}$ receptor levels (Orr, Orr, Li, Gross, & Traynelis, 2009). Elevated A$_{2A}$ receptors are implicated in a number of microglial reactions including process dynamics, morphological changes and cytokine release and immune responses in different conditions. Antagonizing microglial A$_{2A}$ receptors has been shown to be beneficial in a number of disease states.

As described above, surveilling microglia are highly responsive to ATP stimulation and undergo rapid morphological changes with elongated processes towards ATP-releasing sources. However, LPS-primed microglia substantially attenuated their responses to ATP with relatively static morphology of swollen soma and shortened processes (Orr et al., 2009; Wollmer et al., 2001). Moreover, microglial process retraction is often seen in reactive microglia that adopt an amoeboid morphology in chronic brain injury and neurodegeneration. The typical amoeboid morphology of microglia has also been considered a hallmark of neuroinflammation in the brain. Microglial process retraction has been shown to be mediated by adenosine A$_{2A}$ receptor, which is upregulated in activated microglia. Adenosine A$_{2A}$ receptor stimulation by adenosine, causes microglia to retract their processes and become less responsive to chemoattractants such as ATP or ADP. This chemotactic reversal is also associated with the downregulation of P2Y12 receptor, which mediates microglial process extension toward sites of CNS injury. Thus, adenosine A$_{2A}$ receptor modulates microglial motility and morphology by switching the signaling from ATP-mediated P2Y12 receptor signaling to adenosine-mediated A$_{2A}$ receptor signaling during chronic brain diseases and neuroinflammation (Gyoneva et al., 2014; Orr et al., 2009).

Several studies have shown that A$_{2A}$R expression is upregulated in microglia under inflammatory conditions induced by lipopolysaccharide (LPS), amyloid-beta (Aβ), or hypoxia (Duan et al., 2009; Ingwersen et al., 2016; Pedata et al., 2016). This upregulation coincides with the downregulation of P2Y12 receptor expression and the retraction of microglial processes. Pharmacological or genetic blockade of A$_{2A}$R prevents microglial process retraction and restores their responsiveness to ATP or ADP. Conversely, pharmacological activation of A$_{2A}$R induces microglial process retraction even in the absence of inflammatory stimuli. These findings indicate that $A_{2A}R$ activation is sufficient and necessary to trigger microglial morphological changes during neuroinflammation.

The molecular mechanisms by which $A_{2A}R$ activation causes microglial process retraction are not fully elucidated, but some possible pathways have been proposed. The activation of protein kinase A (PKA) and cAMP response element-binding protein (CREB), which leads to the transcriptional regulation of genes involved in cytoskeletal remodeling and cell motility (Duan et al., 2018; Wong & Schlichter, 2014). Moreover, elevated $A_{2A}$ receptor in microglia or macrophages facilitates the stabilization of the hypoxia-inducible factor 1α (HIF1α), thereby helping sustaining the inflammasome activation and promoting chronic inflammation (Ouyang et al., 2013). $A_{2A}$ receptor activation also upregulates Cyclooxygenase-2 (COX-2), Prostaglandin E2 (PGE2) and induces Nitric oxide (NO) release by upregulating NO synthase-II expression (Fiebich et al., 1996; Saura et al., 2005). Since COX-2 and NO synthase-II are both neurotoxic (Teismann et al., 2003), $A_{2A}$ signaling may play a neurotoxic role.

## 7. Microglial $A_{2A}$ receptor and neuroinflammation

Neuroinflammation is a complex process that involves the activation of microglia, the resident immune cells of the central nervous system. Microglia can exert both beneficial and detrimental effects on neuronal function and survival, depending on the context and the signals they receive. The $A_{2A}$ receptor has been shown to play a crucial role in regulating microglial inflammatory responses and neurotoxicity in multiple neurodegenerative diseases such as Alzheimer's disease, Parkinson's disease, and multiple sclerosis.

Microglia can adopt different phenotypes and functions depending on the environmental cues (Gutmann & Kettenmann, 2019; Paolicelli et al., 2022; Wolf, Boddeke, & Kettenmann, 2017). Under normal conditions, microglia have a ramified morphology with highly motile processes that constantly scan the brain parenchyma. Upon stimulation by ATP or ADP released from injured cells, microglia extend their processes toward the source of damage via the activation of P2Y12 receptor, a purinergic receptor that mediates chemotaxis. However, under chronic or severe brain injury or disease, microglia undergo a phenotypic switch from a ramified to an amoeboid morphology, characterized by process retraction and cell body enlargement. This morphological change is accompanied by a functional change from a neuroprotective to a proinflammatory state, in which microglia secrete inflammatory cytokines, reactive oxygen species, and nitric oxide that can exacerbate neuronal damage and impair synaptic plasticity. Moreover, microglia lose their responsiveness to ATP and/or ADP and downregulate the expression of P2Y12 receptor. Mechanisms underlying this transition from a ramified to an amoeboid phenotype are not fully understood, but recent evidence suggests that $A_{2A}R$ plays a pivotal role in this process (Orr et al., 2009).

$A_{2A}R$ has been implicated in modulating microglial activation and neuroinflammation through several mechanisms. First, adenosine-$A_{2A}R$ activation prevents the chemotaxis of microglia toward sites of injury. Second, $A_{2A}R$ regulates the production and release of inflammatory mediators by microglia, by enhancing the production and release of proinflammatory elements such as IL-1b, reactive oxygen species (ROS), and nitric oxide (NO), which contribute to neuronal damage and synaptic loss. Third, by interacting with other intracellular signaling pathways, $A_{2A}R$ helps to stabilize the activated inflammasome and sustain chronic inflammation. For example, by acting on $A_{2A}R$, adenosine is able to prolong the duration of the inflammatory responses, which supersedes a tolerogenic state induced by acute LPS stimulation/exposure and drive IL-1b production. This is mainly mediated by the $A_{2A}R$–CAMP–PKA–CREB pathway, which helps to stabilize the HIF-1a, thereby sustaining the inflammasome activity (Ouyang et al., 2013).

The role of microglial $A_{2A}R$ in neuroinflammation has been demonstrated in various animal models of CNS disorders, such as Alzheimer's disease, Parkinson's disease, glaucoma. In general, pharmacological or genetic blockade of $A_{2A}R$ has been shown to reduce neuroinflammation and to improve neuronal survival and function in these models. Conversely, activation of $A_{2A}R$ has been shown to exacerbate neuroinflammation and to worsen neuronal damage and dysfunction. Therefore, targeting $A_{2A}R$ may represent a promising therapeutic strategy for modulating neuroinflammation and its consequences in various CNS diseases.

## 8. Microglial $A_{2A}R$ and AD

Alzheimer's disease (AD) is a neurodegenerative disorder characterized by progressive cognitive decline and memory loss, as well as pathological features such as amyloid-beta (Aβ) plaques and neurofibrillary tangles of tau protein. AD is also associated with chronic neuroinflammation, which involves the activation of microglia. Microglia play a dual role in

AD, as they can both clear Aβ deposits and release proinflammatory cytokines that exacerbate neuronal damage. Therefore, modulating microglial function is a potential therapeutic strategy for AD. $A_{2A}R$ expression is upregulated in microglia surrounding the Aβ-associated plaques, suggesting that $A_{2A}R$ signaling may contribute to the pathophysiology of AD by influencing microglial function.

Several studies have investigated the role of $A_{2A}R$ in AD using pharmacological or genetic approaches. $A_{2A}R$ antagonists have been reported to reduce Aβ production and aggregation, enhance Aβ clearance by microglia, attenuate neuroinflammation and oxidative stress, protect synaptic function and neuronal survival, and improve cognitive performance in animal models of AD. Conversely, $A_{2A}R$ agonists have been shown to have opposite effects, worsening AD pathology and symptoms. Moreover, $A_{2A}R$ knockout mice have been found to be resistant to Aβ-induced neurotoxicity and cognitive impairment. However, since $A_{2A}R$ is upregulated in multiple cell types including both astrocytes and microglia, it is difficult to distinguishing the role of astrocytic or microglial $A_{2A}R$ in the disease pathology. Further studies using selective genetic ablation tools will be helpful to dissect the role of $A_{2A}R$ in different cells in different diseases. In conclusion, microglial adenosine $A_{2A}$ receptor plays an important role in Alzheimer's disease by modulating microglial function and interacting with other receptors that are involved in AD pathogenesis. Targeting $A_{2A}R$ may represent a novel therapeutic approach for AD by enhancing the neuroprotective effects of endogenous or exogenous cannabinoids on microglia.

In endothelial cells, $A_{2A}R$ activation can influence the blood–brain barrier (BBB) integrity and permeability. The BBB is a specialized structure that separates the CNS from the peripheral circulation, regulating the exchange of molecules and cells between them. In AD, the BBB is compromised due to inflammation and oxidative stress, leading to increased leakage of plasma proteins and immune cells into the brain parenchyma. This can further aggravate neuroinflammation and neuronal damage in AD. $A_{2A}R$ activation can either protect or disrupt the BBB integrity depending on the level of receptor stimulation.

## 9. Microglial adenosine $A_{2A}R$ and PD

PD is a neurodegenerative disorder characterized by the loss of dopaminergic neurons in the substantia nigra and the accumulation of alpha-synuclein aggregates in the brain. PD causes motor symptoms such as bradykinesia, rigidity, tremor and postural instability, as well as non-motor symptoms such as cognitive impairment, depression and sleep disorders. The current pharmacological treatment of PD is based on dopamine replacement therapy, which can improve motor function but has limited efficacy and causes side effects such as dyskinesia and psychosis. The approval of $A_{2A}R$ antagonist by FDA reflects the importance of $A_{2A}R$ in PD pathogenesis.

Microglial $A_{2A}R$ also contributes to PD pathology. CD73, an ectonucleotidase that hydrolyzes AMP into adenosine is enriched in the striatum. Interestingly, CD39, an ectonucleotidase catalyzes the hydrolysis ATP to AMP, is restrictively expressed in microglia. CD39–CD73 coupling therefore drives the catabolism of extracellular ATP and provides an important source of extracellular adenosine in the striatum. Since $A_{2A}R$ is also highly expressed in the striatum. CD73-mediated adenosine formation provides an important input to activate $A_{2A}R$. Interestingly, CD73 and microglial $A_{2A}R$ are upregulated in the striatum in MPTP-induced cytotoxic models, which coordinatively contribute to the elevated adenosine $A_{2A}R$ signaling in PD models. Inhibition of CD73 robustly attenuated lipopolysaccharide-induced proinflammatory responses in both primary microglial cultures and microglia in vivo. However, microglial process extension, movement and chemotaxis towards the injury site was enhanced in the absence of CD73, which is opposite to the $A_{2A}R$ activation-induced microglial process retraction. These studies suggest that CD73-derived adenosine provides an important source to activate microglial $A_{2A}$, thereby regulating the dynamics and immune responses of microglia (Meng et al., 2019). Removing CD73 to restrict the adenosine input substantially suppressed microglia-mediated neuroinflammation and played a neuroprotective role for dopaminergic neurons in Parkinson's disease models. Moreover, CD73 removal suppressed microglial $A_{2A}R$ induction and $A_{2A}R$-mediated proinflammatory responses, whereas adenosine replenishment restored these effects, suggesting that CD73 produces a self-regulating feed-forward adenosine formation to activate $A_{2A}R$ and promote neuroinflammation (Meng et al., 2019). These studies suggest that targeting nucleotide metabolic pathways to limit adenosine production is also an effective way to reduce the effects of $A_{2A}R$ signaling and inhibit neuroinflammation in Parkinson's disease.

## 10. Microglial adenosine $A_{2A}$ receptor and stroke

Ischemic stroke is a leading cause of death and disability worldwide. Microglia are activated by ischemic injury and release proinflammatory cytokines, chemokines, ROS and nitric oxide, which exacerbate the inflammatory response and neuronal damage. However, microglia also have neuroprotective functions, such as phagocytosis of debris and apoptotic cells, secretion of neurotrophic factors and modulation of synaptic plasticity. Therefore, the balance between the proinflammatory and neuroprotective roles of microglia is crucial for the outcome of ischemic stroke.

The sterile inflammatory responses in ischemic brains are primarily initiated by the danger-associate molecular patterns (DAMP). ATP, released by the dying to the extracellular DAMP. Extracellular ATP rapidly activates microglial P2 signaling, including both P2X and P2Y receptor-mediated signaling, leading to microglial chemotaxis, microglial micropinocytosis, promoting cytokine and chemokine production and subsequent immune cell recruitment from the periphery which further amplifies post-stroke inflammation (Koles, Furst, & Illes, 2005; Li et al., 2013; Schadlich et al., 2023). The ectonucleotidases CD39 and CD73 help to balance the inflammatory environment by hydrolyzing extracellular ATP to adenosine. Adenosine elicit neuroprotective effects by acting on $A_1$ receptors on neurons, thereby suppressing glutamatergic excitotoxicity. On the other hand, adenosine may have anti-inflammatory effect through A(2A) receptor activation on infiltrating immune cells in the subacute phase after stroke (Schadlich et al., 2023).

Studies have shown that $A_{2A}R$ activation by low doses of adenosine or selective agonists suppresses the proinflammatory phenotype of microglia and shifts them towards a neuroprotective phenotype in a rat model of transient medial cerebral artery occlusion (MCAo) (Melani, Corti, Cellai, Vannucchi, & Pedata, 2014). $A_{2A}R$ activation may inhibit the production of proinflammatory cytokines including TNF-α, IL-1β and IL-6, and ROS by microglia, and enhance their phagocytic activity and secretion of neurotrophic factors (Schadlich et al., 2023). $A_{2A}R$ activation also modulates microglial interactions with other cell types in the ischemic brain, such as neurons, astrocytes and endothelial cells. For example, $A_{2A}R$ activation promotes microglial–neuronal communication via gap junctions and reduces microglial-mediated neuronal death in ischemic models. $A_{2A}R$ activation also attenuates microglial–astrocytic crosstalk via purinergic receptors and reduces astrocytic activation and gliosis. Moreover, $A_{2A}R$ activation improves microglial–endothelial interactions via adhesion molecules and tight junction proteins and enhances blood–brain barrier integrity and cerebral blood flow.

Several preclinical studies have demonstrated that pharmacological or genetic manipulation of $A_{2A}R$ can reduce infarct size, improve neurological function and promote neurogenesis and angiogenesis after ischemic stroke in animal models. However, these studies in microglia are in conflict with the anti-inflammatory role of using $A_{2A}R$ antagonist in other diseases including both AD and PD diseases. These contradictory results may reflect the diverse roles of microglia in different contexts with different phases of disease progression at different brain regions. Further studies characterizing the heterogeneity and plasticity of microglia in different brain regions and stages of diseases are needed to understand the role of $A_{2A}R$ in microglia-mediated responses.

## 11. Microglial $A_{2A}R$ in other diseases

Synaptic pruning eliminates excess or inappropriate synapses to facilitate synaptic maturation during brain development. It is essential for the refinement of neural circuits and the optimization of brain function. Adenosine $A_{2A}$ receptor has been implicated in the synaptic pruning of the retinogeniculate pathway, which connects the retina to the dorsal lateral geniculate nucleus (dLGN) in the thalamus (Miao et al., 2021). Antagonizing $A_{2A}R$ at postnatal stages by pharmacologic blockade or genetic ablation reduces facilitate the eye-specific segregation of the dLGN, as a result of synaptic pruning of retinal ganglion cell (RGC) inputs, in a microglia-dependent manner (Miao et al., 2021). $A_{2A}R$ antagonism at postnatal phases appears to promote microglial activation and enhance lysosomal activity, thereby facilitating microglial engulfment of RGC inputs. As well, it also affects components of postsynaptic proteins and synaptic density in the dLGN, suggesting that $A_{2A}R$ controls synaptic pruning by multiple mechanisms involving both microglial phagocytosis and synaptic remodeling at both presynaptic and postsynaptic areas.

Glaucoma is a leading cause of vision loss and blindness characterized by progressive loss of retinal neurons. Microglial reactions contribute to the pathology by driving local inflammation and retinal neurodegeneration. Pharmacologically blocking adenosine $A_{2A}$ receptor ($A_{2A}R$) by its antagonist (SCH 58261) prevents the neuroinflammatory responses of microglia and play a protective role for retinal ganglion cells (RGCs) after elevated intraocular pressure (IOP), the main risk factor for glaucoma development. In a high IOP-induced transient ischemia (ischemia-reperfusion, I-R) animal model, $A_{2A}R$ blockade reduced retinal neuroinflammation and neurodegeneration

(Ahmad et al., 2013; Aires, Boia et al., 2019). SCH 58261 treatment significantly attenuated microglial reactivity and the increased expression and release of proinflammatory cytokines in primary microglial cell cultures challenged with lipopolysaccharide or with elevated eye pressure (Ahmad et al., 2013; Aires, Boia et al., 2019; Aires, Madeira et al., 2019). Moreover, intravitreal administration of SCH 58261 prevented ischemia-reperfusion I-R-induced cell death and RGC loss, via reduced microglial-mediated neuroinflammatory responses (Aires, Madeira et al., 2019), suggesting that $A_{2A}R$ blockade may have great potential in the management of retinal neurodegenerative diseases characterized by microglia reactivity and RGC death, such as glaucoma and ischemic diseases.

## 12. Role of other microglial adenosine $A_1$ and $A_3$ receptors

In addition to $A_{2A}R$, other adenosine receptors are also involved in the regulation of microglial activity. Adenosine $A_1$ receptor, a Gi-coupled receptor, is upregulated after ATP treatment and activation of $A_1$ receptor inhibits the morphological activation of microglia (Luongo et al., 2014). $A_1$ receptor is also increased in microglia adjacent to the glioblastoma in both mouse models and human patients, and adenosine may attenuate glioblastoma growth acting on microglial $A_1$ receptor (Synowitz et al., 2006). In contrast to the proinflammatory role of $A_{2A}R$, $A_1R$ may be involved in suppressing neuroinflammation as knocking out $A_1$ receptor promotes neuroinflammation in a multiple sclerosis model (Tsutsui et al., 2004). Moreover, agonizing $A_{3A}$ receptor reduces neuropathic pain by suppressing spinal microglia activation (Terayama, Tabata, Maruhama, & Iida, 2018), suggesting anti-inflammatory/neuroprotective roles for both $A_1$ and $A_{3A}$ receptors. Moreover, $A_{2B}$ receptor may also affect microglial immune responses by modulating the expression of adenosine-induced HIF1α target genes, such as glucose transporter-1 (GLUT-1) and inducible nitric-oxide synthase (iNOS) (Merighi et al., 2015). By activating $A_{2B}$ receptor, adenosine simultaneously raised vascular endothelial growth factor (VEGF) and decreased TNF-α levels (Merighi et al., 2015).

Adenosine-P1 and ATP-P2 signaling coordinately contribute to microglial responses and activity. On the one hand, there may be differences in temporal dynamics of P2 and P1 signaling as CD39–CD73 convert ATP into adenosine, with ATP-triggered P2 receptors initiating the signaling, followed by adenosine-mediated P1 signaling. In addition, there might be antagonistic effects between different types of P1 receptors. For instance, the Gs-coupled $A_{2A}$ and Gi-coupled $A_1$ receptors may antagonize each other, and the rivalry may help explain that $A_{2A}$ receptor antagonists, by potentiating the $A_1$ receptor-mediated effects, play a neuroprotective role in disease pathology (Pedata, Corsi, Melani, Bordoni, & Latini, 2001). Moreover, blocking $A_3$ receptor also appears to facilitate $A_{2A}$-mediated signaling, suggesting that $A_{2A}R$ and $A_3R$ may antagonize each other by default (van der Putten et al., 2009). Intriguingly, $A_1$ and $A_3$ receptors may cooperate with $P2Y_{12}$ receptor to regulate ATP/ADP-induced microglial migration (Ohsawa et al., 2012).

## 13. $A_{2A}R$ and oligodendrocytes

Oligodendrocytes are responsible for producing and maintaining the myelin sheath that insulates axons in the central nervous system. Several studies have shown that $A_{2A}R$ stimulation can affect oligodendrocyte development, survival, and function in different contexts (Genovese et al., 2009; Melani et al., 2009). For example, $A_{2A}R$ activation can promote oligodendrocyte differentiation and myelin formation in a pharmacological model of Niemann–Pick type C disease, a rare neurovisceral disorder characterized by dysmyelination and maturational arrest of oligodendrocyte progenitors (Genovese et al., 2009; Melani et al., 2009). $A_{2A}R$ stimulation also appears to protect oligodendrocytes from ischemic injury by reducing microglial activation and inflammation, as well as by attenuating mitochondrial dysfunction and oxidative stress (Genovese et al., 2009; Melani et al., 2009). Moreover, $A_{2A}R$ antagonism can reduce JNK activation and caspase-3 cleavage in oligodendrocytes after cerebral ischemia, suggesting a dual role of $A_{2A}R$ signaling in modulating cell death pathways (Genovese et al., 2009; Melani et al., 2009).

Multiple sclerosis (MS) is a common autoimmune disease characterized by the repeated axon demyelination and remyelination. Chronic treatment with caffeine, showed a protective role against myelin oligodendrocyte glycoprotein (MOG)-induced experimental autoimmune encephalomyelitis (EAE), an animal model of MS. Administration of caffeine during the effector phase, a stage corresponding to phenotypic appearance, significantly reduced EAE-induced behavioral deficits, inflammatory cell infiltration into the spinal cord and demyelination. $A_{2A}R$ is upregulated in the brain in murine experimental autoimmune encephalomyelitis (EAE) models, predominantly on T cells and macrophages/microglia (Ingwersen et al., 2016). Interestingly, genetic deletion of $A_{2A}R$ potentiated MOG-induced brain damage, whereas chronic caffeine administration to $A_{2A}R$ knockout mice alleviated EAE pathology, suggesting that caffeine may act on non-$A_{2A}R$ target to elicit protection. Further studies show that the protective effects of caffeine treatment were associated with $A_1R$ antagonism (Wang et al., 2014). Similar to effects of caffeine treatment at different disease phases, blocking $A_{2A}R$ at different disease phases also led to different disease consequences. Preventive treatment with a specific $A_{2A}R$ agonist before the EAE onset inhibited myelin-specific T cell proliferation and disease progression, whereas application of the same drug after disease onset exacerbated behavioral deficits. Knocking out $A_{2A}R$ accelerated disease progression and exacerbated disease phenotype with increased inflammatory lesions at early stages. However, at later stage, $A_{2A}R$-deficient mice appeared to recover more quickly with less myelin debris accumulation (Ingwersen et al., 2016). At the onset of motor symptoms in EAE mdoels, a single intrathecal administration of $A_{2A}R$ agonists significantly attenuated progression of motor symptoms (Loram et al., 2015). Moreover, $A_{2A}R$ administration also attenuated microglial activation in the lumbar spinal cord following $A_{2A}R$ administration compared to vehicle. Therefore, $A_{2A}R$ agonists attenuate motor symptoms of EAE by acting on $A_{2A}R$ in the spinal cord. Conversely, activating $A_{2A}R$ with agonist, CGS21680 (CGS) significantly suppressed specific lymphocyte proliferation, reduced infiltration of CD4(+) T lymphocytes, and attenuated the expression of inflammatory cytokines, which in turn inhibited the EAE progression (Liu et al., 2016). These findings indicate that $A_{2A}R$ is a potential therapeutic target for enhancing oligodendrocyte viability and function in various neurological disorders involving demyelination and white matter damage.

## 14. Perspective

Adenosine $A_{2A}R$ signaling plays a wide range of roles in modulating both neuronal and glial activity under physiological and pathological condition. Sincere adenosine can be released from both neurons and astrocytes, and $A_{2A}$ receptors are expressed in both neuron and glial cells, adenosine $A_{2A}R$ signaling serves as a critical link for neuron-glial interactions. As adenosine release from neurons is induced by neuronal activity and $A_{2A}R$ levels are susceptible to external stimuli, adenosine $A_{2A}R$ signaling provides an important mechanism for fine-tuning neural function and for multitude effects of $A_{2A}Rs$ in regulating neuronal and glial activity.

Studies have well-characterized the roles of adenosine $A_{2A}$ receptors in physiology and pathology in different cell types. The prominent elevation of $A_{2A}$ receptors in reactive glial cells in different brain diseases have made them a potential therapeutic target in both neurodegenerative and psychiatric diseases. While multiple $A_{2A}R$ agonists and antagonists are extensively used in the laboratory, few of them are currently used in clinical and preclinical trials. There are still challenges and limitations in optimizing the doses, duration and effects of these drug in in vivo disease models. As well, to selectively achieve cell type-specific targeting with precise time window is still a big hurdle on the way of developing new therapeutic drugs for adenosine $A_{2A}$ (Chen, Eltzschig, & Fredholm, 2013; Jacobson & Gao, 2006). Further characterization of adenosine signaling and their interaction with other signaling may reveal additional targets for better drug designs to specifically manipulate $A_{2A}$ activities, thereby providing new therapeutic strategies to fine-tune microglia and astrocyte-mediated immune responses and neuroinflammation, and to facilitate potential pharmacological tools to treat brain diseases.

## References

Ahmad, S., Fatteh, N., El-Sherbiny, N. M., Naime, M., Ibrahim, A. S., El-Sherbini, A. M., ... Gonzales, J. (2013). Potential role of A2A adenosine receptor in traumatic optic neuropathy. *Journal of Neuroimmunology, 264*, 54–64.

Aires, I. D., Boia, R., Rodrigues-Neves, A. C., Madeira, M. H., Marques, C., Ambrosio, A. F., ... Santiago, A. R. (2019). Blockade of microglial adenosine A(2A) receptor suppresses elevated pressure-induced inflammation, oxidative stress, and cell death in retinal cells. *Glia, 67*, 896–914.

Aires, I. D., Madeira, M. H., Boia, R., Rodrigues-Neves, A. C., Martins, J. M., Ambrosio, A. F., ... Santiago, A. R. (2019). Intravitreal injection of adenosine A(2A) receptor antagonist reduces neuroinflammation, vascular leakage and cell death in the retina of diabetic mice. *Scientific Reports, 9*, 17207.

Butovsky, O., Jedrychowski, M. P., Moore, C. S., Cialic, R., Lanser, A. J., Gabriely, G., ... Doykan, C. E. (2014). Identification of a unique TGF-beta-dependent molecular and functional signature in microglia. *Nature Neuroscience, 17*, 131–143.

Cervetto, C., Venturini, A., Guidolin, D., Maura, G., Passalacqua, M., Tacchetti, C., ... Ramoino, P. (2018). Homocysteine and A2A-D2 receptor-receptor interaction at striatal astrocyte Processes. *Journal of Molecular Neuroscience: MN, 65*, 456–466.

Cervetto, C., Venturini, A., Passalacqua, M., Guidolin, D., Genedani, S., Fuxe, K., ... Maura, G. (2017). A2A-D2 receptor-receptor interaction modulates gliotransmitter release from striatal astrocyte processes. *Journal of Neurochemistry, 140*, 268–279.

Chen, J. F., Eltzschig, H. K., & Fredholm, B. B. (2013). Adenosine receptors as drug targets—What are the challenges? *Nature Reviews. Drug Discovery, 12*, 265–286.

Cieslak, M., & Wojtczak, A. (2018). Role of purinergic receptors in the Alzheimer's disease. *Purinergic Signalling, 14*, 331–344.

Cserep, C., Posfai, B., Lenart, N., Fekete, R., Laszlo, Z. I., Lele, Z., ... Schwarcz, A. D. (2020). Microglia monitor and protect neuronal function through specialized somatic purinergic junctions. *Science (New York, N. Y.), 367*, 528–537.

Davalos, D., Grutzendler, J., Yang, G., Kim, J. V., Zuo, Y., Jung, S., ... Gan, W. B. (2005). ATP mediates rapid microglial response to local brain injury in vivo. *Nature Neuroscience, 8*, 752–758.

Dissing-Olesen, L., LeDue, J. M., Rungta, R. L., Hefendehl, J. K., Choi, H. B., & MacVicar, B. A. (2014). Activation of neuronal NMDA receptors triggers transient ATP-mediated microglial process outgrowth. *The Journal of Neuroscience: The Official Journal of the Society for Neuroscience, 34*, 10511–10527.

Duan, W., Gui, L., Zhou, Z., Liu, Y., Tian, H., Chen, J. F., ... Zheng, J. (2009). Adenosine A2A receptor deficiency exacerbates white matter lesions and cognitive deficits induced by chronic cerebral hypoperfusion in mice. *Journal of the Neurological Sciences, 285*, 39–45.

Duan, W., Wang, H., Fan, Q., Chen, L., Huang, H., & Ran, H. (2018). Cystatin F involvement in adenosine A(2A) receptor-mediated neuroinflammation in BV2 microglial cells. *Scientific Reports, 8*, 6820.

Espinosa, J., Rocha, A., Nunes, F., Costa, M. S., Schein, V., Kazlauckas, V., ... Porciuncula, L. O. (2013). Caffeine consumption prevents memory impairment, neuronal damage, and adenosine A2A receptors upregulation in the hippocampus of a rat model of sporadic dementia. *Journal of Alzheimer's Disease: JAD, 34*, 509–518.

Eyo, U. B., Peng, J., Swiatkowski, P., Mukherjee, A., Bispo, A., & Wu, L. J. (2014). Neuronal hyperactivity recruits microglial processes via neuronal NMDA receptors and microglial P2Y12 receptors after status epilepticus. *The Journal of Neuroscience: The Official Journal of the Society for Neuroscience, 34*, 10528–10540.

Fiebich, B. L., Biber, K., Lieb, K., vanCalker, D., Berger, M., Bauer, J., ... GebickeHaerter, P. J. (1996). Cyclooxygenase-2 expression in rat microglia is induced by adenosine A(2a)-receptors. *Glia, 18*, 152–160.

Genovese, T., Melani, A., Esposito, E., Mazzon, E., Di Paola, R., Bramanti, P., ... Cuzzocrea, S. (2009). The selective adenosine A2A receptor agonist CGS 21680 reduces JNK MAPK activation in oligodendrocytes in injured spinal cord. *Shock (Augusta, Ga.), 32*, 578–585.

Gutmann, D. H., & Kettenmann, H. (2019). Microglia/brain macrophages as central drivers of brain tumor pathobiology. *Neuron, 104*, 442–449.

Gyoneva, S., Shapiro, L., Lazo, C., Garnier-Amblard, E., Smith, Y., Miller, G. W., ... Traynelis, S. F. (2014). Adenosine A2A receptor antagonism reverses inflammation-induced impairment of microglial process extension in a model of Parkinson's disease. *Neurobiology of Disease, 67*, 191–202.

Haynes, S. E., Hollopeter, G., Yang, G., Kurpius, D., Dailey, M. E., Gan, W. B., ... Julius, D. (2006). The P2Y12 receptor regulates microglial activation by extracellular nucleotides. *Nature Neuroscience, 9*, 1512–1519.

Hickman, S. E., Kingery, N. D., Ohsumi, T. K., Borowsky, M. L., Wang, L. C., Means, T. K., ... El Khoury, J. (2013). The microglial sensome revealed by direct RNA sequencing. *Nature Neuroscience, 16*, 1896–1905.

Ingwersen, J., Wingerath, B., Graf, J., Lepka, K., Hofrichter, M., Schroter, F., ... Hartung, H. P. (2016). Dual roles of the adenosine A2a receptor in autoimmune neuroinflammation. *Journal of Neuroinflammation, 13*, 48.

Jacobson, K. A., & Gao, Z. G. (2006). Adenosine receptors as therapeutic targets. *Nature Reviews. Drug Discovery, 5*, 247–264.

Koles, L., Furst, S., & Illes, P. (2005). P2X and P2Y receptors as possible targets of therapeutic manipulations in CNS illnesses. *Drug News & Perspectives, 18*, 85–101.

Li, H. Q., Chen, C., Dou, Y., Wu, H. J., Liu, Y. J., Lou, H. F., ... Duan, S. (2013). P2Y4 receptor-mediated pinocytosis contributes to amyloid beta-induced self-uptake by microglia. *Molecular and Cellular Biology, 33*, 4282–4293.

Luongo, L., Guida, F., Imperatore, R., Napolitano, F., Gatta, L., Cristino, L., ... Bellini, G. (2014). The A1 adenosine receptor as a new player in microglia physiology. *Glia, 62*, 122–132.

Madry, C., Kyrargyri, V., Arancibia-Carcamo, I. L., Jolivet, R., Kohsaka, S., Bryan, R. M., ... Attwell, D. (2018). Microglial ramification, surveillance, and interleukin-1beta release are regulated by the two-pore domain K(+) channel THIK-1. *Neuron, 97*, 299–312.e296.

Matos, M., Shen, H. Y., Augusto, E., Wang, Y., Wei, C. J., Wang, Y. T., ... Chen, J. F. (2015). Deletion of adenosine A2a receptors from astrocytes disrupts glutamate homeostasis leading to psychomotor and cognitive impairment: Relevance to schizophrenia. *Biological Psychiatry, 78*, 763–774.

Melani, A., Cipriani, S., Vannucchi, M. G., Nosi, D., Donati, C., Bruni, P., ... Pedata, F. (2009). Selective adenosine A2A receptor antagonism reduces JNK activation in oligodendrocytes after cerebral ischaemia. *Brain, 132*, 1480–1495.

Melani, A., Corti, F., Cellai, L., Vannucchi, M. G., & Pedata, F. (2014). Low doses of the selective adenosine A2A receptor agonist CGS21680 are protective in a rat model of transient cerebral ischemia. *Brain Research, 1551*, 59–72.

Meng, F., Guo, Z., Hu, Y., Mai, W., Zhang, Z., Zhang, B., ... Chen, J. (2019). CD73-derived adenosine controls inflammation and neurodegeneration by modulating dopamine signalling. *Brain, 142*, 700–718.

Merighi, S., Borea, P. A., Stefanelli, A., Bencivenni, S., Castillo, C. A., Varani, K., ... Gessi, S. (2015). A2a and a2b adenosine receptors affect HIF-1alpha signaling in activated primary microglial cells. *Glia, 63*, 1933–1952.

Merighi, S., Borea, P. A., Varani, K., Vincenzi, F., Jacobson, K. A., & Gessi, S. (2022). A(2A) adenosine receptor antagonists in neurodegenerative diseases. *Current Medicinal Chemistry, 29*, 4138–4151.

Miao, Y., Chen, X., You, F., Jia, M., Li, T., Tang, P., ... Chen, J. F. (2021). Adenosine A(2A) receptor modulates microglia-mediated synaptic pruning of the retinogeniculate pathway during postnatal development. *Neuropharmacology, 200*, 108806.

Ohsawa, K., Sanagi, T., Nakamura, Y., Suzuki, E., Inoue, K., & Kohsaka, S. (2012). Adenosine A3 receptor is involved in ADP-induced microglial process extension and migration. *Journal of Neurochemistry, 121*, 217–227.

Orr, A. G., Hsiao, E. C., Wang, M. M., Ho, K., Kim, D. H., Wang, X., ... Adame, A. (2015). Astrocytic adenosine receptor A2A and Gs-coupled signaling regulate memory. *Nature Neuroscience, 18*, 423–434.

Orr, A. G., Orr, A. L., Li, X. J., Gross, R. E., & Traynelis, S. F. (2009). Adenosine A(2A) receptor mediates microglial process retraction. *Nature Neuroscience, 12*, 872–878.

Ouyang, X., Ghani, A., Malik, A., Wilder, T., Colegio, O. R., Flavell, R. A., ... Mehal, W. Z. (2013). Adenosine is required for sustained inflammasome activation via the A(2)A receptor and the HIF-1alpha pathway. *Nature Communications, 4*, 2909.

Paiva, I., Carvalho, K., Santos, P., Cellai, L., Pavlou, M. A. S., Jain, G., ... Fischer, A. (2019). A(2A) R-induced transcriptional deregulation in astrocytes: An in vitro study. *Glia, 67*, 2329–2342.

Paolicelli, R. C., Sierra, A., Stevens, B., Tremblay, M. E., Aguzzi, A., Ajami, B., ... Bennett, M. (2022). Microglia states and nomenclature: A field at its crossroads. *Neuron, 110*, 3458–3483.

Pedata, F., Corsi, C., Melani, A., Bordoni, F., & Latini, S. (2001). Adenosine extracellular brain concentrations and role of A2A receptors in ischemia. *Annals of the New York Academy of Sciences, 939*, 74–84.

Pedata, F., Dettori, I., Coppi, E., Melani, A., Fusco, I., Corradetti, R., ... Pugliese, A. M. (2016). Purinergic signalling in brain ischemia. *Neuropharmacology, 104*, 105–130.

Phatnani, H., & Maniatis, T. (2015). Astrocytes in neurodegenerative disease. *Cold Spring Harbor Perspectives in Biology, 7*, a020628.

Preeti, K., Sood, A., & Fernandes, V. (2022). Metabolic regulation of glia and their neuroinflammatory role in Alzheimer's disease. *Cellular and Molecular Neurobiology, 42*, 2527–2551.

Qian, K., Jiang, X., Liu, Z. Q., Zhang, J., Fu, P., Su, Y., ... Zhu, L. Q. (2023). Revisiting the critical roles of reactive astrocytes in neurodegeneration. *Molecular Psychiatry*. https://doi.org/10.1038/s41380-023-02061-8.

Rahman, A. (2009). The role of adenosine in Alzheimer's disease. *Current Neuropharmacology, 7*, 207–216.

Sasaki, Y., Hoshi, M., Akazawa, C., Nakamura, Y., Tsuzuki, H., Inoue, K., ... Kohsaka, S. (2003). Selective expression of Gi/o-coupled ATP receptor P2Y12 in microglia in rat brain. *Glia, 44*, 242–250.

Saura, J., Angulo, E., Ejarque, A., Casado, V., Tusell, J. M., Moratalla, R., ... Franco, R. (2005). Adenosine A(2A) receptor stimulation potentiates nitric oxide release by activated microglia. *Journal of Neurochemistry, 95*, 919–929.

Schadlich, I. S., Winzer, R., Stabernack, J., Tolosa, E., Magnus, T., & Rissiek, B. (2023). The role of the ATP-adenosine axis in ischemic stroke. *Seminars in Immunopathology, 45*, 347–365.

Silva, A. C., Lemos, C., Goncalves, F. Q., Pliassova, A. V., Machado, N. J., Silva, H. B., ... Agostinho, P. (2018). Blockade of adenosine A(2A) receptors recovers early deficits of memory and plasticity in the triple transgenic mouse model of Alzheimer's disease. *Neurobiology of Disease, 117*, 72–81.

Swiatkowski, P., Murugan, M., Eyo, U. B., Wang, Y., Rangaraju, S., Oh, S. B., ... Wu, L. J. (2016). Activation of microglial P2Y12 receptor is required for outward potassium currents in response to neuronal injury. *Neuroscience, 318*, 22–33.

Synowitz, M., Glass, R., Farber, K., Markovic, D., Kronenberg, G., Herrmann, K., ... Kiwit, J. (2006). A1 adenosine receptors in microglia control glioblastoma-host interaction. *Cancer Research, 66*, 8550–8557.

Teismann, P., Vila, M., Choi, D. K., Tieu, K., Wu, D. C., Jackson-Lewis, V., ... Przedborski, S. (2003). COX-2 and neurodegeneration in Parkinson's disease. *Annals of the New York Academy of Sciences, 991*, 272–277.

Terayama, R., Tabata, M., Maruhama, K., & Iida, S. (2018). A(3) adenosine receptor agonist attenuates neuropathic pain by suppressing activation of microglia and convergence of nociceptive inputs in the spinal dorsal horn. *Experimental Brain Research, 236*, 3203–3213.

Trujillo-Estrada, L., Gomez-Arboledas, A., Forner, S., Martini, A. C., Gutierrez, A., Baglietto-Vargas, D., ... LaFerla, F. M. (2019). Astrocytes: From the physiology to the disease. *Current Alzheimer Research, 16*, 675–698.

Tsutsui, S., Schnermann, J., Noorbakhsh, F., Henry, S., Yong, V. W., Winston, B. W., ... Power, C. (2004). A1 adenosine receptor upregulation and activation attenuates neuroinflammation and demyelination in a model of multiple sclerosis. *The Journal of Neuroscience, 24*, 1521–1529.

van der Putten, C., Zuiderwijk-Sick, E. A., van Straalen, L., de Geus, E. D., Boven, L. A., Kondova, I., ... Bajramovic, J. J. (2009). Differential expression of adenosine A3 receptors controls adenosine A2A receptor-mediated inhibition of TLR responses in microglia. *Journal of Immunology, 182*, 7603–7612.

Wolf, S. A., Boddeke, H. W., & Kettenmann, H. (2017). Microglia in physiology and disease. *Annual Review of Physiology, 79*, 619–643.

Wollmer, M. A., Lucius, R., Wilms, H., Held-Feindt, J., Sievers, J., & Mentlein, R. (2001). ATP and adenosine induce ramification of microglia in vitro. *Journal of Neuroimmunology, 115*, 19–27.

Wong, R., & Schlichter, L. C. (2014). PKA reduces the rat and human KCa3.1 current, CaM binding, and Ca2+ signaling, which requires Ser332/334 in the CaM-binding C terminus. *The Journal of Neuroscience, 34*, 13371–13383.

Xu, K., Di Luca, D. G., Orru, M., Xu, Y., Chen, J. F., & Schwarzschild, M. A. (2016). Neuroprotection by caffeine in the MPTP model of parkinson's disease and its dependence on adenosine A2A receptors. *Neuroscience, 322*, 129–137.

# CHAPTER THREE

# The adenosine $A_{2A}$ receptor in the basal ganglia: Expression, heteromerization, functional selectivity and signalling

**Rafael Franco[a,b,c,*], Gemma Navarro[b,d,e], and Eva Martínez-Pinilla[f,g,h]**

[a]Molecular Neurobiology laboratory, Department of Biochemistry and Molecular Biomedicine, Faculty of Biology, Universitat de Barcelona, Barcelona, Spain
[b]CiberNed, Network Center for Neurodegenerative diseases, National Spanish Health Institute Carlos III, Madrid, Spain
[c]School of Chemistry, Universitat de Barcelona, Barcelona, Spain
[d]Department of Biochemistry and Physiology, School of Pharmacy and Food Science Universitat de Barcelona, Barcelona, Spain
[e]Institute of Neurosciences, Universitat de Barcelona, Barcelona, Spain
[f]Department of Morphology and Cell Biology, Faculty of Medicine, University of Oviedo, Asturias, Spain
[g]Instituto de Neurociencias del Principado de Asturias (INEUROPA), Asturias, Spain
[h]Instituto de Investigación Sanitaria del Principado de Asturias (ISPA), Asturias, Spain
*Corresponding author. e-mail address: rfranco123@gmail.com

## Contents

| | |
|---|---|
| 1. Introduction | 50 |
| 2. Expression of $A_{2A}R$ in the basal ganglia | 51 |
| 3. Functional selectivity and signalling of $A_{2A}R$ | 53 |
|    3.1 $A_{2A}Rs$ as monomers | 53 |
|    3.2 $A_{2A}Rs$ as heteromers | 54 |
|    3.3 $A_{2A}R$-containing heteromers with relevance in basal ganglia circuits | 55 |
|    3.4 Multiple mechanisms of action of $A_{2A}R$ antagonists | 58 |
|    3.5 $A_{2A}R$ mediates the adenosine-dopamine functional interactions in the basal ganglia | 59 |
| 4. Heteromer expression in neurological diseases affecting the basal ganglia and in alterations of the reward circuits by drugs of abuse | 61 |
| 5. Concluding remarks | 62 |
| Conflict of interests | 63 |
| Author contribution | 63 |
| Funding | 63 |
| References | 63 |

## Abstract

Adenosine is a neuroregulatory nucleoside that acts through four G protein-coupled receptors (GPCRs), $A_1$, $A_{2A}$, $A_{2B}$ and $A_3$, which are widely expressed in cells of the nervous system. The $A_{2A}$ receptor ($A_{2A}R$), the GPCR with the highest expression in the striatum, has a similar role to that of receptors for dopamine, one of the main neurotransmitters. Neuronal and glial $A_{2A}Rs$ participate in the modulation of dopaminergic transmission and act in almost any action in which the basal ganglia is involved. This chapter revisits the expression of the $A_{2A}R$ in the basal ganglia in health and disease, and describes the diversity of signalling depending on whether the receptors are expressed as monomer or as heteromer. The $A_{2A}R$ can interact with other receptors as adenosine $A_1$, dopamine $D_2$, or cannabinoid $CB_1$ to form heteromers with relevant functions in the basal ganglia. Heteromerization, with these and other GPCRs, provides diversity to $A_{2A}R$-mediated signalling and to the modulation of neurotransmission. Thus, selective $A_{2A}R$ antagonists have neuroprotective potential acting directly on neurons, but also through modulation of glial cell activation, for example, by decreasing neuroinflammatory events that accompany neurodegenerative diseases. In fact, $A_{2A}R$ antagonists are safe and their potential in the therapy of Parkinson's disease has already led to the approval of one of them, istradefylline, in Japan and United States. The receptor also has a key role in reward circuits and, again, heteromers with dopamine receptors, but also with cannabinoid $CB_1$ receptors, participate in the events triggered by drugs of abuse.

## Abbreviations

| | |
|---|---|
| **GPCR** | G protein-coupled receptor |
| **$A_{2A}R$** | Adenosine $A_{2A}$ receptor |
| **CNS** | Central nervous system |
| **GABA** | Gamma aminobutyric acid |

## 1. Introduction

Approximately, 34% of all drugs approved by the Food and Drug Administration target 108 G protein-coupled receptors (GPCRs) (Hauser, Attwood, Rask-Andersen, Schiöth, & Gloriam, 2017). GPCRs constitute the most populated family of the human proteome with approximately 1000 members, many of which are olfactory receptors. This chapter is devoted to the adenosine $A_{2A}$ receptor ($A_{2A}R$) in the basal ganglia, more specifically to the $A_{2A}R$ in the largest basal ganglia component, the corpus striatum.

The nucleoside adenosine is relevant in metabolic processes in eukaryotic cells, but also in bacteria. In mammals, adenosine is also a modulator of several physiological processes. This molecule may be released into the extracellular milieu or may result from the action of phosphatases acting on extracellular ATP. Adenosine can interact with four cell surface GPCRs, $A_1$, $A_{2A}$, $A_{2B}$ and $A_3$, widely expressed and that belong to class A rhodopsin-type GPCRs; the difference between class A and class C GPCRs is the N-terminal domain, which is much larger in the case of the latter (Alexander et al., 2021). Almost any mammalian tissue/organ contains cells that express one or various adenosine receptors. In neurons, adenosine receptors appear in different locations and may be both presynaptic and postsynaptic. By acting on presynaptic receptors, adenosine was thought to inhibit neurotransmitter release, but this opinion has changed as the nucleoside can inhibit and facilitate neurotransmitter release into the synaptic cleft.

The striatum is the region where the $A_{2A}$ receptor is most abundant, both with respect to other brain areas and the rest of the tissues/organs. The receptor is key in regulating all functions involved in the basal ganglia circuits and has been for long considered a target in the therapy of Parkinson's disease (PD). Currently, a first-class $A_{2A}R$ antagonist, istradefylline, is available in Japan and the United States as an adjunctive medication in PD therapy.

## 2. Expression of $A_{2A}R$ in the basal ganglia

Unlike other adenosine receptors, $A_{2A}R$ shows an unique distribution in the mammalian central nervous system (CNS). Early pioneering radioligand binding and immunohistochemical studies in mammalian brain had already observed enrichment of receptor expression in the areas receiving dopaminergic projections in the basal ganglia i.e., striatum, nucleus accumbens (rostral pole, core and shell), external globus pallidus, olfactory tubercles, areas of extended amygdala and, even, the nucleus of the solitary tract but with a low labelling (Calon et al., 2004; Martinez-Mir, Probst, & Palacios, 1991; Palmer, Jacobson, & Stiles, 1992; Rosin, Robeva, Woodard, Guyenet, & Linden, 1998). The progress in techniques as in situ hybridisation, double-labelling immunohistochemistry or neuroanatomical tract-tracing enabled the identification of the exact neuronal populations in which $A_{2A}R$ is expressed in the striatum. In this sense, it has been demonstrated that labelling for specific $A_{2A}R$ mRNA preferentially appears in the medium-sized γ-aminobutyric acid (GABA)-expressing striatopallidal neurons (indirect pathway), colocalizing in rat striatum with enkephalin and dopamine $D_2$ receptor, but not with dopamine $D_1$ receptor, substance P or somatostatin (Augood & Emson, 1994; Bogenpohl, Ritter, Hall, & Smith, 2012; Fink et al., 1992; Quiroz et al., 2009; Schiffmann & Vanderhaeghen, 1993; Schiffmann, Jacobs, & Vanderhaeghen, 1991). Furthermore, a detailed ultrastructural analysis by

Hettinger et al., in 2001 revealed that $A_{2A}R$ is mainly located in dendrites and dendritic spines of GABA-labelled cells (in pre- and postsynaptic elements) in dorsolateral rat striatum, although it can also be found associated with plasma membrane or membrane-bound organelles in axons, axon terminals and somas (Hettinger, Lee, Linden, & Rosin, 2001).

A similar pattern of expression has been described in monkey and human striatum. Autoradiographic mapping (Martinez-Mir et al., 1991), in situ hybridisation (Svenningsson et al., 1998) or indirect analysis by positron emission tomography (PET) (Mishina et al., 2007, 2011) revealed the $A_{2A}R$ expression in the medium-sized spiny neurons (at pre- and postsynaptic level) of the caudate nucleus, putamen, accumbens, olfactory tubercle and the lateral segment of the globus pallidus, and that these cells also express the mRNA specific for $D_2$ receptor and preproenkephalin A but no for $D_1$ receptor or substance P (Bogenpohl et al., 2012).

At the single cell level, we made a discovery in basal ganglia output neurons of *Macaca fascicularis* whose relevance is not yet deciphered due to technical challenges (Luquin et al., 2012). Despite expression of mRNA for the $A_{2A}R$, the receptor was not detectable by a specific antibody that detected it in in other cells. Indeed, the $A_{2A}R$ was present in the output nuclei of basal ganglia, but it was only detectable by the specific antibody after unmasking using cholera toxin subunit B. This interesting observation may reflect an interaction with another protein, eventually with another GPCR, that hides the epitope recognised by the antibody.

In light with the proven distribution of $A_{2A}R$ in the brain, the receptor appears to be crucial in the maintenance of striatal neurotransmission architecture. Changes in receptor expression could be related to pathological situations as it has been demonstrated both in animal models of epilepsy (Rebola et al., 2011) or dementia (Espinosa et al., 2013; Moreira-de-Sá, Lourenço, Canas, & Cunha, 2021), and in patients of neurological diseases (Moreira-de-Sá et al., 2021; Stockwell, Jakova, & Cayabyab, 2017; Temido-Ferreira et al., 2018). In fact, some studies have described decreased densities of $A_{2A}R$ in striatum of Huntington's disease patients compared with controls (Martinez-Mir et al., 1991). Noteworthy, Calon et al., (2005) demonstrated increases of $A_{2A}R$ mRNA levels in the external globus pallidus of PD subjects versus controls and, more importantly, in the putamen of dyskinetic PD patients after treatment with levodopa, the most common dopamine-replacement therapy (Calon et al., 2004).

## 3. Functional selectivity and signalling of $A_{2A}R$
### 3.1 $A_{2A}Rs$ as monomers

G proteins to which GPCRs couple are heterotrimeric because they consist of three subunits α, β and γ. Following agonist binding, receptor-mediated signal transduction is initiated by opening the GPCR transmembrane helix 6, altering the structure of the coupled G proteins and allowing their activation through GTP/GDP exchange; subsequent events lead to changes in the cytosolic level of second messengers, mainly cAMP and calcium ions, until effect(s)/response(s) is(are) achieved (Oldham & Hamm, 2008). Considering the α subunit, human cells express up to 16 G proteins grouped into 4 families: $G_s$, $G_{i/o}$, $G_{q/11}$, and $G_{12/13}$ and which are encoded by 16 genes (Avet et al., 2022). The $G_s$ or activator family has two members: $G_s$ and $G_{olf}$, and is characterised by activating adenylate cyclase, increasing cAMP levels and stimulating protein kinase A (PKA). Conversely, the $G_i$ or inhibitor family has eight members: $G_{i1}$ $G_{i2}$, $G_{i3}$, rod transducin, cone transducin, $G_{\alpha oA}$, $G_{\alpha oB}$ and $G_{\alpha z}$. They block adenylate cyclase and/or activate phosphodiesterases, thus decreasing the cytosolic levels of cAMP and/or cGMP (Alexander et al., 2021; Neubert & Hurley, 1998; Neubert, Johnson, Hurley, & Walsh, 1992; Schertler, Villa, & Henderson, 1993). The $G_q$ family has 4 members in humans: $G_q$, $G_{11}$, $G_{14}$ and $G_{16}$; they mediate the regulation of the conversion of phosphatidylinositol 4,5-bisphosphate to inositol triphosphate and diacylglycerol, that in turns leads to activating protein kinase C (PKC) and increasing cytosolic calcium ion levels. Finally, the $G_{12/13}$ family has two members: $G_{12}$ and $G_{13}$ that have been linked to the progression of various types of malignant tumours (Rasheed et al., 2021; Suzuki, Hajicek, & Kozasa, 2009; Syrovatkina, Alegre, Dey, & Huang, 2016; Yang, Kuen, Chung, Kurose, & Kim, 2020). It is widely accepted that, upon activation of GPCRs, heterotrimeric G proteins may also provide signal transduction via β and γ subunits. The β subunit comprises five members, and the γ subunit includes 12 subtypes (Syrovatkina et al., 2016). In summary, the signal transduction machinery allows what is known as functional selectivity. For example, activation of a given receptor may lead to a myriad of responses depending on the context of the receptor and the cell type. Similarly, a given response would depend at the very least on the type of subunit (α, β and γ) of the coupled G protein. Also, if synthetic agonists are compared, the effect may vary because different binding modes, in the orthosteric site, may guide a differential coupling to the signal transduction machinery and, accordingly, to different (qualitative and/or quantitative) effects. This "property" of synthetic agonists is known as "biased agonism" and has potential in drug discovery. G protein-independent effects are also gaining interest, although its detailed description after $A_{2A}R$ activation is beyond the scope of this chapter.

Since $A_{2A}R$ couples to $G_s$ proteins, agonist binding leads to activation of adenylate cyclase, increases of cytosolic cAMP levels and protein kinase A activation (Alexander et al., 2021). On the one hand, activation of protein kinase A in the striatal neurons leads to the regulation of the activity of one of the main intracellular players in dopaminergic neurotransmission, 32 kDa dopamine- and cAMP-regulated phosphoprotein (DARPP-32). On the other hand, activation of the receptor in glial cells of the basal ganglia is relevant for regulating the action of activated astrocytes and activated microglia (see below).

Receptor activation produced by endogenous adenosine or synthetic $A_{2A}R$ agonists can lead to different signalling outputs as detailed above. Advances in the understanding of functional diversity mediated by a single receptor have occurred assuming both that the receptor is a monomer and that the receptor establishes direct interactions with other GPCRs, as already predicted four decades ago (Agnati, Fuxe, Zoli, Rondanini, & Ogren, 1982).

## 3.2 $A_{2A}Rs$ as heteromers

Early assays on GPCR-mediated signalling were performed in heterologous expression systems and assuming that receptors were expressed on the cell surface as monomers. More than two decades ago the prediction of receptor–receptor interactions started to be sustained by experimental evidence (Agnati et al., 1982); for rhodopsin-like class A receptors both homomers and heteromers were identified (Cvejic & Devi, 1997; Gines et al., 2000; Jordan & Devi, 1999). In these early times, heterodimers for class C GPCRs, $GABA_B$ receptors, were also discovered (White et al., 1998). Whereas class C receptors interact via the N-terminal domain, class A receptors interact via transmembrane domains. The first crystal structure of a dimer formed by the extracellular domain of a class C GPCR was obtained in the year 2000 (Kunishima et al., 2000). Interaction among class A receptors has been always questioned despite convergent evidence coming from biophysical, biochemical and pharmacological approaches (Franco, Martínez-Pinilla, Lanciego, & Navarro, 2016). It is a matter of time to know the first structure of a class A receptor heteromer, after the recent resolution of the apelin receptor homodimer (Yue et al., 2022). This recent report has not only shown that class A receptors may interact but that the overall structure of the functional GPCR unit is important for G protein-mediated signalling. When a receptor is part of a heteromer, its properties change, thus constituting a unique functional unit.

The $A_{2A}R$ may form homodimers (Canals et al., 2004) and, according to the recent data of the homodimer of the apelin receptor, the receptor-G protein stoichiometry may affect agonist-mediated effects (Yue et al., 2022). However, the breakthrough in understanding the role of $A_{2A}R$ occurs when complexes formed by $A_{2A}R$ and other GPCRs are considered.

The $A_{2A}R$ may interact with several other class A GPCRs and also with metabotropic glutamate class C GPCRs (see http://gpcr-hetnet.com/, accessed on December 30, 2022; Borroto-Escuela et al., 2014). Based on experimental evidence, it is suggested that the effect of $A_{2A}R$ agonists and/or $A_{2A}R$ antagonists may be quantitative or qualitatively different depending on the receptor that is interacting with the $A_{2A}R$. In the adenosine field, it has been demonstrated the existence of some $A_{2A}R$-containing heteromers with relevance in basal ganglia circuits. The most accepted possibility is the formation of heteroreceptor complexes between $A_{2A}R$ and $A_1R$ (Cristóvão-Ferreira et al., 2013; Schicker et al., 2009), and $A_{2A}R$ and $D_2$ receptor (Borroto-Escuela et al., 2011; Franco et al., 2000; Navarro et al., 2008), which have been proved both in vitro and in vivo assays, and in heterologous expression systems as well as in natural sources (primary cells and/or brains sections). Moreover, it has been observed that $A_{2A}Rs$ could physically interact with adenosine $A_3$ (Lillo, Martínez-Pinilla, Reyes-Resina, Navarro, & Franco, 2020), dopamine $D_3$ (Hillefors, Hedlund, & von Euler, 1999; Torvinen et al., 2005), group I metabotropic glutamate mGlu$_5$ (Cabello et al., 2009), cannabinoid $CB_1$ (Carriba et al., 2007; Navarro et al., 2008) and prosaposin GPR37 (Dunham, Meyer, Garcia, & Hall, 2009) receptors. Also, it may interact with the $P2Y_1$ purinoceptor (Nakata, Suzuki, Namba, & Oyanagi, 2010; Schicker et al., 2009), $P2Y_2$ purinoceptor (Schicker et al., 2009), the $P2Y_{12}$ purinoceptor (Nakata et al., 2010; Schicker et al., 2009) and the $P2Y_{13}$ purinoceptor (Schicker et al., 2009). There is indirect but strong evidence that the $A_{2A}R$ may simultaneously interact with two GPCRs, for instance with the $D_2$ and the $CB_1$ receptors (Bonaventura et al., 2014; Navarro et al., 2010).

## 3.3 $A_{2A}R$-containing heteromers with relevance in basal ganglia circuits

The physiological role of heteromer formation is to provide functional diversity, something that is especially relevant in the most complex structure of the mammalian body, the brain. There are allosteric interactions between the interacting receptors, in such a way that the structure of the receptor depends on the heteromer context. The allosteric modulation may automatically arise from the interaction; it may be triggered by binding of ligands, agonists or antagonists, or both. The discovery of $A_{2A}$–$D_2$ receptor heteromers was first interpreted as if the affinity of adenosine and/or dopamine for the cognate receptor in the heteromeric context varies, e.g., adenosine binding leads to lower affinity of dopamine binding to the $D_2$, thus reducing to $G_i$-mediated signalling (Hillion et al., 2002). To the best of our knowledge, there are no substantial changes in the affinity of endogenous ligands for most receptor heteromers, although some exceptions may occur. Alternatively, heteromer formation leads to differential functional selectivity and the exact nature of it depends on the heteromer. Every heteromer has its own signature, i.e., its own "heteromer imprint". The three main consequences of GPCR heteromerization are (i) differential coupling to G proteins, (ii) altered/biased signalling, and (iii) cross-antagonism (Franco et al., 2016). In drug discovery, the properties of the heteromer can be boosted because synthetic ligands may trigger novel allosteric interactions of therapeutic interest. Below are some examples of the consequences of the heteromerization of $A_{2A}$Rs.

At present, it is unclear whether the $A_{2A}$R may interact together with the $D_1$–$D_2$ receptor heteromer in those neurons in the basal ganglia in which the two dopamine receptors are expressed at the same time (Hasbi et al., 2020; Rico et al., 2016). This possibility should be explored as dopaminergic neurotransmission, in cells co-expressing these two dopamine receptors (for instance in dynorphin/enkephalin neurons), may proceed via calcium ions instead of cAMP. The dopamine receptor heteromer was the first example in which a change in G-protein coupling was demonstrated. Whereas GABAergic neurons containing $D_1$ receptors respond to dopamine via $G_i$ proteins and neurons containing $D_2$ receptors respond via $G_s$ proteins, the $D_1$–$D_2$ receptor heteromer is coupled to $G_q$ (Hasbi et al., 2009; Hasbi, O'Dowd, & George, 2011; Lee et al., 2004; Perreault et al., 2010).

Another GPCR-containing macromolecular complex, of interest in the basal ganglia, is that formed by adenosine $A_{2A}$ and $A_1$ receptors. The heteromer is present in nerve terminals arriving to the striatum from cortical neurons. Structurally, it is likely that the heteromer consists of a receptor heterotetramer in complex with one $G_i$ protein and one $G_s$ protein (Navarro et al., 2016, 2018). The heteromer affords something that cannot be achieved by monomeric (even by homodimeric) receptors; $G_i$ coupling occurs at low adenosine concentrations like the ones in the synaptic extracellular space, whereas $G_s$ coupling takes place at higher adenosine concentrations. Therefore, the heteromer imprint consists of differential G protein coupling depending on the concentration of adenosine. In striatal nerve terminals, the heteromer is involved in regulating up or down the release of neurotransmitters. In astrocytes, the same $A_1$–$A_{2A}$ receptor heteromer mediates, by a similar mechanism, the regulation of GABA transport by extracellular adenosine (Cristóvão-Ferreira et al., 2013).

Discovery of the interaction of the $A_{2A}R$ with the $CB_1$ receptor led to the hypothesis that a functional unit consisting of the three $A_{2A}$, $D_2$ and $CB_1$ receptors could occur in the striatum. Convergent evidence supports the idea that heteromers containing the three receptors are key components in the dendritic spines of GABAergic striatal neurons (see (Ferré et al., 2010, 2011) for review). In 2007, the term "striatal spine module" was coined and defined as: "*comprised of the dendritic spine of the medium spiny neuron (MSN), its glutamatergic and dopaminergic terminals and astroglial processes*" (Ferré et al., 2007). This module would be one of the many local models that operates as a unit that integrates information, processes it and delivers ad hoc outputs within the CNS. The metabotropic glutamate mGlu$_5$ receptor must be considered in the context of local modules as it is also able to interact with the $A_{2A}R$ and its imprint consists of an adenosine/glutamate synergistic effect (Ciruela et al., 2001; Ferré et al., 2002).

In the basal ganglia, the $A_{2A}R$ may be expressed in glial cells. PD courses with activation of microglia in which the receptor is upregulated, therefore, the receptor is also a target to afford neuroprotection by reducing neuroinflammation. $A_{2A}R$ antagonists would reduce the production of pro-inflammatory cytokines and/or boost the release of neuroprotective factors by activated microglia. One example is provided by the $A_{2A}R$ antagonist SCH-58261 that limits the noxious effect of lipopolysaccharide in the brain (Rebola et al., 2011). Similarly, $A_{2A}R$ antagonists reverse neuroinflammation occurring after the lesion in a rodent model of PD (Frau et al., 2011) or of striatal neurodegeneration (Minghetti et al., 2007). Recent evidence from our laboratory confirms heteromerization of the $A_{2A}R$ and another adenosine receptor, the $A_3$, in microglia. The $A_{2A}$–$A_3$ receptor heteromer imprint consists of blockade of $A_3R$ activation and its reversal by $A_{2A}R$ antagonists (Lillo et al., 2020). This imprint is consistent with reports on the anti-inflammatory potential of $A_3R$ activation in retinal subarachnoid haemorrhage (Li et al., 2020), neurodegeneration (Galvao et al., 2015), and post-ischaemic brain damage (Choi et al., 2011). Although the $A_{2A}$–$A_3$ receptor heteromer appears to be relevant in the field of Alzheimer's disease (Lillo et al., 2022), it may not have a significant role in physiological performance and/or pathophysiological mechanisms underlying neurological diseases affecting the basal ganglia or reward circuits.

We wonder whether the heteromer resulting from the interaction of $A_{2A}R$ with the adenosine $A_{2B}R$ has relevance for basal ganglia function. The $A_{2A}$–$A_{2B}$ heteromer has been already reported and has interesting properties. One of the imprints of the heteromer, namely a very low affinity of selective $A_{2A}R$ agonists, makes it difficult to understand its physiological role (De Filippo et al., 2020; Hinz et al., 2018). Interestingly though, data from in vivo models has shown that the heteromer participates in the control of obesity, also improving some of the effects of aging (Gnad et al., 2020). The reason to mention the $A_{2A}$–$A_{2B}$ heteromer in this chapter is because it has long been known that the $A_{2B}R$ is also a receptor for netrin-1 and is involved in the outgrowth of dorsal spinal cord axons (Corset et al., 2000). The future will tell whether the $A_{2A}$–$A_{2B}$ heteromer is expressed in the basal ganglia and, if so, whether it mediates some of the actions of netrin-1 in the development of the CNS and in the maturation of the brain in adolescence (Hoops & Flores, 2017).

## 3.4 Multiple mechanisms of action of $A_{2A}R$ antagonists

$A_{2A}R$ antagonists are generally safe and have potential in a variety of disease including those of the CNS (Armentero et al., 2011; Lopes et al., 2011; Willingham, Hotson, & Miller, 2020). The consequences of $A_{2A}R$ blockade by antagonists may vary from cell to cell. On the one hand, targeting cells that express a given heteromer would reduce the adverse effects of the medication. On the other hand, the affinity and/or potency of a given antagonist depends on the heteromeric context of the $A_{2A}R$. Several years ago, we discovered that a given antagonist could have differential effects when tested on postsynaptic versus presynaptic $A_{2A}R$-containing heteromers (Orru et al., 2011). In fact, the potency of the antagonists to modulate motor activity and inhibit striatal glutamate release, measured in vivo, suggested that the two selective antagonists, istradefylline (KW-6002) and SCH-442416, targeted postsynaptic and presynaptic $A_{2A}R$, respectively. These results correlate with in vitro assays showing differential binding affinity depending on the drug and the heteromer: SCH-442416 binds with high affinity to the $A_{2A}R$ when interacting with the $A_1R$, while it is istradefylline that has high affinity for binding to the

$A_{2A}R$ when forming heteromers with the $D_2$ receptor. As above indicated, the $A_1$–$A_{2A}$ heteromer is presynaptic whereas the $A_{2A}$–$D_2$ is postsynaptic/extrasynaptic. Other selective antagonists did not show any particular preference for the binding to the $A_{2A}R$ in any specific heteromeric context (Orru et al., 2011). These findings are a solid indication that antagonists, even selective, can have a stronger or milder effect depending on the heteromer. Furthermore, the results indicate that the target is often a GPCR in a heteromer rather than a "monomeric" receptor. Therefore, GPCR-focused drug discovery programs may consider conducting evaluations to select heteromer-selective drugs.

## 3.5 $A_{2A}R$ mediates the adenosine-dopamine functional interactions in the basal ganglia

The approval of istradefylline for anti-parkinsonian therapy is consequence of decades of intense research leading to decipher the mechanisms by which adenosine regulates dopaminergic neurotransmission in the basal ganglia. It was first reported that adenosine antagonised dopaminergic neurotransmission, and that the effect was mediated by cell surface receptors on neurons. Classically, GABAergic medium spiny neurons in the striatum of adult mammalian brains have been considered to segregate into cells containing $D_1$ and into cells containing $D_2$ receptors. This happens in a high percentage of these neurons, but there is a non-negligible percentage of striatal neurons that express both dopamine receptors (see above for details of the properties -$G_q$ coupling- arising from the interaction of these two dopamine receptors). Instrumental to advance in the knowledge of the central mechanisms of motor control, has been the indirect and direct pathway nomenclature; the difference lies in the number of intermediate brain regions needed to convey information from the cortex and striatum to the thalamus. The direct pathway (excitatory) consists of neurons expressing the $D_1$ receptor, and the indirect pathway (inhibitory) consists of neurons expressing the $D_2$ receptor. With this scenario in mind, another relevant discovery was related to a similar segregation of $A_1$ and $A_{2A}$ receptors; indeed, striatal medium spiny neurons that express the $D_1$ receptor but not the $D_2$ receptor, express the $A_1R$ and not the $A_{2A}R$, while those that express the $D_2$ receptor but not the $D_1$ receptor contain the $A_{2A}R$ and not the $A_1R$. Thus, the most obvious functional consequence was an adenosine–dopamine antagonism at the cAMP level. The $D_1$ receptor couples to $G_s$ and the $A_1R$ couples to $G_i$, therefore, activation of the $D_1$ receptor produces an increase of cytoplasmic cAMP levels that is counteracted by adenosine activation of $A_1R$. In the medium spiny neurons of the indirect pathway, the opposite occurs, activation of the $G_s$-coupled $A_{2A}R$ by adenosine counteracts the effect of dopamine on the $G_i$-coupled $D_2$ receptor. This antagonism and the substantial evidence of benefit of adenosine receptors blockade in models of PD prompted the proposal of targeting these receptors to combat this neurodegenerative disease. Interestingly, if not enough dopamine reaches the striatum of the patients, dopaminergic signalling may be boosted by adenosine receptor antagonists. In fact, due to the excellent safety record of $A_{2A}R$ antagonists several of them entered clinical trials in which, by ethical reasons, the dopamine replacement therapy could not be abandoned. Finally, a selective $A_{2A}R$ antagonist, istradefylline, was approved in Japan and United States to be taken by patients together with dopamine replacement therapy (marketed as Nourianz® in US and Nouriast® in Japan) (Jenner et al., 2009; Jenner, 2014; Mori et al., 2022; Saki, Yamada, Koshimura, Sasaki, & Kanda, 2013).

There have been further discoveries such as the heteromerization of $A_1$ and $D_1$ receptors on GABAergic neurons of the direct pathway, and of $A_{2A}$ and $D_2$ receptors on GABAergic neurons of the indirect pathway. Taken together, the findings suggest that the actual targets of istradefylline in striatal medium spiny neurons are $A_{2A}Rs$ that do not form heteromers plus the $A_{2A}R$-containing heteromers. Another consideration is the interest in evaluating the expression of adenosine receptors and adenosine–dopamine receptor heteromers in the striatum of both healthy and PD brains (or brains of animal models of PD) before and after dopamine replacement therapy, and when dyskinesias appear as a side effect of chronic levodopa medication. It is also relevant to consider the $A_{2A}$–$D_3$ receptor heteromer (Hillefors et al., 1999; Torvinen et al., 2005) since it is expressed in the basal ganglia and the $D_3$ receptor is now considered a therapeutic target for dyskinesias (Farré et al., 2015; Fiorentini et al., 2008; Franco et al., 2015; Marcellino et al., 2008; Scarselli et al., 2001; Solís, Garcia-Montes, González-Granillo, Xu, & Moratalla, 2017).

Findings in animals lacking the $D_2$ receptor have led to the demonstration that, in addition to dopaminergic modulation, $A_{2A}R$ ligands may exert functions in the CNS that are independent of regulation of $D_2$ receptor-mediated effects (Chen et al., 2001). The motor impairment in animals that are knockout for the $D_2$ receptor are rescued by $A_{2A}R$ antagonists, which were also able to restore enkephalin and substance P levels to normal (Aoyama, Kase, & Borrelli, 2000). These similar results obtained on two different laboratories suggest that *"selective $A_{2A}R$ antagonists can exhibit their anti-parkinsonian activities through a nondopaminergic mechanism"* (Aoyama et al., 2000).

The approval of istradefylline is a milestone in the search for neuroprotective interventions. At present, there is still no reliable marker for neuroprotection, so the neuroprotective efficacy of istradefylline could be evaluated in longitudinal studies comparing disease progression in patients taking or not the $A_{2A}R$ antagonist.

## 4. Heteromer expression in neurological diseases affecting the basal ganglia and in alterations of the reward circuits by drugs of abuse

The first study providing reliable data on how the expression of a given $A_{2A}R$-containing GPCR heteromer is altered in a neurological disease was performed in the *Macaca fascicularis* non-human primate model of PD. Results were obtained using a technique instrumental to detect complexes formed by two proteins in brain sections, the in situ proximity ligation assay (PLA). Data from PLAs addressing (two by two) the interactions between the $A_{2A}$, the $D_1$ and/or the cannabinoid $CB_1$ receptors, showed that heteromers were abundant in the caudate and putamen nuclei. Such marked expression was also found in animals with nigral degeneration produced by 1-methyl-4-phenyl-1,2,3,6-tetrahydropyridine (MPTP). Administration of levodopa to MPTP-induced parkinsonian animals lead to decreased expression of heteromers and disappearance of heteromer imprints, as detected by biochemical approaches (Bonaventura et al., 2014). Fully equivalent results were obtained in the 6-hydroxydopamine-based hemilesioned rodent model of PD (Pinna et al., 2014). Since a side effect of levodopa medication is the appearance of involuntary movements, the participation of the heteromer disruption in the development of dyskinesias is a real possibility.

According to the relevance of dopamine in reward circuits, which include the basal ganglia, and due to the high expression and seminal role of $A_{2A}R$ in this brain area, this receptor and the $A_{2A}$–$D_2$ heteromer have been investigated in models of drug addiction. In this chapter, we will briefly focus on cocaine addition (Borroto-Escuela et al., 2016; Lopes et al., 2011; Wydra et al., 2020). The $A_{2A}$ and the $D_2$ receptors are involved in sensitising and locomotor effects of cocaine (Filip et al., 2006). Both cocaine self-administration and withdrawal lead to significant changes in the expression of $A_{2A}$ and of dopamine receptors in the *nucleus accumbens* (Marcellino et al., 2007). Concerning the $A_{2A}$–$D_2$ heteromer, cocaine provokes conformational changes that may be detected by biophysical and pharmacological approaches (Marcellino et al., 2010). Also, recent data indicates that the expression and the imprint of the heteromers are altered in *"extinction from cocaine use, lost in cue induced reinstatement of cocaine seeking"* (Romero-Fernandez et al., 2022).

Considering that the cannabinoid $CB_1$ receptor mediates the psychotropic effect of (−)-trans-$\Delta^9$tetrahydrocannabinol, an interesting piece of evidence is the cocaine milieu signal mediated by the $CB_1$ receptor expressed in $A_{2A}R$-containing neurons (Turner et al., 2021). Another is the finding that $A_{2A}R$ antagonists alter some effects of cocaine via $CB_1$ receptor activation (Tozzi et al., 2012). Such crosstalk could be explained by the participation of $A_{2A}$–$CB_1$ receptor heteromers, although experimental proof of this possibility is lacking.

## 5. Concluding remarks

Consistent with the enormous amount of $A_{2A}Rs$ in striatum, expressed in both neurons and glia, the receptor has a key role in all events in which basal ganglia participate; in health but, also, in diseases affecting this brain area and in events affecting the plasticity of reward circuits such as in drug addiction. The huge amount of data on the potential of the $A_{2A}R$ in PD has led to the approval of the first in class antagonist, istradefylline, for anti-parkinsonian therapy. The current knowledge on the $A_{2A}R$ indicates that drugs aimed at targeting this receptor in the basal ganglia are, in fact, targeting the $A_{2A}R$ in a heteromeric context. Remarkably, the different structure of the $A_{2A}R$ acquired depending on the heteromeric environment makes that not all selective $A_{2A}R$ antagonists are equal; advantages are that different $A_{2A}R$ antagonists may serve to combat different diseases. This theoretical benefit is supported by the demonstration that of two different antagonists one has shown preferential action on the $A_1$–$A_{2A}$ heteromer (presynaptic), whereas the other has shown preferential action on the $A_{2A}$–$D_2$ heteromer (postsynaptic). In summary, drug screening should be done considering the various heteromers, whose targeting would provide advantage in a given disease.

In neurological diseases or in drug addiction, it would be convenient to address how the expression of $A_{2A}R$-containing heteromers change. This would help in select the right target, i.e., the one that is expressed, and the right antagonist, i.e., the one that shows preference for the heteromer to be targeted. The disruption of $A_{2A}R$-containing heteromers by the dopamine replacement therapy should be considered to either look for more efficacious adjunctive medications or to select a heteromer to target to avoid levodopa-induced dyskinesias. In this sense, comparing the progression of the disease and/or the appearance of dyskinesias induced by levodopa in PD patients, treated or not with istradefylline, will provide substantial benefit to understand the mechanisms of dyskinesias appearance and, eventually, to prevent its acquisition.

## Conflict of interests

Authors declare no conflict of interest.

## Author contribution

All the authors participated in the design and the writing, and have agreed on the submitted version.

## Funding

This work was supported by grants PID2020–113430RB-I00 and PID2021–126600OB-I00 funded by Spanish MCIN/AEI/10.13039/501100011033 and, as appropriate, by "ERDF A way of making Europe", by the "European Union" or by the "European Union Next Generation EU/PRTR". The research group of the University of Barcelona is considered of excellence (grup consolidat #2017 SGR 1497) by the Regional Catalonian Government, which does not provide any specific funding for reagents or for payment of services or Open Access fees.

## References

Agnati, L. F., Fuxe, K., Zoli, M., Rondanini, C., & Ogren, S. O. (1982). New vistas on synaptic plasticity: The receptor mosaic hypothesis of the engram. *Medical Biology, 60*(4), 183–190. ⟨https://pubmed.ncbi.nlm.nih.gov/6128444/⟩.

Alexander, S. P., Christopoulos, A., Davenport, A. P., Kelly, E., Mathie, A., Peters, J. A., ... Ye, R. D. (2021). The concise guide to pharmacology 2021/22: G protein-coupled receptors. *British Journal of Pharmacology, 178*(S1), S27–S156. https://doi.org/10.1111/BPH.15538

Aoyama, S., Kase, H., & Borrelli, E. (2000). Rescue of locomotor impairment in dopamine D2 receptor-deficient mice by an adenosine A2A receptor antagonist. *The Journal of Neuroscience: The Official Journal of the Society for Neuroscience, 20*(15), 5848–5852. https://doi.org/10.1523/JNEUROSCI.20-15-05848.2000

Armentero, M. T., Pinna, A., Ferré, S., Lanciego, J. L., Müller, C. E., & Franco, R. (2011). Past, present and future of A2A adenosine receptor antagonists in the therapy of Parkinson's disease. *Pharmacology & Therapeutics, 132*(3), 280–299. https://doi.org/10.1016/j.pharmthera.2011.07.004

Augood, S. J., & Emson, P. C. (1994). Adenosine A2a receptor mRNA is expressed by enkephalin cells but not by somatostatin cells in rat striatum: A co-expression study. *Brain, 22*(1–4), 204–210. https://doi.org/10.1016/0169-328X(94)90048-5

Avet, C., Mancini, A., Breton, B., Le Gouill, C., Hauser, A. S., ... Bouvier, M. (2022). Effector membrane translocation biosensors reveal G protein and Parrestin coupling profiles of 100 therapeutically relevant GPCRs. *ELife, 11*. https://doi.org/10.7554/ELIFE.74101

Bogenpohl, J. W., Ritter, S. L., Hall, R. A., & Smith, Y. (2012). Adenosine A 2A receptor in the monkey basal ganglia: Ultrastructural localization and colocalization with the metabotropic glutamate receptor 5 in the striatum. *Journal of Comparative Neurology, 520*(3), 570–589. https://doi.org/10.1002/cne.22751

Bonaventura, J., Rico, A. J., Moreno, E., Sierra, S., Sánchez, M., Luquin, N., & Franco, R. (2014). L-DOPA-treatment in primates disrupts the expression of A(2A) adenosine-CB (1) cannabinoid-D(2) dopamine receptor heteromers in the caudate nucleus. *Neuropharmacology, 79*, 90–100. https://doi.org/10.1016/j.neuropharm.2013.10.036

Borroto-Escuela, D. O., Brito, I., Romero-Fernandez, W., Di Palma, M., Oflijan, J., Skieterska, K., ... Fuxe, K. (2014). The G protein-coupled receptor heterodimer network (GPCR-HetNet) and its hub components. *International Journal of Molecular Sciences, 15*(5), 8570–8590. https://doi.org/10.3390/ijms15058570

Borroto-Escuela, D. O., Romero-Fernandez, W., Tarakanov, A. O., Ciruela, F., Agnati, L. F., & Fuxe, K. (2011). On the existence of a possible A2A-D2-β- arrestin2 complex: A2A agonist modulation of D2 agonist-induced β-arrestin2 recruitment. *Journal of Molecular Biology, 406*(5), 687–699. https://doi.org/10.1016/j.jmb.2011.01.022

Borroto-Escuela, D. O., Wydra, K., Pintsuk, J., Narvaez, M., Corrales, F., Zaniewska, M., ... Fuxe, K. (2016). Understanding the functional plasticity in neural networks of the basal ganglia in cocaine use disorder: A role for allosteric receptor-receptor interactions in A2A-D2 heteroreceptor complexes. *Neural Plasticity, 2016*, 1–12. https://doi.org/10.1155/2016/4827268

Cabello, N., Gandía, J., Bertarelli, D. C. G., Watanabe, M., Lluís, C., Franco, R., ... Ciruela, F. (2009). Metabotropic glutamate type 5, dopamine D2 and adenosine A 2a receptors form higher-order oligomers in living cells. *Journal of Neurochemistry, 109*(5), 1497–1507. https://doi.org/10.1111/j.1471-4159.2009.06078.x

Calon, F., Dridi, M., Hornykiewicz, O., Bédard, P. J., Rajput, A. H., & Di Paolo, T. (2004). Increased adenosine A2A receptors in the brain of Parkinson's disease patients with dyskinesias. *Brain: A Journal of Neurology, 127*(Pt 5), 1075–1084. https://doi.org/10.1093/BRAIN/AWH128

Canals, M., Burgueño, J., Marcellino, D., Cabello, N., Canela, E. I., Mallol, J., ... Franco, R. (2004). Homodimerization of adenosine A2A receptors: Qualitative and quantitative assessment by fluorescence and bioluminescence energy transfer. *Journal of Neurochemistry, 88*(3), 726–734. https://doi.org/10.1046/j.1471-4159.2003.02200.x

Carriba, P., Ortiz, O., Patkar, K., Justinova, Z., Stroik, J., Themann, A., ... Ferré, S. (2007). Striatal adenosine A2A and cannabinoid CB1 receptors form functional heteromeric complexes that mediate the motor effects of cannabinoids. *Neuropsychopharmacology: Official Publication of the American College of Neuropsychopharmacology, 32*(11), 2249–2259. https://doi.org/10.1038/sj.npp.1301375

Chen, J. F., Moratalla, R., Impagnatiello, F., Grandy, D. K., Cuellar, B., Rubinstein, M., ... Schwarzschild, M. A. (2001). The role of the D(2) dopamine receptor (D(2)R) in A(2A) adenosine receptor (A(2A)R)-mediated behavioral and cellular responses as revealed by A(2A) and D(2) receptor knockout mice. *Proceedings of the National Academy of Sciences of the United States of America, 98*(4), 1970–1975. https://doi.org/10.1073/PNAS.98.4.1970

Choi, I.-Y., Lee, J.-C., Ju, C., Hwang, S., Cho, G.-S., Lee, H. W., ... Kim, W.-K. (2011). A3 adenosine receptor agonist reduces brain ischemic injury and inhibits inflammatory cell migration in rats. *The American Journal of Pathology, 179*(4), 2042–2052. https://doi.org/10.1016/j.ajpath.2011.07.006

Ciruela, F., Escriche, M., Burgueno, J., Angulo, E., Casado, V., Soloviev, M. M., ... Franco, R. (2001). Metabotropic glutamate 1alpha and adenosine A1 receptors assemble into functionally interacting complexes. *Journal of Biological Chemistry, 276*(21), 18345–18351. https://doi.org/10.1074/jbc.M006960200

Corset, V., Nguyen-Ba-Charvet, K. T., Forcet, C., Moyse, E., Chédotal, A., & Mehlen, P. (2000). Netrin-1-mediated axon outgrowth and cAMP production requires interaction with adenosine A2b receptor. *Nature, 407*(6805), 747–750. https://doi.org/10.1038/35037600

Cristóvão-Ferreira, S., Navarro, G., Brugarolas, M., Pérez-Capote, K., Vaz, S. H., Fattorini, G., ... Sebastião, A. M. (2013). A1R-A2AR heteromers coupled to Gs and G i/o proteins modulate GABA transport into astrocytes. *Purinergic Signalling, 9*(3), 433–449. https://doi.org/10.1007/s11302-013-9364-5

Cvejic, S., & Devi, L. A. (1997). Dimerization of the delta opioid receptor: Implication for a role in receptor internalization. *The Journal of Biological Chemistry, 272*(43), 26959–26964.

De Filippo, E., Hinz, S., Pellizzari, V., Deganutti, G., El-Tayeb, A., Navarro, G., ... Müller, C. E. (2020). A2A and A2B adenosine receptors: The extracellular loop 2 determines high (A2A) or low affinity (A2B) for adenosine. *Biochemical Pharmacology, 172*, 113718. https://doi.org/10.1016/j.bcp.2019.113718

Dunham, J. H., Meyer, R. C., Garcia, E. L., & Hall, R. A. (2009). GPR37 surface expression enhancement via N-terminal truncation or protein-protein interactions. *Biochemistry, 48*(43), 10286–10297. https://doi.org/10.1021/bi9013775

Espinosa, J., Rocha, A., Nunes, F., Costa, M. S., Schein, V., Kazlauckas, V., ... Porciúncula, L. O. (2013). Caffeine consumption prevents memory impairment, neuronal damage, and adenosine A2A receptors upregulation in the hippocampus of a rat model of sporadic dementia. *Journal of Alzheimer's Disease: JAD, 34*(2), 509–518. https://doi.org/10.3233/JAD-111982

Farré, D., Muñoz, A., Moreno, E., Reyes-Resina, I., Canet-Pons, J., Dopeso-Reyes, I. G., ... Franco, R. (2015). Stronger dopamine D1 receptor-mediated neurotransmission in dyskinesia. *Molecular Neurobiology, 52*(3), 1408–1420. https://doi.org/10.1007/s12035-014-8936-x

Ferré, S., Agnati, L. F., Ciruela, F., Lluis, C., Woods, A. S., Fuxe, K., ... Franco, R. (2007). Neurotransmitter receptor heteromers and their integrative role in "local modules": The striatal spine module. *Brain Research Reviews, 55*(1), 55–67. https://doi.org/10.1016/j.brainresrev.2007.01.007

Ferré, S., Karcz-Kubicha, M., Hope, B. T., Popoli, P., Burgueño, J., Gutiérrez, M. A., ... Ciruela, F. (2002). Synergistic interaction between adenosine A2A and glutamate mGlu5 receptors: Implications for striatal neuronal function. *Proceedings of the National Academy of Sciences of the United States of America, 99*(18), 11940–11945. https://doi.org/10.1073/pnas.172393799

Ferré, S., Lluís, C., Justinova, Z., Quiroz, C., Orru, M., Navarro, G., ... Goldberg, S. R. (2010). Adenosine-cannabinoid receptor interactions. Implications for striatal function. *British Journal of Pharmacology, 160*(3), 443–453. https://doi.org/10.1111/j.1476-5381.2010.00723.x

Ferré, S., Quiroz, C., Orru, M., Guitart, X., Navarro, G., Cortés, A., ... Franco, R. (2011). Adenosine A2A and dopamine D2 receptor heteromers as key players in striatal function. *Frontiers in Neuroanatomy, 5*, 36. https://doi.org/10.3389/fnana.2011.00036

Filip, M., Frankowska, M., Zaniewska, M., Przegaliński, E., Muller, C. E., Agnati, L., ... Fuxe, K. (2006). Involvement of adenosine A2A and dopamine receptors in the locomotor and sensitizing effects of cocaine. *Brain Research, 1077*(1), 67–80. https://doi.org/10.1016/j.brainres.2006.01.038

Fink, J. S., Weaver, D. R., Rivkees, S. A., Peterfreund, R. A., Pollack, A. E., Adler, E. M., ... Reppert, S. M. (1992). Molecular cloning of the rat A2 adenosine receptor: Selective co-expression with D2 dopamine receptors in rat striatum. *Brain, 14*(3), 186–195. https://doi.org/10.1016/0169-328X(92)90173-9

Fiorentini, C., Busi, C., Gorruso, E., Gotti, C., Spano, P. F., & Missale, C. (2008). Reciprocal regulation of dopamine D1 and D3 receptor function and trafficking by heterodimerization. *Molecular Pharmacology, 74*(1), 59–69. https://doi.org/10.1124/mol.107.043885

Franco, R., Casadó-Anguera, V., Muñoz, A., Petrovic, M., Navarro, G., Moreno, E., ... Casadó, V. (2015). Hints on the lateralization of dopamine binding to D1 receptors in rat striatum. *Molecular Neurobiology*, *53*(8), 5436–5445. https://doi.org/10.1007/s12035-015-9468-8

Franco, R., Ferré, S., Agnati, L., Torvinen, M., Ginés, S., Hillion, J., ... Fuxe, K. (2000). Evidence for adenosine/dopamine receptor interactions: Indications for heteromerization. *Neuropsychopharmacology: Official Publication of the American College of Neuropsychopharmacology*, *23*(4), S50–S59. https://doi.org/10.1016/S0893-133X(00)00144-5

Franco, R., Martínez-Pinilla, E., Lanciego, J. L., & Navarro, G. (2016). Basic pharmacological and structural evidence for class A G-protein-coupled receptor heteromerization. *Frontiers in Pharmacology*, *7*, 76. https://doi.org/10.3389/fphar.2016.00076

Frau, L., Borsini, F., Wardas, J., Khairnar, A. S., Schintu, N., & Morelli, M. (2011). Neuroprotective and anti-inflammatory effects of the adenosine A(2A) receptor antagonist ST1535 in a MPTP mouse model of Parkinson's disease. *Synapse (New York, N. Y.)*, *65*(3), 181–188. https://doi.org/10.1002/SYN.20833

Galvao, J., Elvas, F., Martins, T., Cordeiro, M. F., Ambrósio, A. F., & Santiago, A. R. (2015). Adenosine A3 receptor activation is neuroprotective against retinal neurodegeneration. *Experimental Eye Research*, *140*, 65–74. https://doi.org/10.1016/J.EXER.2015.08.009

Gines, S., Hillion, J., Torvinen, M., Le Crom, S., Casado, V., Canela, E. I., ... Franco, R. (2000). Dopamine D1 and adenosine A1 receptors form functionally interacting heteromeric complexes. *Proceedings of the National Academy of Sciences*, *97*(15), 8606–8611. https://doi.org/10.1073/pnas.150241097

Gnad, T., Navarro, G., Lahesmaa, M., Reverte-Salisa, L., Copperi, F., Cordomi, A., ... Pfeifer, A. (2020). Adenosine/A2B receptor signaling ameliorates the effects of aging and counteracts obesity. *Cell Metabolism*, *32*(1), 56–70.e7. https://doi.org/10.1016/j.cmet.2020.06.006

Hasbi, A., Fan, T., Alijaniaram, M., Nguyen, T., Perreault, M. L., O'Dowd, B. F., ... George, S. R. (2009). Calcium signaling cascade links dopamine D1-D2 receptor heteromer to striatal BDNF production and neuronal growth. *Proceedings of the National Academy of Sciences of the United States of America*, *106*(50), 21377–21382. https://doi.org/10.1073/pnas.0903676106

Hasbi, A., O'Dowd, B. F., & George, S. R. (2011). Dopamine D1-D2 receptor heteromer signaling pathway in the brain: Emerging physiological relevance. *Molecular Brain*, *4*, 26. https://doi.org/10.1186/1756-6606-4-26

Hasbi, A., Sivasubramanian, M., Milenkovic, M., Komarek, K., Madras, B. K., & George, S. R. (2020). Dopamine D1-D2 receptor heteromer expression in key brain regions of rat and higher species: Upregulation in rat striatum after cocaine administration. *Neurobiology of Disease*, *143*, 105017. https://doi.org/10.1016/J.NBD.2020.105017

Hauser, A. S., Attwood, M. M., Rask-Andersen, M., Schiöth, H. B., & Gloriam, D. E. (2017). Trends in GPCR drug discovery: New agents, targets and indications. *Nature Reviews. Drug Discovery*, *16*(12), 829–842. https://doi.org/10.1038/nrd.2017.178

Hettinger, B. D., Lee, A., Linden, J., & Rosin, D. L. (2001). Ultrastructural localization of adenosine A2A receptors suggests multiple cellular sites for modulation of GABAergic neurons in rat striatum. *Journal of Comparative Neurology*, *431*(3), 331–346. https://doi.org/10.1002/1096-9861(20010312)431:3<331::AID-CNE1074>3.0.CO;2-W

Hillefors, M., Hedlund, P. B., & von Euler, G. (1999). Effects of adenosine A(2A) receptor stimulation in vivo on dopamine D3 receptor agonist binding in the rat brain. *Biochemical Pharmacology*, *58*(12), 1961–1964. ⟨http://www.ncbi.nlm.nih.gov/pubmed/10591151⟩.

Hillion, J., Canals, M., Torvinen, M., Casado, V., Scott, R., Terasmaa, A., ... Fuxe, K. (2002). Coaggregation, cointernalization, and codesensitization of adenosine A2A receptors and dopamine D2 receptors. *The Journal of Biological Chemistry*, *277*(20), 18091–18097. https://doi.org/10.1074/jbc.M107731200

Hinz, S., Navarro, G., Borroto-Escuela, D., Seibt, B. F., Ammon, C., De Filippo, E., ... Müller, C. E. (2018). Adenosine A2A receptor ligand recognition and signaling is blocked by A2B receptors. *Oncotarget, 9*(17), 13593–13611. https://doi.org/10.18632/oncotarget.24423

Hoops, D., & Flores, C. (2017). Making dopamine connections in adolescence. *Trends in Neurosciences, 40*(12), 709–719. https://doi.org/10.1016/J.TINS.2017.09.004

Jenner, P., Mori, A., Hauser, R., Morelli, M., Fredholm, B. B., & Chen, J. F. (2009). Adenosine, adenosine A 2A antagonists, and Parkinson's disease. *Parkinsonism & Related Disorders, 15*(6), 406–413. https://doi.org/10.1016/j.parkreldis.2008.12.006

Jenner, P. (2014). An overview of adenosine A2A receptor antagonists in Parkinson's disease. *International Review of Neurobiology, 119*, 71–86. https://doi.org/10.1016/B978-0-12-801022-8.00003-9

Jordan, B. A., & Devi, L. A. (1999). G-protein-coupled receptor heterodimerization modulates receptor function. *Nature, 399*(6737), 697–700. https://doi.org/10.1038/21441

Kunishima, N., Shimada, Y., Tsuji, Y., Sato, T., Yamamoto, M., Kumasaka, T., ... Morikawa, K. (2000). Structural basis of glutamate recognition by a dimeric metabotropic glutamate receptor. *Nature, 407*(6807), 971–977. https://doi.org/10.1038/35039564

Lee, S. P., So, C. H., Rashid, A. J., Varghese, G., Cheng, R., Lança, A. J., ... George, S. R. (2004). Dopamine D1 and D2 receptor co-activation generates a novel phospholipase C-mediated calcium signal. *Journal of Biological Chemistry, 279*(34), 35671–35678. https://doi.org/10.1074/jbc.M401923200

Li, P., Li, X., Deng, P., Wang, D., Bai, X., Li, Y., ... Yi, B. (2020). Activation of adenosine A3 receptor reduces early brain injury by alleviating neuroinflammation after subarachnoid hemorrhage in elderly rats. *Aging, 13*(1), 694–713. https://doi.org/10.18632/AGING.202178

Lillo, A., Martínez-Pinilla, E., Reyes-Resina, I., Navarro, G., & Franco, R. (2020). Adenosine A2a and A3 receptors are able to interact with each other. A further piece in the puzzle of adenosine receptor-mediated signaling. *International Journal of Molecular Sciences, 21*(14), 5070. https://doi.org/10.3390/ijms21145070

Lillo, A., Raïch, I., Lillo, J., Pérez-Olives, C., Navarro, G., & Franco, R. (2022). Expression of the adenosine $A_{2A-A3}$ receptor heteromer in different brain regions and marked upregulation in the microglia of the transgenic APPSw,Ind Alzheimer's disease model. *Biomedicines, 10*(2), 214. https://doi.org/10.3390/BIOMEDICINES10020214

Lopes, L. V., Sebastião, A. M., Ribeiro, J. A., V. Lopes, L. M., Sebastiao, A., A. Ribeiro, J., ... Ribeiro, J. A. (2011). Adenosine and related drugs in brain diseases: Present and future in clinical trials. *Current Topics in Medicinal Chemistry, 11*(8), 1087–1101. https://doi.org/10.2174/156802611795347591

Luquin, N., Sierra, S., Rico, A. J., Gómez-Bautista, V., Roda, E., Conte-Perales, L., ... Lanciego, J. L. (2012). Unmasking adenosine 2A receptors (A2ARs) in monkey basal ganglia output neurons using cholera toxin subunit B (CTB). *Neurobiology of Disease, 47*(3), 347–357. https://doi.org/10.1016/j.nbd.2012.05.006

Marcellino, D., Ferré, S., Casadó, V., Cortés, A., Le Foll, B., Mazzola, C., ... Franco, R. (2008). Identification of dopamine D1-D3 receptor heteromers: Indications for a role of synergistic D1-D3 receptor interactions in the striatum. *Journal of Biological Chemistry, 283*(38), 26016–26025. https://doi.org/10.1074/jbc.M710349200

Marcellino, D., Navarro, G., Sahlholm, K., Nilsson, J., Agnati, L. F., Canela, E. I., ... Fuxe, K. (2010). Cocaine produces D2R-mediated conformational changes in the adenosine A(2A)R-dopamine D2R heteromer. *Biochemical and Biophysical Research Communications, 394*(4), 988–992. https://doi.org/10.1016/j.bbrc.2010.03.104

Marcellino, D., Roberts, D. C. S. S., Navarro, G., Filip, M., Agnati, L., Lluís, C., ... Fuxe, K. (2007). Increase in A2A receptors in the nucleus accumbens after extended cocaine self-administration and its disappearance after cocaine withdrawal. *Brain Research, 1143*(1), 208–220. https://doi.org/10.1016/j.brainres.2007.01.079

Martinez-Mir, M. I., Probst, A., & Palacios, J. M. (1991). Adenosine A2 receptors: Selective localization in the human basal ganglia and alterations with disease. *Neuroscience, 42*(3), 697–706. https://doi.org/10.1016/0306-4522(91)90038-P

Minghetti, L., Greco, A., Potenza, R. L., Pezzola, A., Blum, D., Bantubungi, K., ... Popoli, P. (2007). Effects of the adenosine A2A receptor antagonist SCH 58621 on cyclooxygenase-2 expression, glial activation, and brain-derived neurotrophic factor availability in a rat model of striatal neurodegeneration. *Journal of Neuropathology and Experimental Neurology, 66*(5), 363–371. https://doi.org/10.1097/nen.0b013e3180517477

Mishina, M., Ishiwata, K., Kimura, Y., Naganawa, M., Oda, K., Kobayashi, S., ... Ishii, K. (2007). Evaluation of distribution of adenosine A2A receptors in normal human brain measured with [11C]TMSX PET. *Synapse (New York, N. Y.), 61*(9), 778–784. https://doi.org/10.1002/SYN.20423

Mishina, M., Ishiwata, K., Naganawa, M., Kimura, Y., Kitamura, S., Suzuki, M., ... Ishii, K. (2011). Adenosine A2A receptors measured with [11C]TMSX PET in the striata of Parkinson's disease patients. *PLoS One, 6*(2), https://doi.org/10.1371/JOURNAL.PONE.0017338

Moreira-de-Sá, A., Lourenço, V. S., Canas, P. M., & Cunha, R. A. (2021). Adenosine A2A receptors as biomarkers of brain diseases. *Frontiers in Neuroscience, 15*, 702581. https://doi.org/10.3389/fnins.2021.702581

Mori, A., Chen, J.-F., Uchida, S., Durlach, C., King, S. M., & Jenner, P. (2022). The pharmacological potential of adenosine A 2A receptor antagonists for treating Parkinson's disease. *Molecules (Basel, Switzerland), 27*(7), 2366. https://doi.org/10.3390/MOLECULES27072366

Nakata, H., Suzuki, T., Namba, K., & Oyanagi, K. (2010). Dimerization of G protein-coupled purinergic receptors: Increasing the diversity of purinergic receptor signal responses and receptor functions. *Journal of Receptors and Signal Transduction, 30*(5), 337–346. https://doi.org/10.3109/10799893.2010.509729

Navarro, G., Carriba, P., Gandía, J., Ciruela, F., Casadó, V., Cortés, A., ... Franco, R. (2008). Detection of heteromers formed by cannabinoid CB1, dopamine D2, and adenosine A2A G-protein-coupled receptors by combining bimolecular fluorescence complementation and bioluminescence energy transfer. *The Scientific World Journal, 8*, 1088–1097. https://doi.org/10.1100/tsw.2008.136

Navarro, G., Cordomí, A., Brugarolas, M., Moreno, E., Aguinaga, D., Pérez-Benito, L., ... Franco, R. (2018). Cross-communication between Gi and Gs in a G-protein-coupled receptor heterotetramer guided by a receptor C-terminal domain. *BMC Biology, 16*(1), 24. https://doi.org/10.1186/s12915-018-0491-x

Navarro, G., Cordomí, A., Zelman-Femiak, M., Brugarolas, M., Moreno, E., Aguinaga, D., ... Franco, R. (2016). Quaternary structure of a G-protein-coupled receptor heterotetramer in complex with Gi and Gs. *BMC Biology, 14*(1), 26. https://doi.org/10.1186/s12915-016-0247-4

Navarro, G., Ferré, S., Cordomi, A., Moreno, E., Mallol, J., Casadó, V., ... Woods, A. S. (2010). Interactions between intracellular domains as key determinants of the quaternary structure and function of receptor heteromers. *Journal of Biological Chemistry, 285*(35), 27346–27359. https://doi.org/10.1074/jbc.M110.115634

Neubert, T. A., Johnson, R. S., Hurley, J. B., & Walsh, K. A. (1992). The rod transducin alpha subunit amino terminus is heterogeneously fatty acylated. *Journal of Biological Chemistry, 267*(26), 18274–18277. https://doi.org/10.1016/S0021-9258(19)36955-8

Neubert, T. A., & Hurley, J. B. (1998). Functional heterogeneity of transducin α subunits. *FEBS Letters, 422*(3), 343–345. https://doi.org/10.1016/S0014-5793(98)00037-4

Oldham, W. M., & Hamm, H. E. (2008). Heterotrimeric G protein activation by G-protein-coupled receptors. *Nature Reviews Molecular Cell Biology, 9*(1), 60–71. https://doi.org/10.1038/nrm2299

Orru, M., Bakešová, J., Brugarolas, M., Quiroz, C., Beaumont, V., Goldberg, S. R., ... Ferré, S. (2011). Striatal pre- and postsynaptic profile of adenosine A2A receptor antagonists. *PLoS One, 6*(1), e16088. https://doi.org/10.1371/journal.pone.0016088

Palmer, T. M., Jacobson, K. A., & Stiles, G. L. (1992). Immunological identification of A2 adenosine receptors by two antipeptide antibody preparations. *Molecular Pharmacology, 42*(3), 391–397.

Perreault, M. L., Hasbi, A., Alijaniaram, M., Fan, T., Varghese, G., Fletcher, P. J., & George, S. R. (2010). The dopamine D1-D2 receptor heteromer localizes in dynorphin/enkephalin neurons: Increased high affinity state following amphetamine and in schizophrenia. *The Journal of Biological Chemistry, 285*(47), 36625–36634. https://doi.org/10.1074/jbc.M110.159954

Pinna, A., Bonaventura, J., Farré, D., Sánchez, M., Simola, N., Mallol, J., & Franco, R. (2014). L-DOPA disrupts adenosine A(2A)-cannabinoid CB(1)-dopamine D(2) receptor heteromer cross-talk in the striatum of hemiparkinsonian rats: Biochemical and behavioral studies. *Experimental Neurology, 253*, 180–191. https://doi.org/10.1016/j.expneurol.2013.12.021

Quiroz, C., Luján, R., Uchigashima, M., Simoes, A. P., Lerner, T. N., Borycz, J., ... Ferré, S. (2009). Key modulatory role of presynaptic adenosine A 2A receptors in cortical neurotransmission to the striatal direct pathway. *The Scientific World Journal, 9*, 1321–1344. https://doi.org/10.1100/tsw.2009.143

Rasheed, S. A. K., Subramanyan, L. V., Lim, W. K., Udayappan, U. K., Wang, M., ... Casey, P. J. (2021). The emerging roles of Gα12/13 proteins on the hallmarks of cancer in solid tumors. *Oncogene, 41*(2), 147–158. https://doi.org/10.1038/s41388-021-02069-w

Rebola, N., Simões, A. P., Canas, P. M., Tomé, A. R., Andrade, G. M., Barry, C. E., ... Cunha, R. A. (2011). Adenosine A2A receptors control neuroinflammation and consequent hippocampal neuronal dysfunction. *Journal of Neurochemistry, 117*(1), 100–111. https://doi.org/10.1111/j.1471-4159.2011.07178.x

Rico, A. J., Dopeso-Reyes, I. G., Martínez-Pinilla, E., Sucunza, D., Pignataro, D., Roda, E., ... Lanciego, J. L. (2016). Neurochemical evidence supporting dopamine D1–D2 receptor heteromers in the striatum of the long-tailed macaque: Changes following dopaminergic manipulation. *Brain Structure and Function, 222*(4), 1–18. https://doi.org/10.1007/s00429-016-1306-x

Romero-Fernandez, W., Wydra, K., Borroto-Escuela, D. O., Jastrzębska, J., Zhou, Z., Frankowska, M., ... Fuxe, K. (2022). Increased density and antagonistic allosteric interactions in A2AR-D2R heterocomplexes in extinction from cocaine use, lost in cue induced reinstatement of cocaine seeking. *Pharmacology, Biochemistry, and Behavior, 215*, 173375. https://doi.org/10.1016/J.PBB.2022.173375

Rosin, D. L., Robeva, A., Woodard, R. L., Guyenet, P. G., & Linden, J. (1998). Immunohistochemical localization of adenosine A(2A) receptors in the rat central nervous system. *Journal of Comparative Neurology, 401*(2), 163–186. https://doi.org/10.1002/(SICI)1096-9861(19981116)401:2<163::AID-CNE2>3.0.CO;2-D

Saki, M., Yamada, K., Koshimura, E., Sasaki, K., & Kanda, T. (2013). In vitro pharmacological profile of the A2A receptor antagonist istradefylline. *Naunyn-Schmiedeberg's Archives of Pharmacology, 386*(11), 963–972. https://doi.org/10.1007/s00210-013-0897-5

Scarselli, M., Novi, F., Schallmach, E., Lin, R., Baragli, A., Colzi, A., ... Maggio, R. (2001). D2/D3 dopamine receptor heterodimers exhibit unique functional properties. *Journal of Biological Chemistry, 276*(32), 30308–30314. https://doi.org/10.1074/jbc.M102297200

Schertler, G. F., Villa, C., & Henderson, R. (1993). Projection structure of rhodopsin. *Nature, 362*(6422), 770–772. https://doi.org/10.1038/362770a0

Schicker, K., Hussl, S., Chandaka, G. K., Kosenburger, K., Yang, J. W., Waldhoer, M., ... Boehm, S. (2009). A membrane network of receptors and enzymes for adenine nucleotides and nucleosides. *Biochimica et Biophysica Acta, 1793*(2), 325–334. https://doi.org/10.1016/J.BBAMCR.2008.09.014

Schiffmann, S. N., & Vanderhaeghen, J. J. (1993). Adenosine A2 receptors regulate the gene expression of striatopallidal and striatonigral neurons. *The Journal of Neuroscience, 13*(3), 1080. https://doi.org/10.1523/JNEUROSCI.13-03-01080.1993

Schiffmann, S. N., Jacobs, O., & Vanderhaeghen, J.-J. (1991). Striatal restricted adenosine A2 receptor (RDC8) is expressed by enkephalin but not by substance P neurons: An in situ hybridization histochemistry study. *Journal of Neurochemistry, 57*(3), 1062–1067. https://doi.org/10.1111/J.1471-4159.1991.TB08257.X

Solís, O., Garcia-Montes, J. R., González-Granillo, A., Xu, M., & Moratalla, R. (2017). Dopamine D3 receptor modulates l-DOPA-induced dyskinesia by targeting D1 receptor-mediated striatal signaling. *Cerebral Cortex, 27*(1), 435–446. https://doi.org/10.1093/cercor/bhv231

Stockwell, J., Jakova, E., & Cayabyab, F. S. (2017). Adenosine A1 and A2A receptors in the brain: Current research and their role in neurodegeneration. *Molecules (Basel, Switzerland), 22*(4), 676. https://doi.org/10.3390/molecules22040676

Suzuki, N., Hajicek, N., & Kozasa, T. (2009). Regulation and physiological functions of G12/13-mediated signaling Pathways. *Neuro-Signals, 17*(1), 55–70. https://doi.org/10.1159/000186690

Svenningsson, P., Le Moine, C., Aubert, I., Burbaud, P., Fredholm, B. B., & Bloch, B. (1998). Cellular distribution of adenosine A2A receptor mRNA in the primate striatum. *The Journal of Comparative Neurology, 399*(2), 229–240. https://doi.org/10.1002/(sici)1096-9861(19980921)399:2<229::aid-cne6>3.0.co;2-2

Syrovatkina, V., Alegre, K. O., Dey, R., & Huang, X. Y. (2016). Regulation, signaling, and physiological functions of G-proteins. *Journal of Molecular Biology, 428*(19), 3850–3868. https://doi.org/10.1016/J.JMB.2016.08.002

Temido-Ferreira, M., Ferreira, D. G., Batalha, V. L., Marques-Morgado, I., Coelho, J. E., Pereira, P., ... Lopes, L. V. (2018). Age-related shift in LTD is dependent on neuronal adenosine A2A receptors interplay with mGluR5 and NMDA receptors. *Molecular Psychiatry*, 1–25. https://doi.org/10.1038/s41380-018-0110-9

Torvinen, M., Marcellino, D., Canals, M., Agnati, L. F., Lluis, C., Franco, R., ... Fuxe, K. (2005). Adenosine A2A receptor and dopamine D3 receptor interactions: Evidence of functional A2A/D3 heteromeric complexes. *Molecular Pharmacology, 67*(2), 400–407. https://doi.org/10.1124/mol.104.003376

Tozzi, A., de Iure, A., Marsili, V., Romano, R., Tantucci, M., Filippo, M., ... Calabresi, P. (2012). A2A adenosine receptor antagonism enhances synaptic and motor effects of cocaine via CB1 cannabinoid receptor activation. *PLoS One, 7*(6), https://doi.org/10.1371/JOURNAL.PONE.0038312

Turner, B. D., Smith, N. K., Manz, K. M., Chang, B. T., Delpire, E., Grueter, C. A., ... Grueter, B. A. (2021). Cannabinoid type 1 receptors in A2a neurons contribute to cocaine-environment association. *Psychopharmacology, 238*(4), 1121–1131. https://doi.org/10.1007/S00213-021-05759-1

White, J. H., Wise, A., Main, M. J., Green, A., Fraser, N. J., Disney, G. H., ... Marshall, F. H. (1998). Heterodimerization is required for the formation of a functional GABA(B) receptor. *Nature, 396*(6712), 679–682. https://doi.org/10.1038/25354

Willingham, S. B., Hotson, A. N., & Miller, R. A. (2020). Targeting the A2AR in cancer; early lessons from the clinic. *Current Opinion in Pharmacology, 53*, 126–133. https://doi.org/10.1016/j.coph.2020.08.003

Wydra, K., Gawliński, D., Gawlińska, K., Frankowska, M., Borroto-Escuela, D. O., Fuxe, K., & Filip, M. (2020). Adenosine A2AReceptors in substance use disorders: A focus on cocaine. *Cells, 9*(6), https://doi.org/10.3390/CELLS9061372

Yang, Y. M., Kuen, D. S., Chung, Y., Kurose, H., & Kim, S. G. (2020). Gα12/13 signaling in metabolic diseases. *Experimental & Molecular Medicine, 52*(6), 896–910. https://doi.org/10.1038/S12276-020-0454-5

Yue, Y., Liu, L., Wu, L. J., Wu, Y., Wang, L., Li, F., ... Xu, F. (2022). Structural insight into apelin receptor-G protein stoichiometry. *Nature Structural & Molecular Biology, 29*(7), 688–697. https://doi.org/10.1038/s41594-022-00797-5

CHAPTER FOUR

# How and why the adenosine $A_{2A}$ receptor became a target for Parkinson's disease therapy

Peter Jenner[a,*], Tomoyuki Kanda[b], and Akihisa Mori[c]
[a]Institute of Pharmaceutical Sciences, King's College London, London, United Kingdom
[b]Kyowa Kirin Co., Ltd., Otemachi. Chiyoda-ku, Tokyo, Japan
[c]SNLD Ltd., Akashi-sho, Chuo-ku, Tokyo, Japan
*Corresponding author e-mail address: peter.jenner@kcl.ac.uk

## Contents

| | |
|---|---|
| 1. Introduction | 74 |
| 2. The relevance of the adenosinergic system and the $A_{2A}$ adenosine receptor | 77 |
| 3. Adenosine $A_{2A}$ receptors and motor function | 80 |
| 4. Adenosine $A_{2A}$ antagonists and dyskinesia | 84 |
| 5. Adenosine $A_{2A}$ antagonists and motor function in Parkinson's disease | 85 |
| 6. Other potential mechanisms of action of $A_{2A}$ adenosine antagonists | 88 |
| 7. Adenosine antagonists and non-motor symptoms of Parkinson's disease | 90 |
| 8. Exploring the future potential of adenosine $A_{2A}$ antagonists in Parkinson's disease | 92 |
| References | 95 |

## Abstract

Dopaminergic therapy for Parkinson's disease has revolutionised the treatment of the motor symptoms of the illness. However, it does not alleviate all components of the motor deficits and has only limited effects on non-motor symptoms. For this reason, alternative non-dopaminergic approaches to treatment have been sought and the adenosine $A_{2A}$ receptor provided a novel target for symptomatic therapy both within the basal ganglia and elsewhere in the brain. Despite an impressive preclinical profile that would indicate a clear role for adenosine $A_{2A}$ antagonists in the treatment of Parkinson's disease, the road to clinical use has been long and full of difficulties. Some aspects of the drugs preclinical profile have not translated into clinical effectiveness and not all the clinical studies undertaken have had a positive outcome. The reasons for this will be explored and suggestions made for the further development of this drug class in the treatment of Parkinson's disease. However, one adenosine $A_{2A}$ antagonist, namely istradefylline has been introduced successfully for the treatment of late-stage Parkinson's disease in two major areas of the world and has become a commercial success through offering the first non-dopaminergic approach to the treatment of unmet need to be introduced in several decades.

## 1. Introduction

There is no doubt that dopaminergic replacement therapy in Parkinson's disease has had a major impact on the control of motor dysfunction, improved clinical outcomes and quality of life (Mao, Qin, Zhang, & Ye, 2020). This primarily centres on levodopa and the range of enzyme inhibitors used to enhance levodopa's actions (peripheral dopa decarboxylase inhibitors, DDCi's; catechol-O-methyltransferase inhibitors, COMTi's and monoamine oxidase B inhibitors, MAO-Bi's), the oral dopamine agonist drugs (ropinirole, pramipexole), a range of formulations of levodopa and dopamine agonists and the technologies used to deliver both levodopa and dopamine agonist drugs by non-oral routes of administration in early and late disease (rotigotine, apomorphine) (Lees et al., 2023).

However, dopamine replacement therapy is commonly associated with a range short term side effects and long-term treatment related complications (Kulisevsky, 2022). In addition, some components of motor disability in Parkinson's disease, such as postural instability and gait, are not responsive to dopaminergic medications. This suggests that their aetiology relates to biochemical or pathological change in the basal ganglia unrelated to the loss of dopaminergic input or to the loss of non-dopaminergic neuronal systems in other parts of the brain. Even phenomena that appear to be primarily dopaminergic in nature, for example 'wearing off' appears to have non-dopaminergic components as it cannot be fully controlled through alterations in dopaminergic medication (Rota et al., 2022).

In this context, it is perhaps surprising that there are few non-dopaminergic approaches to treating the motor symptoms of Parkinson's disease. Those in general use are non-selective anti-cholinergic drugs, and amantadine which is a low affinity antagonist for NMDA receptors among many other actions (Rascol, Fabbri, & Poewe, 2021) and in Japan, repurposing of the anti-epileptic drug zonisamide which has multi-mechanisms of action (i.e., MAO-Bi, blockade for T-type $Ca^{++}$ channels) (Murata, 2010). There have been multiple attempts to introduce novel therapies that are non-dopaminergic in nature, and which act through a variety of other neuronal systems including noradrenergic, serotoninergic and glutamatergic receptor systems (Cenci, Skovgard, & Odin, 2022; Gonzalez-Latapi, Bhowmick, Saranza, & Fox, 2020). Overall, these have shown potential efficacy in preclinical studies but failed to translate into clinical effect in Parkinson's disease due to a lack of effect or a narrow therapeutic window or unexpected adverse events. In addition, most have been aimed at the control or prevention of established levodopa-induced dyskinesia rather than addressing a primary role in the control of motor abnormalities. Recently, drugs exhibiting a multimodal activity have also been shown to be effective—for example safinamide which acts as both as a MAO-Bi and a sodium channel blocker that inhibits excessive glutamate release (Pagonabarraga, Tinazzi, Caccia, & Jost, 2021).

Another fact of Parkinson's disease that in recent years has been recognised as being more clinically relevant than even the treatment of motor deficits, are the non-motor components of the illness. These are wide and diverse in nature affecting both the peripheral and central nervous systems and which can be a greater contributor to poor quality of life than motor symptoms and which are proving more difficult to combat through pharmacological manipulation (Chaudhuri & Schapira, 2009; Schapira, Chaudhuri, & Jenner, 2017). A range of studies have shown that dopaminergic therapy can have some effect on some non-motor symptoms but do not exert complete control and that some non-motor symptoms are unresponsive to the available dopaminergic medications. Consequently, non-motor symptoms and their treatment is a key unmet need in Parkinson's disease. While some non-motor symptoms may arise from alterations in basal ganglia function in Parkinson's disease, it is likely that most reflect the widespread biochemical and pathological change that occurs in this illness. Indeed, non-motor symptoms of the illness may appear many years before any motor manifestations perhaps showing that the early pathology of Parkinson's disease may start distant from the basal ganglia and then progressively spread through the brain during the illness (Sauerbier, Jenner, Todorova, & Chaudhuri, 2016) Unfortunately, there has not been the investment in preclinical studies of non-motor symptoms and the pharmacological manipulation which the unmet medical need would warrant. At present, it remains unclear what the pathophysiological basis of individual symptoms might be. Consequently, non-motor symptoms are treated with drugs aimed at controlling the expression of each symptom as it appears leading to a situation in which polypharmacy becomes dominant (Csoti, Herbst, Urban, Woitalla, & Wullner, 2019).

The conclusion at present must be that while dopaminergic drugs are a major asset to the treatment of Parkinson's disease and that their use will continue to be necessary in all patients developing the illness. However, given that they do not control all aspects of the illness and given the widespread pathological and progressive biochemical change that takes place in Parkinson's disease, it seems only sensible that novel treatments affecting key neuronal systems involved in the expression and control of motor function and in the development and expression of non-motor symptoms should be sought. Such targets might be within the strio-thalamo-cortical loops that control motor function within the basal ganglia or in other components of the basal ganglia involved in sensory or cognitive control. More likely, they will be found in other brain areas in which non-dopaminergic mechanisms contribute to both non-motor and motor symptoms. Whatever the reality, the need is for non-dopaminergic targets for future drug treatment that would either control motor and non-motor symptoms by themselves or synergise with the currently used dopaminergic approaches to therapy.

This is why exploring adenosine $A_{2A}$ antagonists as a novel therapeutic approach to Parkinson's disease appears to have much to offer in overcoming the deficits of current symptomatic treatments. In this chapter, we will explain the potential of this drug class and the obstacles that have had to overcome during development and in the future. To set the scene for the reader and to layout the case for the use of adenosine $A_{2A}$ antagonists, the key reasons for focussing on this class of molecule are summarised below and each of these will be fully explored in the following sections:

1. Adenosine $A_{2A}$ receptors have a selective localisation to the brain and to specific areas of the brain, most notably the basal ganglia.
2. Within the basal ganglia, adenosine $A_{2A}$ receptors are selectively found on the cell bodies and terminals of the indirect striatal GABAergic output pathway that is intimately involved in the control of motor function.
3. Adenosine exerts an excitatory effect on GABAergic transmission via adenosine $A_{2A}$ receptors that is overactive in Parkinson's disease.
4. Adenosine $A_{2A}$ receptors are also present in other brain regions associated with the motor and non-motor symptoms of Parkinson's disease, most notably limbic areas of the brain.
5. Adenosine $A_{2A}$ receptors interact with other important neurotransmitter systems relevant to Parkinson's disease most notably the serotoninergic and glutamatergic systems.
6. Selective adenosine $A_{2A}$ receptor antagonists are effective in reversing motor symptoms in experimental models of Parkinson's disease predictive of clinical efficacy.
7. Selective adenosine $A_{2A}$ receptor antagonists are also effective in alleviating some non-motor symptoms of Parkinson's disease in preclinical models of the illness.

8. In clinical trial, adenosine $A_{2A}$ antagonists, most notably istradefylline, have been shown to be effective in reducing 'off' time and increasing 'on' time without troublesome dyskinesia in late-stage patients already receiving optimal dopaminergic therapy.

## 2. The relevance of the adenosinergic system and the $A_{2A}$ adenosine receptor

The identification of adenosine as an important neuromodulator opened a new chapter in the biology of the peripheral and central nervous system and this has been documented extensively elsewhere (Chen, Lee, & Chern, 2014; Fredholm, Chern, Franco, & Sitkovsky, 2007). For those in the Parkinson's disease area, most of whom are still unfamiliar with adenosine, it is important to point out that it does not exist as a classical neurotransmitter in that there are no adenosinergic neurons identified as such and the classical machinery associated with synaptic transmission is absent – with the possible exception of striatal GABAergic neurons. Rather, adenosine is produced as a product of adenine nucleotide metabolism in which neurons are bathed and which then acts through adenosine receptors to modulate neuronal activity. This simple summary is explored in extensive detail in previous chapters of this volume and extensively in other publications.

Key to the interest in adenosine receptors for the treatment of Parkinson's disease has been the classification of receptor subtypes into $A_1$, $A_{2A}$, $A_{2B}$ and $A_3$ receptors (Fredholm, AP, Jacobson, Linden, & Müller, 2011). A detailed description of the discovery and classification and characteristics of the different subtypes is presented elsewhere in this volume and in previous publications. From the perspective of Parkinson's disease, there is one key discovery that led to the evaluation of adenosine $A_{2A}$ antagonists as a potential treatment for Parkinson's disease. Namely that the $A_{2A}$ receptor has an almost specific localisation to brain and within the brain has a selective localisation to the basal ganglia (Ishiwata et al., 2005; Jarvis & Williams, 1989; Svenningsson, Hall, Sedvall, & Fredholm, 1997). It is important to point out that relevant to later sections of this paper, $A_{2A}$ receptors are also found in other brain regions in low concentrations most notably limbic brain areas, and this will be discussed in more detail in relation to the potential effects of these drugs on non-motor symptoms of Parkinson's disease.

The selective localisation of adenosine $A_{2A}$ receptors to the basal ganglia and to the striatum would be sufficient reason for this to become an interesting target for the treatment of Parkinson's disease. However, this interest is increased by the impressive demonstration of the predominant localisation of the $A_{2A}$ adenosine receptor to the cell bodies and terminals of the indirect striatal GABAergic output pathway to the external segment of the globus pallidus (Mori & Shindou, 2003; Mori, 2020; Mori, Shindou, Ichimura, Nonaka, & Kase, 1996; Shindou et al., 2002; Shindou, Mori, Kase, & Ichimura, 2001; Shindou, Richardson, Mori, Kase, & Ichimura, 2003). This has particular relevance to Parkinson's disease as the balance between the indirect (NO-GO pathway) and direct (GO pathway) striatal output pathways is thought to underlie the deficits of voluntary movement in the illness. In particular, the indirect output pathway is thought to have a major influence on the control of motor function that occur in Parkinson's disease. Whereas the direct output pathway is thought to have a more relevant role to the onset and expression of dyskinesia. Sophisticated electrophysiological and biochemical studies have shown that the manipulation of adenosine $A_{2A}$ receptor function leads to changes in GABAergic transmission in the indirect output pathway and in the pallido-subthalamic GABAergic pathway with which it makes synaptic connectivity. Adenosine acts to enhance GABAergic transmission whereas adenosine $A_{2A}$ antagonists reduce GABAergic function. This would be key to a role in the treatment of Parkinson's disease as the indirect output pathway becomes overactive so increasing the NO-GO influence on the expression of movement. All of this evidence places the adenosine $A_{2A}$ receptor in a position that is thought to be essential for controlling striatal output and identifies it as a pharmacological target that would allow manipulation of this key pathway in the control of movement (Mori et al., 2022; Mori, 2020) (see Fig. 1).

It needs to be pointed out that there is an alternative view that the adenosine $A_{2A}$ receptor coexist dimerised with dopamine D2 receptors in the striatum and that it is the action of adenosine at this site that serves to manipulate dopaminergic transmission and striatal output (Ferre et al., 2002; Ferre et al., 2008; Ferre et al., 2018). This view is also backed by a wealth of preclinical evidence and controversy has raged over where drugs acting at adenosine $A_{2A}$ receptors would function in Parkinson's disease. The prevailing view is that in Parkinson's disease it is the $A_{2A}$ adenosine receptors located on the indirect output pathway that are key to the potential actions of drugs treating motor dysfunction. However, there could be several different scenarios in which adenosine $A_{2A}$ receptors are

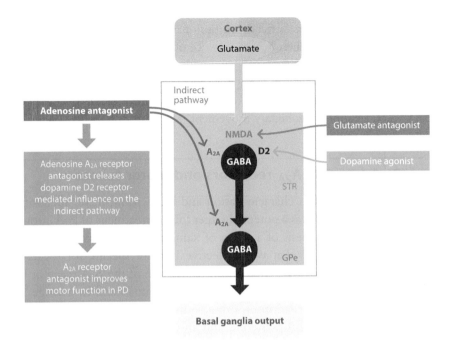

**Fig. 1** A schematic illustration of the ability of an adenosine $A_{2A}$ antagonist to alter the activity of the indirect striatal output pathway and to modulate motor function. *(This figure is reproduced from the cover of 'Parkinsonism and Related Disorders' volume 80, Suppl. 1, 2020, edited by Jenner P and LeWitt P, 'Adenosine $A_{2A}$ receptor antagonism in the treatment paradigm of Parkinson's disease: The why and the how? With permission from Elsevier.)*

involved in the basal ganglia control of movement. These are as follows: (1) Adenosine $A_{2A}$ receptors colocalised with dopamine D2 receptors on the cell bodies of the indirect output pathway are functionally important in the modification of dopaminergic transmission and that adenosinergic drugs exert their action at this level; (2) Adenosine $A_{2A}$ receptors are functionally localised on the cell bodies and terminals of the indirect striatal output pathway and it is at the level of the stratum and the external globus pallidus that manipulation of adenosine receptor function alters motor activity; (3) Both concepts of adenosine $A_{2A}$ receptor function are correct in the normal brain but in Parkinson's disease where dopaminergic transmission is increasingly lost, it is the $A_{2A}$ receptors located on striatal output neurons that becomes the key target for manipulation of adenosinergic activity and the control of motor function.

At this point it is also necessary to make mention of the fact that while there is a dominance of the views that adenosine $A_{2A}$ receptors are either found on the indirect striatal output pathway or colocalised with D2 dopamine receptors there is also evidence of the presence of $A_{2A}$ receptors on cholinergic and glutamatergic inputs within the striatum suggesting that manipulation of adenosine $A_{2A}$ receptor function also alters transmission in these pathways which are themselves key to the control of the overall basal ganglia contribution to voluntary movement (Sebastiao & Ribeiro, 1996).

## 3. Adenosine $A_{2A}$ receptors and motor function

The identification characterisation and localisation of the adenosine $A_{2A}$ adenosine receptor as a potential target for the treatment of Parkinson's disease led to the synthesis of a range of xanthine and non-xanthine adenosine antagonist drugs that were selective but not specific for the $A_{2A}$ receptor. This then allowed exploration of the role of the $A_{2A}$ receptor in the control of motor function both in normal animals and in a range of experimental models of Parkinson's disease using classical pharmacological techniques based on both systemic and intra-cerebral administration of agonist and antagonist drugs (Kanda & Jenner, 2020). Using the classical preclinical models of Parkinson's disease, the potential anti parkinsonian activities of antagonists such as preladenant, tozadenant, ST1535 and istradefylline were explored in rodents and primates (Kanda & Jenner, 2020; Michel et al., 2017; Michel, Downey, Nicolas, & Scheller, 2014). For example, istradefylline exerted effects on motor function in (1) haloperidol-treated mice; (2) reserpine-treated mice; (3) 6-hydroxydopamine lesioned rats; (4) MPTP-treated rats and mice; (5) MPTP-treated common marmosets and (6) MPTP-treated cynomolgus monkeys (Kanda & Jenner, 2020) (see Table 1).

There has been general agreement over the effects on motor function observed with the range of $A_{2A}$ antagonist drugs so far evaluated in rodent models (Collins et al., 2012; Correa et al., 2004; Fenu, Pinna, Ongini, & Morelli, 1997; Koga, Kurokawa, Ochi, Nakamura, & Kuwana, 2000; Rose, Ramsay Croft, & Jenner, 2007; Salamone, 2010) and in MPTP-treated primates (Grondin et al., 1999; Hodgson et al., 2010; Kanda, Jackson, et al., 1998; Kanda, Tashiro, Kuwana, & Jenner, 1998; Rose et al., 2006; Uchida et al., 2014; Uchida et al., 2015a, 2015b). The effects can be summarised as follows:

1. Administration of $A_{2A}$ antagonist drugs alone produces a mild symptomatic improvement in motor function.

**Table 1** Pharmacological activity of selective adenosine $A_{2A}$ antagonists in experimental models used to evaluate antiparkinsonian activity.

| $A_{2A}$ antagonist | Model | Pharmacological activities |
|---|---|---|
| Preladenant | • Rodents | • Potentiated levodopa-induced contralateral rotation in 6-OHDA-lesioned rats<br>• Attenuated haloperidol-induced catalepsy in rats |
| | • Non-human primates | • Improved motor ability without causing dopaminergic-mediated dyskinetic or motor complications in MPTP-treated cynomolgus monkeys<br>• Significantly improved the parkinsonian scores and increased locomotor activity in combination with L-DOPA in MPTP-treated cynomolgus monkeys |
| Istradefylline | • Rodents | • Antagonised haloperidol-induced catalepsy in mice and rats<br>• Antagonised reserpine-induced catalepsy in mice<br>• Antagonised CGS21680-induced catalepsy in mice<br>• Enhanced MPTP- or reserpine-induced hypo-locomotion in mice<br>• Enhanced the contralateral rotational response induced by an optimal dose of apomorphine and a suboptimal dose of levodopa in 6-OHDA-lesioned rats |
| | • Non-human primates | • Increased locomotor activity and improved motor disability in MPTP-treated cynomolgus monkeys<br>• Little or no dyskinesia produced by the drug alone in MPTP-treated cynomolgus monkeys<br>• In combination with levodopa, produced an additive effect in increasing locomotor activity without affecting dyskinesia expression in MPTP-treated cynomolgus monkeys<br>• Reversed motor disability but prevented the induction of dyskinesia by repeated apomorphine administration in MPTP-treated cynomolgus monkeys<br>• Drug alone Improved motor disability and increased locomotor activity in MPTP-treated common marmosets |

(*continued*)

**Table 1** Pharmacological activity of selective adenosine $A_{2A}$ antagonists in experimental models used to evaluate antiparkinsonian activity. (cont'd)

| $A_{2A}$ antagonist | Model | Pharmacological activities |
|---|---|---|
| | | • Drug alone produced little or no dyskinesia in MPTP-treated common marmosets |
| | | • On repeated administration, antiparkinsonian effects were maintained for ≥ 21 days in MPTP-treated common marmosets |
| | | • In combination with levodopa, produced an additive antiparkinsonian effect without worsening levodopa-induced dyskinesia in MPTP-treated common marmosets |
| | | • Did not worsen the severity of existing dyskinesia when administered in combination with suboptimal dose of levodopa for 21days in MPTP-treated common marmosets |
| | | • In combination with dopamine agonists, enhanced the reversal of motor disability and increased 'on' time in MPTP-treated common marmosets |
| | | • Enhanced the antiparkinsonian response produced by a threshold dose of a dopamine agonist or a suboptimal dose of levodopa in MPTP-treated common marmosets |
| Tozadenant | • Rodents | • Potentiated levodopa-induced contralateral rotation in 6-OHDA-lesioned rats |
| | | • Produced rearing behaviour in 6-OHDA-lesioned rats |
| | • Non-human primates | • Produced an increase in locomotor activity and improvement in motor disability in MPTP-treated common marmosets. |
| ST1535 | • Rodents | • Increased spontaneous locomotor activity and reversed haloperidol-induced catalepsy in mice |
| | | • Enhanced contralateral rotational response induced by a threshold dose of levodopa in 6-OHDA-lesioned rats |
| | • Non-human primates | • Produced a dose-related increase in locomotor activity and improvement in motor disability in MPTP-treated common marmosets. |
| | | • Increased 'on' time induced by levodopa in MPTP-treated common marmosets |

Data for the table were reconstructed from Kanda and Jenner (2020), Michel et al. (2014, 2017).

2. $A_{2A}$ antagonist drugs showed additive effects in improving motor function when co-administered with levodopa suggesting a role as adjunct therapy in Parkinson's disease.
3. Activity was maintained on repeated administration with no suggestion of tolerance to the effects of adenosine $A_{2A}$ antagonist drugs. The dose-response curve was shallow, and the effect maintained over a wide range of doses once maximal affect was achieved.
4. $A_{2A}$ antagonist drugs also showed additive effects on motor function when administered in conjunction with a dopamine agonist drug. Again, suggesting a role as adjunct therapy in Parkinson's disease.

A number of other observations were also relevant to the potential use of the drugs. First, the effects of $A_{2A}$ antagonist drugs when administered with levodopa where were most marked when low or threshold doses of levodopa were employed. The effects of the adenosine antagonist compounds were less obvious and less marked when combined with optimal or maximal levodopa dosage (Rose et al., 2006; Uchida et al., 2014). This suggests that the use of these compounds would be in the earlier stages of Parkinson's disease or as levodopa sparing therapy. Second, an interesting but unexplained observation was that the effect of a single dose of levodopa was enhanced whether administered with an $A_{2A}$ antagonist drug or whether administered 24 or 48 h following adenosine antagonist administration to levodopa-naïve MPTP-treated primates which had no motor complications (i.e., 'wearing off' or dyskinesia) and when the $A_{2A}$ antagonist would have been cleared from blood and brain (Kanda et al., 2000). This suggests an adaptive change in the responsiveness to levodopa is induced by $A_{2A}$ antagonist administration, but this phenomenon has never been further explored either in a preclinical model or of Parkinson's disease or in clinical evaluation. Third, Pharmacological analysis showed that the effects of the antagonists examined is mediated through the $A_{2A}$ receptor and not through the $A_1$ receptor for which these compounds also have some affinity (Kanda, Tashiro, et al., 1998). For compounds such as istradefylline, there are no other pharmacological sites at which these drugs are known to interact clearly suggesting their activity is mediated through $A_{2A}$ receptor antagonism alone (Saki, Yamada, Koshimura, Sasaki, & Kanda, 2013). One last relevant point is that the models of Parkinson's disease used are largely a reflection of the later stages of the illness in terms of the extent of nigral denervation but are a model of nigro-striatal denervation rather than reflecting the widespread pathology and biochemistry of the illness as it affects man (El-Gamal et al., 2021; Smeyne & Jackson-Lewis, 2005).

## 4. Adenosine $A_{2A}$ antagonists and dyskinesia

A continuing concern over the use of dopaminergic medications for the treatment of the motor symptoms of Parkinson's disease is the onset of involuntary movements (dyskinesia). This is particularly true with the use of levodopa in the long-term control of motor symptoms where both troublesome and non-troublesome dyskinesia commonly occurs. Consequently, a non-dopaminergic approach to the treatment of motor symptoms could be viewed as a possible means of either reducing the risk of dyskinesia or of decreasing the intensity of established involuntary movements. The adenosine $A_{2A}$ antagonists would appear on mechanistic grounds to be such an opportunity to not only control motor function but also to avoid treatment complications related to dyskinesia.

The general view of the aetiology of dyskinesia is that like motor dysfunction in Parkinson's disease, it relates to an imbalance between the activity of the D-1 mediated direct and D-2 mediated indirect striatal output pathways. Whereas motor function is considered as being controlled largely by the activity of the indirect output pathway, it is alterations in the activity of the direct output pathway that are thought to be responsible for dyskinesia (Bastide et al., 2015). This may certainly be the case for the expression of the complex involuntary movements that makeup dyskinesia but in it is unlikely to underlie the priming process that lays down the motor memory for the persistence of dyskinesia once established. Whatever the circumstances, a drug which specifically interacts with the indirect output pathway and modulates its activity could be viewed as a candidate for improving motor function while avoiding dyskinesia induction or expression. This would seem to be one potential clinical advantage of the use of adenosine $A_{2A}$ antagonists both in the early stages of Parkinson's disease for the avoidance of dyskinesia induction and for the later stages of Parkinson's disease where improvements in motor function would not necessarily be accompanied by any increase in the intensity or severity of involuntary movements. This view is enhanced by the finding from post-mortem and PET studies that the density of $A_{2A}$ adenosine receptors in the striatum changes over the course of the disease (Calon et al., 2004; Mishina & Ishiwata, 2014; Mishina et al., 2011; Mishina et al., 2017; Morelli et al., 2007; Ramlackhansingh et al., 2011; Waggan et al., 2023). In early Parkinson's disease, the density of $A_{2A}$ adenosine receptors may be reduced (Hurley, Mash, & Jenner, 2000); Waggan et al., 2013) but as the disease progresses there is an increase in $A_{2A}$ receptor density and this is particularly marked in those patients with advanced disease exhibiting dyskinesia.

There has been some exploration of the effects of selective adenosine $A_{2A}$ antagonist drugs on dyskinesia in experimental models of Parkinson's disease, largely the 6-hydroxydopamine lesioned rat (Jones, Bleickardt, Mullins, Parker, & Hodgson, 2013; Lundblad, Vaudano, & Cenci, 2003; Tronci et al., 2007; Winkler, Kirik, Björklund, & Cenci, 2002) and MPTP-treated primate (Grondin et al., 1999; Kanda, Jackson, et al., 1998; Rose et al., 2006; Uchida et al., 2014; Uchida et al., 2015a) where repeated levodopa treatment induces involuntary movements that are similar to those occurring in patients with Parkinson's disease receiving chronic levodopa administration. Many of such studies were undertaken using istradefylline and the results show that administration of istradefylline as monotherapy improved motor function but did not precipitate dyskinesia in animals previously exposed to levodopa. When administered in conjunction with levodopa, dyskinesia was not enhanced and on repeated istradefylline administration the intensity of dyskinesia decreased. This suggests that this is a functional expression of the selective interaction of adenosine $A_{2A}$ antagonists with the indirect striatal output pathway but not the direct output pathway. However, it is more difficult to explain why this would lead to a decrease in dyskinesia intensity. Very recently, the mainstream view that the D-1 mediated direct output pathway is responsible for dyskinesia has been challenged and major role shown for the D-2 mediated indirect pathway has been suggested for the expression of dyskinesia. In all probability, both output pathways are involved in the genesis and expression of dyskinesia but in an interrelated and complex manner (Castela et al., 2023).

## 5. Adenosine $A_{2A}$ antagonists and motor function in Parkinson's disease

The preclinical evidence shows the anatomical and functional localisation of adenosine $A_{2A}$ receptors at the level of the indirect striatal output pathway. In predictive experimental models of Parkinson's disease, this translated to the ability of adenosine $A_{2A}$ antagonists to improve motor function coupled to limiting the extent of dyskinesia. Extrapolated to clinical use in Parkinson's disease, the expectation would be that this class of drug would produce a mild symptomatic improvement as monotherapy in early disease and an additive affect when used in combination with levodopa or dopamine agonist drugs in later illness.

Two molecules, vipadenant and BIIB 014 were briefly examined in man with vipadenant discontinued because of toxicity issues (Brooks et al., 2010; Pawsey et al., 2016; Pinna, 2014). Three selective adenosine $A_{2A}$ antagonists, namely preladenant (Hauser et al., 2011; Hauser et al., 2015), tozadenant (Hauser et al., 2014) and istradefylline (Fernandez et al., 2010; Hauser et al., 2008; Hauser, Hubble, Truong, & Istradefylline, 2003; LeWitt et al., 2008; Mizuno et al., 2010; Pourcher et al., 2012; Stacy et al., 2008) have been extensively evaluated in clinical trials for the symptomatic improvement of motor function in Parkinson's disease. Analysis of these clinical studies has been reported elsewhere and in a subsequent chapter in this volume and it is not the purpose of this paper to re-analyse these investigations in detail (Hauser et al., 2021; Jenner, 2014; Jenner, Mori, Aradi, & Hauser, 2021; LeWitt, Aradi, Hauser, & Rascol, 2020).

An overall summary of the clinical evaluation of adenosine $A_{2A}$ antagonists is as follows:

1. In a limited number of monotherapy studies in the early stages of Parkinson's disease, the expected symptomatic affect predicted by the preclinical studies was not observed for preladenant or istradefylline.
2. In phase IIA proof-of-concept studies in patients with Parkinson's disease receiving suboptimal levodopa treatment, an improvement in motor function occurred without a worsening of dyskinesia and a reduction in 'off' time was observed (Bara-Jimenez et al., 2003).
3. In phase IIB/III studies in populations with late-stage Parkinson's disease receiving levodopa and optimal dopaminergic therapy but with remaining significant 'off' time aimed at showing a therapeutic effect on off episode with an increase 'on' time without troublesome dyskinesia, the results have been mixed with some trials showing the expected changes in motor markers but others failing to detect any significant drug effect.

The latter requires some explanation. Questions have been raised about trial design, drug dosage, patient demographics, clinical evaluation and placebo effect size among others. We and others have commented on these issues elsewhere (Jenner, 2014; LeWitt et al., 2020). However, the outcomes of these studies have influenced the subsequent development of adenosine $A_{2A}$ antagonists for Parkinson's disease. Tozadenant was not developed further because of toxicity issues while the failure of preladenant in some phase II and all phase III studies led to the termination of the development programme. Istradefylline was examined in 8 phase II/III investigations of which 5 met the primary outcome measures but where an overall analysis of drug effect in all 8 showed a positive effect. In Japan, 2 positive phase II/III clinical trials led to the registration and approval of the product for the treatment of 'off' periods in Parkinson's disease (Dungo & Deeks, 2013). In the United States, the FDA granted approval for marketing of istradefylline based on the results of the phase II/III investigations including Japan (Chen & Cunha, 2020). In the EU, approval was refused based on evidence of efficacy after consideration of all 8 clinical investigations and a lack of drug effect in studies that involved European centres (https://www.ema.europa.eu/en/medicines/human/EPAR/nouryant). EMA considered that the results of the studies were inconsistent and did not satisfactorily show that istradefylline was effective at reducing 'off' time.

The rationale for the phase III development of adenosine $A_{2A}$ antagonists was undertaken in line with regulatory requirements to show an effect of the drugs in later stage patients with Parkinson's disease optimised on dopaminergic medication but where significant 'off' time remained. In other words, it was necessary to show an affect over and above that which could be achieved using already available dopaminergic medication. Consequently, many of the patients recruited to these studies were receiving not only levodopa but also a dopamine agonist drug and in addition an MAO-B inhibitor or COMT inhibitor. This makes showing a further improvement a difficult challenge as the window of opportunity for significant change in standard rating scales derived from the evaluation of dopaminergic drug effect is small.

Another significant problem may relate to the overall strategy for the clinical evaluation of this drug class. The regulatory requirements for evaluation did not fit with the scientific evidence for the potential use of adenosine $A_{2A}$ antagonists in Parkinson's disease. The available evidence from the use of adenosine $A_{2A}$ antagonists, most notably istradefylline, in experimental models of a functional effect of these drugs in Parkinson's disease, showed that when administered with high doses of levodopa little or no additional benefit on motor function was observed. In contrast, when administered with threshold or low doses of levodopa a significant interaction and additive effect with levodopa occurred (see above). This strongly suggests that clinical evaluation in the earlier stages of Parkinson's disease when sub maximal doses of levodopa are employed but where additional benefit is required would have been a more appropriate positioning for this novel non- dopaminergic approach to treatment. The implications of this will be explored in a subsequent paragraph.

## 6. Other potential mechanisms of action of $A_{2A}$ adenosine antagonists

While the emphasis has been on the positioning of $A_{2A}$ adenosine receptors on the indirect striatal output pathway as being the major site of action of selective antagonist drugs, there is evidence that $A_{2A}$ receptors may also modulate the activity of other neurotransmitter systems, including acetylcholine, glutamate, serotonin and endocannabinoids (Pinna, Parekh, & Morelli, 2023; Pinna, Serra, Marongiu, & Morelli, 2020). Two of these interactions seem immediately relevant to the potential use of $A_{2A}$ antagonist drugs in the treatment of the motor symptoms of Parkinson's disease.

There has been extensive interest in the potential for using glutamate antagonists to improve motor function and suppress dyskinesia in Parkinson's disease (Johnson, Conn, & Niswender, 2009). While efficacy has been shown for both ionotropic and metabotropic receptor antagonists in a range of preclinical models (Sebastianutto & Cenci, 2018), this has not reliably translated into clinical effect without the occurrence of adverse events (Montastruc, Rascol, & Senard, 1997). However, interactions between $A_{2A}$ receptor antagonists and glutamate NMDA receptors and metabotropic mGlu4/mGlu5 glutamate receptors have been reported both in vitro and in in vivo animal models of Parkinson's disease which seem of greater interest (Pinna et al., 2020). Recent findings in rodent and non-human primate models of Parkinson's disease have shown that the combined blockade of NMDA receptors and adenosine $A_{2A}$ receptors may improve motor performance without producing dyskinesia (Michel et al., 2015; Michel et al., 2017). How this occurs is debatable but $A_{2A}$ receptors and NMDA receptors located in the striatum may have a common intracellular signalling pathway that reduces the excessive glutamatergic transmission thought to occur in Parkinson's disease and in the expression of levodopa-induced dyskinesia (Nash & Brotchie, 2000). In a similar manner, $A_{2A}$ receptor antagonists may act synergistically to potentiate the effects of both mGlu5 receptor inhibitors and mGlu4 receptor activators. Systemic administration of $A_{2A}$ receptor antagonists and compounds acting on mGlu4/5 receptors improved motor deficits in models of Parkinson's disease, and this may relate to their combined ability to modulate the activity of the striatal indirect output pathway although the exact mechanisms remain unknown (Coccurello, Breysse, & Amalric, 2004; Ferre et al., 2002; Jones et al., 2012; Kachroo et al., 2005; Lopez et al., 2008).

In recent years, there has been considerable interest in the role played by serotoninergic transmission in the genesis and/or expression of levodopa-induced dyskinesia (Corsi, Stancampiano, & Carta, 2021). While the control of serotoninergic transmission is complex with multiple 5-HT receptor subtype, attention has centred on the activity of 5-HT1 agonist drugs as a potential treatment for dyskinesia. This is based on the tight control of serotonin neuron firing and release exerted by both 5HT-1A and 5HT-1B receptors (Pinna et al., 2023). In preclinical studies of the action of selective agonist drugs in experimental models of levodopa-induced dyskinesia, 5HT-1A or 5HT-1B agonists have been shown to reduce the intensity of involuntary movements but also decreasing the ability of levodopa to improve motor function (Iravani, Tayarani-Binazir, Chu, Jackson, & Jenner, 2006). However, the translation of the antidyskinetic effect to Parkinson's disease in man has been difficult with either again a reduction in levodopa's antiparkinsonian activity or a lack of efficacy (Goetz et al., 2007). There may be a more reproducible effect using molecules that are an agonist at both 5HT-1A and 5HT-1B receptors, such as eltoprazine (Bezard et al., 2013; Munoz et al., 2008; Svenningsson et al., 2015).

In this respect, studies using a combination of eltoprazine and an adenosine $A_{2A}$ antagonist, namely preladenant have shown an additive effect in reducing the intensity of established levodopa-induced dyskinesia in both 6-hydroxydopamine lesioned rats and in MPTP-treated primates that occurs without any decrease in the antiparkinsonian effect of levodopa (Ko et al., 2017; Pinna et al., 2016). This is another illustration of the potential use of adenosine $A_{2A}$ antagonists in Parkinson's disease but as an addition to the use of another non-dopaminergic approach to treatment. The use of combinations of drugs for the treatment of motor abnormalities in Parkinson's disease maybe a way forward for future therapy and a significant contributor to such an approach might be using adenosine $A_{2A}$ antagonists. But once more there has been little or no clinical evaluation of this possibility again leading to a lack of exploration of a potentially useful non-dopaminergic approach to therapy.

## 7. Adenosine antagonists and non-motor symptoms of Parkinson's disease

As mentioned earlier in this review, the non-motor symptoms of Parkinson's disease remain a major unmet need (LeWitt & Chaudhuri, 2020). While some non-motor symptoms respond to dopaminergic therapy, most appear to be non-dopaminergic and non-striatal in origin (Chaudhuri & Schapira, 2009). In this respect, it is important to remember that $A_{2A}$ adenosine receptors are also found in areas of the brain outside of the basal ganglia. In particular, the low density localisation in some limbic and cortical regions of the brain may have important implications for treating the neuropsychiatric components of Parkinson's disease as well as an effect on the control of movement (Svenningsson, Le Moine, Fisone, & Fredholm, 1999). Elsewhere, we have reviewed the evidence for an effect of adenosine $A_{2A}$ antagonists and in particular, istradefylline for potential therapeutic use in treating depression, anxiety, cognitive impairment and excessive daytime sleepiness in Parkinson's disease (Jenner, Mori, & Kanda, 2020; Mori et al., 2022). A detailed assessment of the role of adenosine $A_{2A}$ adenosine receptors is presented also in other chapters in this volume.

In brief, extensive studies show that adenosine $A_{2A}$ antagonists are effective in reversing cognitive deficits in a range of experimental models related to the early executive and visuo-spatial effects seen in Parkinson's disease (Chen, 2014). The mechanism through which this effect is mediated is unclear. It could reflect the action of adenosine $A_{2A}$ antagonists in modifying the strio-thalamo-cortical loop responsible for the control of cognition. On the other hand, it has been suggested that $A_{2A}$ antagonists might modify glutamatergic transmission or synaptic plasticity through changes in LTP/LTD although there might also be alterations in trophic factor activity mediated by an action on glial cells (Hu et al., 2016). Apart from some preliminary data from one study of the use of istradefylline as monotherapy in Parkinson's disease, there has been no clinical evaluation of the effect of this drug class.

A similar story emerges when looking at depression in Parkinson's disease. $A_{2A}$ receptor antagonists can reverse depressive symptoms in experimental models of PD including those models with a high predictive value of effect in humans and to the same extent as classical antidepressants (Yamada et al., 2014; Yamada, Kobayashi, Mori, Jenner, & Kanda, 2013). Importantly, $A_{2A}$ antagonists are effective in models of the motivational symptoms of depression which resembles the apathetic and hedonic expression of depression that occurs in the illness (Lopez-Cruz, Salamone,

& Correa, 2018; Nunes et al., 2010; Salamone et al., 2018). $A_{2A}$ antagonists can also be effective in preclinical models of anxiety but the evidence is more mixed (van Calker, Biber, Domschke, & Serchov, 2019). The site of action of $A_{2A}$ antagonists in depression/anxiety is unclear and may again centre around changes in the activity of the strio-thalamo-cortical loops controlling the motor and affective pathways and also from effects in the nucleus accumbens. Equally neuronal activity in the amygdala, hippocampus and limbic brain regions may be altered. So far, only one small open label clinical study as examined the effects of an $A_{2A}$ antagonist on mood disorders in Parkinson's disease. Istradefylline improved rating scales assessing depression including an apathy scale (Nagayama et al., 2019). Despite the considerable evidence of a role for $A_{2A}$ receptors in depression coming from preclinical studies, there has been little clinical evaluation of the potential of this action on depression in Parkinson's disease or on depression in otherwise normal individuals.

Since sleep disturbance and excessive day time sleepiness are common components of Parkinson's disease, it is significant that $A_{2A}$ adenosine receptors play a prominent role in regulating the sleep wake cycle with arousal attributed to $A_{2A}$ receptor antagonism (Jenner et al., 2020). In rodents, $A_{2A}$ receptor antagonists induce arousal in the active part of the daily cycle only and not during the inactive phase (Yuan et al., 2017) suggesting a potential use in excessive day time sleepiness that does not interfere with nocturnal sleep. This was supported by some small clinical studies in Parkinson's disease where $A_{2A}$ antagonism did improve daytime sleepiness without impairing nocturnal sleep (Matsuura et al., 2018; Suzuki et al., 2017). Again, despite the rationale for this use, no other clinical evaluation has ensued.

There may also be uses for adenosine $A_{2A}$ antagonists in other components of the non-motor symptoms of Parkinson's disease. For example, bladder hyper-reflexia occurs in a significant proportion of the Parkinson's disease population and in experimental models of Parkinson's disease (where hyper-reflexia has been shown to occur), the administration of an adenosine $A_{2A}$ antagonist reduced bladder overactivity but again little further clinical evaluation has taken place (Albanese, Jenner, Marsden, & Stephenson, 1988; Kitta et al., 2012; Kitta et al., 2016; Pritchard et al., 2017).

The overall picture is of a non-dopaminergic drug class that may have uses in meeting one of the major unmet needs in Parkinson's disease but where the clinical evaluation of adenosine $A_{2A}$ antagonists has not reflected the potential shown in preclinical studies.

## 8. Exploring the future potential of adenosine $A_{2A}$ antagonists in Parkinson's disease

The objective of this review has been to examine the rationale for the use of adenosine $A_{2A}$ antagonists in the treatment of Parkinson's disease as a novel non-dopaminergic approach to therapy. In fact, this class of compounds represents the first new non-dopaminergic medication for use in Parkinson's disease to be introduced in 40 years. The evidence presented also shows a strong case for the continued evaluation of adenosine $A_{2A}$ antagonists in Parkinson's disease. The main attributes of adenosine $A_{2A}$ antagonists can be summarised as follows:

1. Adenosine $A_{2A}$ antagonists have a unique mechanism and site of action that causes functional change in the activity of the indirect GABAergic striatal output pathway that is accepted to be a key component of the alterations in basal ganglia function that underlie both motor dysfunction in Parkinson's disease and the onset or expression of dyskinesia.
2. Adenosine $A_{2A}$ antagonists show clinically relevant effects in preclinical models of Parkinson's disease that are widely accepted as being predictive of drug action in the illness. However, these models would not be effective in demonstrating an effect on 'wearing off' as this does not occur on repeated levodopa administration in these animals. As a class, adenosine $A_{2A}$ antagonists show a mild symptomatic activity as monotherapy, additive effects on motor function when administered with levodopa with/without a dopamine agonist. The improvement in motor function is not associated with increased dyskinesia in preclinical models of Parkinson's disease but can occur with an exacerbation of established dyskinesia and even increasing frequency of dyskinesia in patient populations.
3. Adenosine $A_{2A}$ antagonists can alter the activity of other key neuronal systems relevant to Parkinson's disease and to the expression of dyskinesia, most notably the glutamatergic and serotoninergic pathways. These actions translate into clinically relevant changes in preclinical models of the motor deficits of Parkinson's disease and the expression of dyskinesia.
4. Adenosine $A_{2A}$ antagonists can act through both striatal and non-striatal $A_{2A}$ receptors - most notably in the limbic regions of the brain- to produce functionally significant effects in experimental models of depression, anxiety and cognition. These suggest a potential role in controlling neuropsychiatric components of Parkinson's disease.

5. One adenosine $A_{2A}$ receptor antagonist, namely istradefylline has been shown to be clinically effective in increasing 'on' time without troublesome dyskinesia in late-stage patients with Parkinson's disease with sufficient evidence of efficacy to be approved in both Japan and in the USA.

However, the development of adenosine $A_{2A}$ antagonists and the path to registration has not being universally successful and some adenosine $A_{2A}$ antagonists have failed during clinical development due to a lack of efficacy (preladenant) or to toxicity (tozadenant). Even istradefylline failed to show efficacy in about half of the phase III clinical studies in which it was examined. The reasons for this require some discussion:

1. The required regulatory pathway to registration for treating 'off' periods in late-stage Parkinson's disease may not be the appropriate for this non-dopaminergic class of drugs.
2. The requirement to use adenosine $A_{2A}$ antagonists as an adjunct to levodopa in patients already optimised to maximal dopaminergic therapy goes against the preclinical evidence that this drug class should be used in conjunction with low or threshold doses of levodopa.
3. The classical trial design for drugs being developed for the treatment of 'off' periods in late-stage Parkinson's disease is designed around adjuncts enhancing dopaminergic based therapy. It does not necessarily adapt for use in the development of non-dopaminergic approaches to treatment.
4. The clinical endpoints required for regulatory registration and approval are based around studies examining the effects of medications acting through dopaminergic mechanisms and have not been adapted to detect the effects of non-dopaminergic approaches to treatment which may affect outcome in a different way.
5. Little attention has been paid to the potential use of adenosine $A_{2A}$ antagonists as a means of limiting the use of levodopa in early Parkinson's disease by keeping doses low and employing a levodopa sparing strategy which would fit with the pharmacological profile of this drug class.
6. The clinical trial programme used to look at a classical indication for adjuncts to the treatment of Parkinson's disease does not take into account that a new drug class might have actions on both motor and non- motor symptoms that lead to an overall improvement in quality of

life and patient experience. Rather indicators of non-motor activity are usually a secondary endpoint or studied in additional clinical evaluations.
7. The effects of non-dopaminergic approaches to treatment of Parkinson's disease are inevitably compared to the powerful effect of dopaminergic medications on motor symptoms. This comparative approach ignores the overall differences between dopaminergic and non-dopaminergic mechanisms, and it continually delays the introduction of non-dopaminergic approaches to treatment which may have a beneficial effect in the patient population.
8. Parkinson's disease is a heterogeneous disorder that may in fact be syndromic in nature and where subtypes of Parkinson's disease are increasingly thought to exist. The unmet need that is satisfied by a non-dopaminergic approach treatment may well be useful in a subset of patients, but this is not taken into account by the regulatory process and receives scant investigation.
9. Expectation that adenosine $A_{2A}$ antagonists would be more than just an adjunct to dopaminergic therapy may have made the outcome of clinical studies appear disappointing and contributed to a negative approach to this drug class.

At this point in time, it is disappointing to note that the further evaluation of adenosine $A_{2A}$ antagonists as a non-dopaminergic approach to Parkinson's disease appears to have largely ceased. This is despite the evidence that this drug class can be effective and used safely in Parkinson's disease that comes from the registration and use of istradefylline in both Japan and the USA.

On a positive note, it is important to remember that istradefylline has provided an additional tool in the toolbox for the treatment of Parkinson's disease. Every clinician who treats patients with Parkinson's disease recognises the value of having as many alternative strategies as possible available for use in individual patients whose response two drug treatment and whose unmet needs vary.

Finally, it may be that the adenosine $A_{2A}$ antagonists so far examined have not been the best in class for modifying the effect of the adenosine $A_{2A}$ receptor on striatal output in Parkinson's disease. While the activity of the $A_{2A}$ receptor is modulated by its interaction with adenosine, this receptor also exerts intrinsic activity on the indirect GABAergic output pathway. Current adenosine $A_{2A}$ antagonists inhibit the stimulation of the receptor produced by adenosine but they do not remove the intrinsic activity. A more effective control of the indirect output pathway in

Parkinson's disease might be achieved by using an inverse agonist drug that blocks both the effect of adenosine and the intrinsic receptor activity so exerting greater modification of GABAergic mediated output (Mori et al., 2022). This has recently been tested using KW-6356 as a potent inverse agonist (Tayama et al., 2023) and where this molecule has been shown to have a significant monotherapy effect in early Parkinson's disease in contrast to an apparent lack of effect of the classical antagonists istradefylline and preladenant (see Abstracts: Maeda, Yamamoto, Kimura, Hattori, 2018; Maeda, Sugiyama, Yamada, Nishi, Hattori, 2022). All of this suggests that the $A_{2A}$ adenosine receptor remains a target for Parkinson's disease but that it requires perseverance with the chemistry, pharmacology and clinical evaluation of this important drug class.

## References

Albanese, A., Jenner, P., Marsden, C. D., & Stephenson, J. D. (1988). Bladder hyperreflexia induced in marmosets by 1-methyl-4-phenyl-1,2,3,6-tetrahydropyridine. *Neuroscience Letters, 87*(1–2), 46–50. https://doi.org/10.1016/0304-3940(88)90143-7

Bara-Jimenez, W., Sherzai, A., Dimitrova, T., Favit, A., Bibbiani, F., Gillespie, M., … Chase, T. N. (2003). Adenosine A(2A) receptor antagonist treatment of Parkinson's disease. *Neurology, 61*(3), 293–296. https://doi.org/10.1212/01.wnl.0000073136.00548.d4

Bastide, M. F., Meissner, W. G., Picconi, B., Fasano, S., Fernagut, P. O., Feyder, M., … Bezard, E. (2015). Pathophysiology of L-dopa-induced motor and non-motor complications in Parkinson's disease. *Progress in Neurobiology, 132*, 96–168. https://doi.org/10.1016/j.pneurobio.2015.07.002

Bezard, E., Tronci, E., Pioli, E. Y., Li, Q., Porras, G., Bjorklund, A., & Carta, M. (2013). Study of the antidyskinetic effect of eltoprazine in animal models of levodopa-induced dyskinesia. *Movement Disorders: Official Journal of the Movement Disorder Society, 28*(8), 1088–1096. https://doi.org/10.1002/mds.25366

Brooks, D. J., Papapetropoulos, S., Vandenhende, F., Tomic, D., He, P., Coppell, A., & O'Neill, G. (2010). An open-label, positron emission tomography study to assess adenosine A2A brain receptor occupancy of vipadenant (BIIB014) at steady-state levels in healthy male volunteers. *Clinical Neuropharmacology, 33*(2), 55–60. https://doi.org/10.1097/WNF.0b013e3181d137d2

Calon, F., Dridi, M., Hornykiewicz, O., Bedard, P. J., Rajput, A. H., & Di Paolo, T. (2004). Increased adenosine A2A receptors in the brain of Parkinson's disease patients with dyskinesias. *Brain, 127*(Pt 5), 1075–1084. https://doi.org/10.1093/brain/awh128

Castela, I., Casado-Polanco, R., Rubio, Y. V., da Silva, J. A., Marquez, R., Pro, B., … Hernandez, L. F. (2023). Selective activation of striatal indirect pathway suppresses levodopa induced-dyskinesias. *Neurobiology of Disease, 176*, 105930. https://doi.org/10.1016/j.nbd.2022.105930

Cenci, M. A., Skovgard, K., & Odin, P. (2022). Non-dopaminergic approaches to the treatment of motor complications in Parkinson's disease. *Neuropharmacology, 210*, 109027. https://doi.org/10.1016/j.neuropharm.2022.109027

Chaudhuri, K. R., & Schapira, A. H. (2009). Non-motor symptoms of Parkinson's disease: Dopaminergic pathophysiology and treatment. *Lancet Neurology, 8*(5), 464–474. https://doi.org/10.1016/S1474-4422(09)70068-7

Chen, J. F. (2014). Adenosine receptor control of cognition in normal and disease. *International Review of Neurobiology, 119*, 257–307. https://doi.org/10.1016/B978-0-12-801022-8.00012-X

Chen, J. F., & Cunha, R. A. (2020). The belated US FDA approval of the adenosine A(2A) receptor antagonist istradefylline for treatment of Parkinson's disease. *Purinergic Signalling, 16*(2), 167–174. https://doi.org/10.1007/s11302-020-09694-2

Chen, J. F., Lee, C. F., & Chern, Y. (2014). Adenosine receptor neurobiology: Overview. *International Review of Neurobiology, 119*, 1–49. https://doi.org/10.1016/B978-0-12-801022-8.00001-5

Coccurello, R., Breysse, N., & Amalric, M. (2004). Simultaneous blockade of adenosine A2A and metabotropic glutamate mGlu5 receptors increase their efficacy in reversing Parkinsonian deficits in rats. *Neuropsychopharmacology: Official Publication of the American College of Neuropsychopharmacology, 29*(8), 1451–1461. https://doi.org/10.1038/sj.npp.1300444

Collins, L. E., Sager, T. N., Sams, A. G., Pennarola, A., Port, R. G., Shahriari, M., & Salamone, J. D. (2012). The novel adenosine A2A antagonist Lu AA47070 reverses the motor and motivational effects produced by dopamine D2 receptor blockade. *Pharmacology, Biochemistry, and Behavior, 100*(3), 498–505. https://doi.org/10.1016/j.pbb.2011.10.015

Correa, M., Wisniecki, A., Betz, A., Dobson, D. R., O'Neill, M. F., O'Neill, M. J., & Salamone, J. D. (2004). The adenosine A2A antagonist KF17837 reverses the locomotor suppression and tremulous jaw movements induced by haloperidol in rats: Possible relevance to parkinsonism. *Behavioural Brain Research, 148*(1–2), 47–54. https://doi.org/10.1016/s0166-4328(03)00178-5

Corsi, S., Stancampiano, R., & Carta, M. (2021). Serotonin/dopamine interaction in the induction and maintenance of L-DOPA-induced dyskinesia: An update. *Progress in Brain Research, 261*, 287–302. https://doi.org/10.1016/bs.pbr.2021.01.032

Csoti, I., Herbst, H., Urban, P., Woitalla, D., & Wullner, U. (2019). Polypharmacy in Parkinson's disease: Risks and benefits with little evidence. *The Journal of Neural Transmission (Vienna), 126*(7), 871–878. https://doi.org/10.1007/s00702-019-02026-8

Dungo, R., & Deeks, E. D. (2013). Istradefylline: First global approval. *Drugs, 73*(8), 875–882. https://doi.org/10.1007/s40265-013-0066-7

El-Gamal, M., Salama, M., Collins-Praino, L. E., Baetu, I., Fathalla, A. M., Soliman, A. M., ... Moustafa, A. A. (2021). Neurotoxin-induced rodent models of Parkinson's disease: Benefits and drawbacks. *Neurotoxicity Research, 39*(3), 897–923. https://doi.org/10.1007/s12640-021-00356-8

Fenu, S., Pinna, A., Ongini, E., & Morelli, M. (1997). Adenosine A2A receptor antagonism potentiates L-DOPA-induced turning behaviour and c-fos expression in 6-hydroxydopamine-lesioned rats. *European Journal of Pharmacology, 321*(2), 143–147. https://doi.org/10.1016/s0014-2999(96)00944-2

Fernandez, H. H., Greeley, D. R., Zweig, R. M., Wojcieszek, J., Mori, A., Sussman, N. M., & Group, U. S. S. (2010). Istradefylline as monotherapy for Parkinson disease: Results of the 6002-US-051 trial. *Parkinsonism & Related Disorders, 16*(1), 16–20. https://doi.org/10.1016/j.parkreldis.2009.06.008

Ferre, S., Bonaventura, J., Zhu, W., Hatcher-Solis, C., Taura, J., Quiroz, C., ... Zwilling, D. (2018). Essential control of the function of the striatopallidal neuron by pre-coupled complexes of adenosine A(2A)-dopamine D(2) receptor heterotetramers and adenylyl cyclase. *Frontiers in Pharmacology, 9*, 243. https://doi.org/10.3389/fphar.2018.00243

Ferre, S., Karcz-Kubicha, M., Hope, B. T., Popoli, P., Burgueno, J., Gutierrez, M. A., ... Ciruela, F. (2002). Synergistic interaction between adenosine A2A and glutamate mGlu5 receptors: implications for striatal neuronal function. *Proceeding of the National Academy of Sciences of the United States of America, 99*(18), 11940–11945. https://doi.org/10.1073/pnas.172393799

Ferre, S., Quiroz, C., Woods, A. S., Cunha, R., Popoli, P., Ciruela, F., ... Schiffmann, S. N. (2008). An update on adenosine A2A-dopamine D2 receptor interactions: implications for the function of G protein-coupled receptors. *Current Pharmaceutical Design, 14*(15), 1468–1474. https://doi.org/10.2174/138161208784480108

Fredholm, B. B., AP, I. J., Jacobson, K. A., Linden, J., & Muller, C. E. (2011). International Union of Basic and Clinical Pharmacology. LXXXI. Nomenclature and classification of adenosine receptors—an update. *Pharmacological Reviews, 63*(1), 1–34. https://doi.org/10.1124/pr.110.003285

Fredholm, B. B., Chern, Y., Franco, R., & Sitkovsky, M. (2007). Aspects of the general biology of adenosine A2A signaling. *Progress in Neurobiology, 83*(5), 263–276. https://doi.org/10.1016/j.pneurobio.2007.07.005

Goetz, C. G., Damier, P., Hicking, C., Laska, E., Muller, T., Olanow, C. W., ... Russ, H. (2007). Sarizotan as a treatment for dyskinesias in Parkinson's disease: A double-blind placebo-controlled trial. *Movement Disorders: Official Journal of the Movement Disorder Society, 22*(2), 179–186. https://doi.org/10.1002/mds.21226

Gonzalez-Latapi, P., Bhowmick, S. S., Saranza, G., & Fox, S. H. (2020). Non-dopaminergic treatments for motor control in Parkinson's disease: An update. *CNS Drugs, 34*(10), 1025–1044. https://doi.org/10.1007/s40263-020-00754-0

Grondin, R., Bédard, P. J., Hadj Tahar, A., Grégoire, L., Mori, A., & Kase, H. (1999). Antiparkinsonian effect of a new selective adenosine A2A receptor antagonist in MPTP-treated monkeys. *Neurology, 52*(8), 1673–1677. https://doi.org/10.1212/wnl.52.8.1673

Hauser, R. A., Cantillon, M., Pourcher, E., Micheli, F., Mok, V., Onofrj, M., ... Wolski, K. (2011). Preladenant in patients with Parkinson's disease and motor fluctuations: A phase 2, double-blind, randomised trial. *Lancet Neurology, 10*(3), 221–229. https://doi.org/10.1016/S1474-4422(11)70012-6

Hauser, R. A., Hattori, N., Fernandez, H., Isaacson, S. H., Mochizuki, H., Rascol, O., ... LeWitt, P. (2021). Efficacy of istradefylline, an adenosine A2A receptor antagonist, as adjunctive therapy to levodopa in Parkinson's disease: A pooled analysis of 8 phase 2b/3 trials. *The Journal of Parkinson's Disease, 11*(4), 1663–1675. https://doi.org/10.3233/JPD-212672

Hauser, R. A., Hubble, J. P., Truong, D. D., & Istradefylline, U. S. S. G. (2003). Randomized trial of the adenosine A(2A) receptor antagonist istradefylline in advanced PD. *Neurology, 61*(3), 297–303. https://doi.org/10.1212/01.wnl.0000081227.84197.0b

Hauser, R. A., Olanow, C. W., Kieburtz, K. D., Pourcher, E., Docu-Axelerad, A., Lew, M., ... Bandak, S. (2014). Tozadenant (SYN115) in patients with Parkinson's disease who have motor fluctuations on levodopa. A phase 2b, double-blind, randomised trial. *Lancet Neurology, 13*(8), 767–776. https://doi.org/10.1016/S1474-4422(14)70148-6

Hauser, R. A., Shulman, L. M., Trugman, J. M., Roberts, J. W., Mori, A., Ballerini, R., ... Istradefylline, U. S. S. G. (2008). Study of istradefylline in patients with Parkinson's disease on levodopa with motor fluctuations. *Movement Disorders: Official Journal of the Movement Disorder Society, 23*(15), 2177–2185. https://doi.org/10.1002/mds.22095

Hauser, R. A., Stocchi, F., Rascol, O., Huyck, S. B., Capece, R., Ho, T. W., ... Hewitt, D. (2015). Preladenant as an adjunctive therapy with levodopa in Parkinson disease: Two randomized clinical trials and lessons learned. *JAMA Neurology, 72*(12), 1491–1500. https://doi.org/10.1001/jamaneurol.2015.2268

Hodgson, R. A., Bedard, P. J., Varty, G. B., Kazdoba, T. M., Di Paolo, T., Grzelak, M. E., ... Hunter, J. C. (2010). Preladenant, a selective A(2A) receptor antagonist, is active in primate models of movement disorders. *Experimental Neurology, 225*(2), 384–390. https://doi.org/10.1016/j.expneurol.2010.07.011

Hu, Q., Ren, X., Liu, Y., Li, Z., Zhang, L., Chen, X., ... Chen, J. F. (2016). Aberrant adenosine A2A receptor signaling contributes to neurodegeneration and cognitive impairments in a mouse model of synucleinopathy. *Experimental Neurology, 283*(Pt A), 213–223. https://doi.org/10.1016/j.expneurol.2016.05.040

Hurley, M. J., Mash, D. C., & Jenner, P. (2000). Adenosine A(2A) receptor mRNA expression in Parkinson's disease. *Neuroscience Letters, 291*(1), 54–58. https://doi.org/10.1016/s0304-3940(00)01371-9

Iravani, M. M., Tayarani-Binazir, K., Chu, W. B., Jackson, M. J., & Jenner, P. (2006). In 1-methyl-4-phenyl-1,2,3,6-tetrahydropyridine-treated primates, the selective 5-hydroxytryptamine 1a agonist (R)-(+)-8-OHDPAT inhibits levodopa-induced dyskinesia but only with\ increased motor disability. *The Journal of Pharmacology and Experimental Therapeutics, 319*(3), 1225–1234. https://doi.org/10.1124/jpet.106.110429

Ishiwata, K., Mishina, M., Kimura, Y., Oda, K., Sasaki, T., & Ishii, K. (2005). First visualization of adenosine A(2A) receptors in the human brain by positron emission tomography with [11C]TMSX. *Synapse (New York, N. Y.), 55*(2), 133–136. https://doi.org/10.1002/syn.20099

Jarvis, M. F., & Williams, M. (1989). Direct autoradiographic localization of adenosine A2 receptors in the rat brain using the A2-selective agonist, [3H]CGS 21680. *European Journal of Pharmacology, 168*(2), 243–246. https://doi.org/10.1016/0014-2999(89)90571-2

Jenner, P. (2014). An overview of adenosine A2A receptor antagonists in Parkinson's disease. *International Review of Neurobiology, 119*, 71–86. https://doi.org/10.1016/B978-0-12-801022-8.00003-9

Jenner, P., Mori, A., & Kanda, T. (2020). Can adenosine A(2A) receptor antagonists be used to treat cognitive impairment, depression or excessive sleepiness in Parkinson's disease? *Parkinsonism & Related Disorders, 80*(Suppl 1), S28–S36. https://doi.org/10.1016/j.parkreldis.2020.09.022

Jenner, P., Mori, A., Aradi, S. D., & Hauser, R. A. (2021). Istradefylline—a first generation adenosine A(2A) antagonist for the treatment of Parkinson's disease. *Expert Review of Neurotherapeutics, 21*(3), 317–333. https://doi.org/10.1080/14737175.2021.1880896

Johnson, K. A., Conn, P. J., & Niswender, C. M. (2009). Glutamate receptors as therapeutic targets for Parkinson's disease. *CNS & Neurological Disorders Drug Targets, 8*(6), 475–491. https://doi.org/10.2174/187152709789824606

Jones, C. K., Bubser, M., Thompson, A. D., Dickerson, J. W., Turle-Lorenzo, N., Amalric, M., ... Niswender, C. M. (2012). The metabotropic glutamate receptor 4-positive allosteric modulator VU0364770 produces efficacy alone and in combination with L-DOPA or an adenosine 2A antagonist in preclinical rodent models of Parkinson's disease. *The Journal of Pharmacology and Experimental Therapeutics, 340*(2), 404–421. https://doi.org/10.1124/jpet.111.187443

Jones, N., Bleickardt, C., Mullins, D., Parker, E., & Hodgson, R. (2013). A2A receptor antagonists do not induce dyskinesias in drug-naive or L-dopa sensitized rats. *Brain Research Bulletin, 98*, 163–169. https://doi.org/10.1016/j.brainresbull.2013.07.001

Kachroo, A., Orlando, L. R., Grandy, D. K., Chen, J. F., Young, A. B., & Schwarzschild, M. A. (2005). Interactions between metabotropic glutamate 5 and adenosine A2A receptors in normal and parkinsonian mice. *The Journal of Neuroscience, 25*(45), 10414–10419. https://doi.org/10.1523/JNEUROSCI.3660-05.2005

Kanda, T., & Jenner, P. (2020). Can adenosine A(2A) receptor antagonists modify motor behavior and dyskinesia in experimental models of Parkinson's disease? *Parkinsonism & Related Disorders, 80*(Suppl 1), S21–S27. https://doi.org/10.1016/j.parkreldis.2020.09.026

Kanda, T., Jackson, M. J., Smith, L. A., Pearce, R. K., Nakamura, J., Kase, H., ... Jenner, P. (1998). Adenosine A2A antagonist: A novel antiparkinsonian agent that does not provoke dyskinesia in parkinsonian monkeys. *Annals of Neurology, 43*(4), 507–513. https://doi.org/10.1002/ana.410430415

Kanda, T., Jackson, M. J., Smith, L. A., Pearce, R. K., Nakamura, J., Kase, H., ... Jenner, P. (2000). Combined use of the adenosine A(2A) antagonist KW-6002 with L-DOPA or with selective D1 or D2 dopamine agonists increases antiparkinsonian activity but not dyskinesia in MPTP-treated monkeys. *Experimental Neurology, 162*(2), 321–327. https://doi.org/10.1006/exnr.2000.7350

Kanda, T., Tashiro, T., Kuwana, Y., & Jenner, P. (1998). Adenosine A2A receptors modify motor function in MPTP-treated common marmosets. *Neuroreport, 9*(12), 2857–2860. https://doi.org/10.1097/00001756-199808240-00032

Kitta, T., Chancellor, M. B., de Groat, W. C., Kuno, S., Nonomura, K., & Yoshimura, N. (2012). Suppression of bladder overactivity by adenosine A2A receptor antagonist in a rat model of Parkinson disease. *The Journal of Urology, 187*(5), 1890–1897. https://doi.org/10.1016/j.juro.2011.12.062

Kitta, T., Yabe, I., Takahashi, I., Matsushima, M., Sasaki, H., & Shinohara, N. (2016). Clinical efficacy of istradefylline on lower urinary tract symptoms in Parkinson's disease. *International Journal of Urology: Official Journal of the Japanese Urological Association, 23*(10), 893–894. https://doi.org/10.1111/iju.13160

Ko, W. K. D., Li, Q., Cheng, L. Y., Morelli, M., Carta, M., & Bezard, E. (2017). A preclinical study on the combined effects of repeated eltoprazine and preladenant treatment for alleviating L-DOPA-induced dyskinesia in Parkinson's disease. *European Journal of Pharmacology, 813*, 10–16. https://doi.org/10.1016/j.ejphar.2017.07.030

Koga, K., Kurokawa, M., Ochi, M., Nakamura, J., & Kuwana, Y. (2000). Adenosine A(2A) receptor antagonists KF17837 and KW-6002 potentiate rotation induced by dopaminergic drugs in hemi-Parkinsonian rats. *European Journal of Pharmacology, 408*(3), 249–255. https://doi.org/10.1016/s0014-2999(00)00745-7

Kulisevsky, J. (2022). Pharmacological management of Parkinson's disease motor symptoms: Update and recommendations from an expert. *Revista de Neurologia, 75*(s04), S1–S10. https://doi.org/10.33588/rn.75s04.2022217

Lees, A., Tolosa, E., Stocchi, F., Ferreira, J. J., Rascol, O., Antonini, A., & Poewe, W. (2023). Optimizing levodopa therapy, when and how? Perspectives on the importance of delivery and the potential for an early combination approach. *Expert Review of Neurotherapeutics, 23*(1), 15–24. https://doi.org/10.1080/14737175.2023.2176220

LeWitt, P. A., & Chaudhuri, K. R. (2020). Unmet needs in Parkinson disease: Motor and non-motor. *Parkinsonism & Related Disorders, 80*(Suppl 1), S7–S12. https://doi.org/10.1016/j.parkreldis.2020.09.024

LeWitt, P. A., Aradi, S. D., Hauser, R. A., & Rascol, O. (2020). The challenge of developing adenosine A(2A) antagonists for Parkinson disease: Istradefylline, preladenant, and tozadenant. *Parkinsonism & Related Disorders, 80*(Suppl 1), S54–S63. https://doi.org/10.1016/j.parkreldis.2020.10.027

LeWitt, P. A., Guttman, M., Tetrud, J. W., Tuite, P. J., Mori, A., Chaikin, P., ... Group, U. S. S. (2008). Adenosine A2A receptor antagonist istradefylline (KW-6002) reduces "off" time in Parkinson's disease: A double-blind, randomized, multicenter clinical trial (6002-US-005). *Annals of Neurology, 63*(3), 295–302. https://doi.org/10.1002/ana.21315

Lopez, S., Turle-Lorenzo, N., Johnston, T. H., Brotchie, J. M., Schann, S., Neuville, P., & Amalric, M. (2008). Functional interaction between adenosine A2A and group III metabotropic glutamate receptors to reduce parkinsonian symptoms in rats. *Neuropharmacology, 55*(4), 483–490. https://doi.org/10.1016/j.neuropharm.2008.06.038

Lopez-Cruz, L., Salamone, J. D., & Correa, M. (2018). Caffeine and selective adenosine receptor antagonists as new therapeutic tools for the motivational symptoms of depression. *Frontiers in Pharmacology, 9*, 526. https://doi.org/10.3389/fphar.2018.00526

Lundblad, M., Vaudano, E., & Cenci, M. A. (2003). Cellular and behavioural effects of the adenosine A2a receptor antagonist KW-6002 in a rat model of l-DOPA-induced dyskinesia. *Journal of Neurochemistry, 84*(6), 1398–1410. https://doi.org/10.1046/j.1471-4159.2003.01632.x

Mao, Q., Qin, W. Z., Zhang, A., & Ye, N. (2020). Recent advances in dopaminergic strategies for the treatment of Parkinson's disease. *Acta Pharmacologica Sinica, 41*(4), 471–482. https://doi.org/10.1038/s41401-020-0365-y

Matsuura, K., Kajikawa, H., Tabei, K. I., Satoh, M., Kida, H., Nakamura, N., & Tomimoto, H. (2018). The effectiveness of istradefylline for the treatment of gait deficits and sleepiness in patients with Parkinson's disease. *Neuroscience Letters, 662,* 158–161. https://doi.org/10.1016/j.neulet.2017.10.018

Maeda, T., Yamamoto, M., Kimura, T., & Hattori, N. (2018). *A novel adenosine A2A receptor antagonist KW-6356 in early Parkinson's disease: A randomised controlled trial for efficacy and safety World congress on Parkinson's disease and related disorders.*

Maeda, T., Sugiyama, K., Yamada, K., Nishi, M., & Hattori, N. (2022). *Effect of KW-6356, a novel adenosine A2A receptor antagonist/inverse agonist, on motor and non-motor symptoms in Parkinson's disease patients as an adjunct to levodopa therapy: results of phase 2b study. International congress of Parkinson's disease and movement disorders.*

Michel, A., Downey, P., Nicolas, J. M., & Scheller, D. (2014). Unprecedented therapeutic potential with a combination of A2A/NR2B receptor antagonists as observed in the 6-OHDA lesioned rat model of Parkinson's disease. *PLoS One, 9*(12), e114086. https://doi.org/10.1371/journal.pone.0114086

Michel, A., Downey, P., Van Damme, X., De Wolf, C., Schwarting, R., & Scheller, D. (2015). Behavioural assessment of the A2a/NR2B combination in the unilateral 6-OHDA-lesioned rat model: A new method to examine the therapeutic potential of non-dopaminergic drugs. *PLoS One, 10*(8), e0135949. https://doi.org/10.1371/journal.pone.0135949

Michel, A., Nicolas, J. M., Rose, S., Jackson, M., Colman, P., Brione, W., ... Downey, P. (2017). Antiparkinsonian effects of the "Radiprodil and Tozadenant" combination in MPTP-treated marmosets. *PLoS One, 12*(8), e0182887. https://doi.org/10.1371/journal.pone.0182887

Mishina, M., & Ishiwata, K. (2014). Adenosine receptor PET imaging in human brain. *International Review of Neurobiology, 119,* 51–69. https://doi.org/10.1016/B978-0-12-801022-8.00002-7

Mishina, M., Ishii, K., Kimura, Y., Suzuki, M., Kitamura, S., Ishibashi, K., ... Ishiwata, K. (2017). Adenosine A(1) receptors measured with (11) C-MPDX PET in early Parkinson's disease. *Synapse (New York, N. Y.), 71*(8), https://doi.org/10.1002/syn.21979

Mishina, M., Ishiwata, K., Naganawa, M., Kimura, Y., Kitamura, S., Suzuki, M., ... Ishii, K. (2011). Adenosine A(2A) receptors measured with [C]TMSX PET in the striata of Parkinson's disease patients. *PLoS One, 6*(2), e17338. https://doi.org/10.1371/journal.pone.0017338

Mizuno, Y., Hasegawa, K., Kondo, T., Kuno, S., Yamamoto, M., & Japanese Istradefylline Study, G. (2010). Clinical efficacy of istradefylline (KW-6002) in Parkinson's disease: A randomized, controlled study. *Movement Disorders: Official Journal of the Movement Disorder Society, 25*(10), 1437–1443. https://doi.org/10.1002/mds.23107

Montastruc, J. L., Rascol, O., & Senard, J. M. (1997). Glutamate antagonists and Parkinson's disease: A review of clinical data. *Neuroscience and Biobehavioral Reviews, 21*(4), 477–480. https://doi.org/10.1016/s0149-7634(96)00035-8

Morelli, M., Di Paolo, T., Wardas, J., Calon, F., Xiao, D., & Schwarzschild, M. A. (2007). Role of adenosine A2A receptors in parkinsonian motor impairment and l-DOPA-induced motor complications. *Progress in Neurobiology, 83*(5), 293–309. https://doi.org/10.1016/j.pneurobio.2007.07.001

Mori, A. (2020). How do adenosine A(2A) receptors regulate motor function? *Parkinsonism & Related Disorders, 80*(Suppl 1), S13–S20. https://doi.org/10.1016/j.parkreldis.2020.09.025

Mori, A., & Shindou, T. (2003). Modulation of GABAergic transmission in the striato-pallidal system by adenosine A2A receptors: A potential mechanism for the antiparkinsonian effects of A2A antagonists. *Neurology, 61*(11 Suppl 6), S44–S48. https://doi.org/10.1212/01.wnl.0000095211.71092.a0

Mori, A., Chen, J. F., Uchida, S., Durlach, C., King, S. M., & Jenner, P. (2022). The pharmacological potential of adenosine A(2A) receptor antagonists for treating Parkinson's disease. *Molecules (Basel, Switzerland), 27*(7), 2366. https://doi.org/10.3390/molecules27072366

Mori, A., Shindou, T., Ichimura, M., Nonaka, H., & Kase, H. (1996). The role of adenosine A2a receptors in regulating GABAergic synaptic transmission in striatal medium spiny neurons. *The Journal of Neuroscience, 16*(2), 605–611. https://doi.org/10.1523/JNEUROSCI.16-02-00605.1996

Munoz, A., Li, Q., Gardoni, F., Marcello, E., Qin, C., Carlsson, T., ... Carta, M. (2008). Combined 5-HT1A and 5-HT1B receptor agonists for the treatment of L-DOPA-induced dyskinesia. *Brain, 131*(Pt 12), 3380–3394. https://doi.org/10.1093/brain/awn235

Murata, M. (2010). Zonisamide: A new drug for Parkinson's disease. *Drugs Today (Barc), 46,* 251–258. https://doi.org/10.1358/dot.2010.46.4.1490077

Nagayama, H., Kano, O., Murakami, H., Ono, K., Hamada, M., Toda, T., ... Hattori, N. (2019). Effect of istradefylline on mood disorders in Parkinson's disease. *Journal of the Neurological Sciences, 396,* 78–83. https://doi.org/10.1016/j.jns.2018.11.005

Nash, J. E., & Brotchie, J. M. (2000). A common signaling pathway for striatal NMDA and adenosine A2a receptors: Implications for the treatment of Parkinson's disease. *The Journal of Neuroscience, 20*(20), 7782–7789. https://doi.org/10.1523/JNEUROSCI.20-20-07782.2000

Nunes, E. J., Randall, P. A., Santerre, J. L., Given, A. B., Sager, T. N., Correa, M., & Salamone, J. D. (2010). Differential effects of selective adenosine antagonists on the effort-related impairments induced by dopamine D1 and D2 antagonism. *Neuroscience, 170*(1), 268–280. https://doi.org/10.1016/j.neuroscience.2010.05.068

Pagonabarraga, J., Tinazzi, M., Caccia, C., & Jost, W. H. (2021). The role of glutamatergic neurotransmission in the motor and non-motor symptoms in Parkinson's disease: Clinical cases and a review of the literature. *Journal of Clinical Neuroscience: Official Journal of the Neurosurgical Society of Australasia, 90,* 178–183. https://doi.org/10.1016/j.jocn.2021.05.056

Pawsey, S., Wood, M., Browne, H., Donaldson, K., Christie, M., & Warrington, S. (2016). Safety, tolerability and pharmacokinetics of FAAH inhibitor V158866: A double-blind, randomised, placebo-controlled phase I study in healthy volunteers, *Drugs in R&D, 16*(2), 181–191. https://doi.org/10.1007/s40268-016-0127-y

Pinna, A. (2014). Adenosine A2A receptor antagonists in Parkinson's disease: Progress in clinical trials from the newly approved istradefylline to drugs in early development and those already discontinued. *CNS Drugs, 28*(5), 455–474. https://doi.org/10.1007/s40263-014-0161-7

Pinna, A., Ko, W. K., Costa, G., Tronci, E., Fidalgo, C., Simola, N., ... Morelli, M. (2016). Antidyskinetic effect of A2A and 5HT1A/1B receptor ligands in two animal models of Parkinson's disease. *Movement Disorders: Official Journal of the Movement Disorder Society, 31*(4), 501–511. https://doi.org/10.1002/mds.26475

Pinna, A., Parekh, P., & Morelli, M. (2023). Serotonin 5-HT(1A) receptors and their interactions with adenosine A(2A) receptors in Parkinson's disease and dyskinesia. *Neuropharmacology, 226,* 109411. https://doi.org/10.1016/j.neuropharm.2023.109411

Pinna, A., Serra, M., Marongiu, J., & Morelli, M. (2020). Pharmacological interactions between adenosine A(2A) receptor antagonists and different neurotransmitter systems. *Parkinsonism & Related Disorders, 80*(Suppl 1), S37–S44. https://doi.org/10.1016/j.parkreldis.2020.10.023

Pourcher, E., Fernandez, H. H., Stacy, M., Mori, A., Ballerini, R., & Chaikin, P. (2012). Istradefylline for Parkinson's disease patients experiencing motor fluctuations: Results of the KW-6002-US-018 study. *Parkinsonism & Related Disorders, 18*(2), 178–184. https://doi.org/10.1016/j.parkreldis.2011.09.023

Pritchard, S., Jackson, M. J., Hikima, A., Lione, L., Benham, C. D., Chaudhuri, K. R., ... Iravani, M. M. (2017). Altered detrusor contractility in MPTP-treated common marmosets with bladder hyperreflexia. *PLoS One, 12*(5), e0175797. https://doi.org/10.1371/journal.pone.0175797

Ramlackhansingh, A. F., Bose, S. K., Ahmed, I., Turkheimer, F. E., Pavese, N., & Brooks, D. J. (2011). Adenosine 2A receptor availability in dyskinetic and nondyskinetic patients with Parkinson disease. *Neurology, 76*(21), 1811–1816. https://doi.org/10.1212/WNL.0b013e31821ccce4

Rascol, O., Fabbri, M., & Poewe, W. (2021). Amantadine in the treatment of Parkinson's disease and other movement disorders. *Lancet Neurology, 20*(12), 1048–1056. https://doi.org/10.1016/S1474-4422(21)00249-0

Rose, S., Jackson, M. J., Smith, L. A., Stockwell, K., Johnson, L., Carminati, P., & Jenner, P. (2006). The novel adenosine A2a receptor antagonist ST1535 potentiates the effects of a threshold dose of L-DOPA in MPTP treated common marmosets. *European Journal of Pharmacology, 546*(1–3), 82–87. https://doi.org/10.1016/j.ejphar.2006.07.017

Rose, S., Ramsay Croft, N., & Jenner, P. (2007). The novel adenosine A2a antagonist ST1535 potentiates the effects of a threshold dose of l-dopa in unilaterally 6-OHDA-lesioned rats. *Brain Research, 1133*(1), 110–114. https://doi.org/10.1016/j.brainres.2006.10.038

Rota, S., Urso, D., van Wamelen, D. J., Leta, V., Boura, I., Odin, P., ... Chaudhuri, K. R. (2022). Why do 'OFF' periods still occur during continuous drug delivery in Parkinson's disease? *Translational Neurodegeneration, 11*(1), 43. https://doi.org/10.1186/s40035-022-00317-x

Saki, M., Yamada, K., Koshimura, E., Sasaki, K., & Kanda, T. (2013). In vitro pharmacological profile of the A2A receptor antagonist istradefylline. *Naunyn-Schmiedeberg's Archives of Pharmacology, 386*(11), 963–972. https://doi.org/10.1007/s00210-013-0897-5

Salamone, J. D. (2010). Preladenant, a novel adenosine A(2A) receptor antagonist for the potential treatment of parkinsonism and other disorders. *IDrugs: The Investigational Drugs Journal, 13*(10), 723–731.

Salamone, J. D., Correa, M., Ferrigno, S., Yang, J. H., Rotolo, R. A., & Presby, R. E. (2018). The psychopharmacology of effort-related decision making: Dopamine, adenosine, and insights into the neurochemistry of motivation. *Pharmacological Reviews, 70*(4), 747–762. https://doi.org/10.1124/pr.117.015107

Sauerbier, A., Jenner, P., Todorova, A., & Chaudhuri, K. R. (2016). Non motor subtypes and Parkinson's disease. *Parkinsonism & Related Disorders, 22*(Suppl 1), S41–S46. https://doi.org/10.1016/j.parkreldis.2015.09.027

Schapira, A. H. V., Chaudhuri, K. R., & Jenner, P. (2017). Non-motor features of Parkinson disease. *Nature Reviews. Neuroscience, 18*(8), 509. https://doi.org/10.1038/nrn.2017.91

Sebastianutto, I., & Cenci, M. A. (2018). mGlu receptors in the treatment of Parkinson's disease and L-DOPA-induced dyskinesia. *Current Opinion in Pharmacology, 38*, 81–89. https://doi.org/10.1016/j.coph.2018.03.003

Sebastiao, A. M., & Ribeiro, J. A. (1996). Adenosine A2 receptor-mediated excitatory actions on the nervous system. *Progress in Neurobiology, 48*(3), 167–189. https://doi.org/10.1016/0301-0082(95)00035-6

Shindou, T., Mori, A., Kase, H., & Ichimura, M. (2001). Adenosine A(2A) receptor enhances GABA(A)-mediated IPSCs in the rat globus pallidus. *The Journal of Physiology, 532*(Pt 2), 423–434. https://doi.org/10.1111/j.1469-7793.2001.0423f.x

Shindou, T., Nonaka, H., Richardson, P. J., Mori, A., Kase, H., & Ichimura, M. (2002). Presynaptic adenosine A2A receptors enhance GABAergic synaptic transmission via a cyclic AMP dependent mechanism in the rat globus pallidus. *British Journal of Pharmacology, 136*(2), 296–302. https://doi.org/10.1038/sj.bjp.0704702

Shindou, T., Richardson, P. J., Mori, A., Kase, H., & Ichimura, M. (2003). Adenosine modulates the striatal GABAergic inputs to the globus pallidus via adenosine A2A receptors in rats. *Neuroscience Letters, 352*(3), 167–170. https://doi.org/10.1016/j.neulet.2003.08.059

Smeyne, R. J., & Jackson-Lewis, V. (2005). The MPTP model of Parkinson's disease. *Brain, 134*(1), 57–66. https://doi.org/10.1016/j.molbrainres.2004.09.017

Stacy, M., Silver, D., Mendis, T., Sutton, J., Mori, A., Chaikin, P., & Sussman, N. M. (2008). A 12-week, placebo-controlled study (6002-US-006) of istradefylline in Parkinson disease. *Neurology, 70*(23), 2233–2240. https://doi.org/10.1212/01.wnl.0000313834.22171.17

Suzuki, K., Miyamoto, M., Miyamoto, T., Uchiyama, T., Watanabe, Y., Suzuki, S., ... Hirata, K. (2017). Istradefylline improves daytime sleepiness in patients with Parkinson's disease: An open-label, 3-month study. *Journal of the Neurological Sciences, 380*, 230–233. https://doi.org/10.1016/j.jns.2017.07.045

Svenningsson, P., Hall, H., Sedvall, G., & Fredholm, B. B. (1997). Distribution of adenosine receptors in the postmortem human brain: an extended autoradiographic study. *Synapse (New York, N. Y.), 27*(4), 322–335. https://doi.org/10.1002/(SICI)1098-2396(199712)27:4<322::AID-SYN6>3.0.CO;2-E

Svenningsson, P., Le Moine, C., Fisone, G., & Fredholm, B. B. (1999). Distribution, biochemistry and function of striatal adenosine A2A receptors. *Progress in Neurobiology, 59*(4), 355–396. https://doi.org/10.1016/s0301-0082(99)00011-8

Svenningsson, P., Rosenblad, C., Af Edholm Arvidsson, K., Wictorin, K., Keywood, C., Shankar, B., ... Widner, H. (2015). Eltoprazine counteracts l-DOPA-induced dyskinesias in Parkinson's disease: A dose-finding study. *Brain, 138*(Pt 4), 963–973. https://doi.org/10.1093/brain/awu409

Tayama, T., Ishiuchi, M., Sugiyama, K., Oka, Y., Maeda, H., Nagata, Y., ... Kagawa, Y. (2023). Safety, tolerability, and pharmacokinetics of the novel adenosine A(2A) antagonist/inverse agonist KW-6356 following single and multiple oral administration in healthy volunteers. *Clinical Pharmacology in Drug Development*. https://doi.org/10.1002/cpdd.1222

Tronci, E., Simola, N., Borsini, F., Schintu, N., Frau, L., Carminati, P., & Morelli, M. (2007). Characterization of the antiparkinsonian effects of the new adenosine A2A receptor antagonist ST1535: Acute and subchronic studies in rats. *European Journal of Pharmacology, 566*(1–3), 94–102. https://doi.org/10.1016/j.ejphar.2007.03.021

Uchida, S., Soshiroda, K., Okita, E., Kawai-Uchida, M., Mori, A., Jenner, P., & Kanda, T. (2015a). The adenosine A2A receptor antagonist, istradefylline enhances anti-parkinsonian activity induced by combined treatment with low doses of L-DOPA and dopamine agonists in MPTP-treated common marmosets. *European Journal of Pharmacology, 766*, 25–30. https://doi.org/10.1016/j.ejphar.2015.09.028

Uchida, S., Soshiroda, K., Okita, E., Kawai-Uchida, M., Mori, A., Jenner, P., & Kanda, T. (2015b). The adenosine A2A receptor antagonist, istradefylline enhances the anti-parkinsonian activity of low doses of dopamine agonists in MPTP-treated common marmosets. *European Journal of Pharmacology, 747*, 160–165. https://doi.org/10.1016/j.ejphar.2014.11.038

Uchida, S., Tashiro, T., Kawai-Uchida, M., Mori, A., Jenner, P., & Kanda, T. (2014). Adenosine A(2)A-receptor antagonist istradefylline enhances the motor response of L-DOPA without worsening dyskinesia in MPTP-treated common marmosets. *Journal of Pharmacological Sciences, 124*(4), 480–485. https://doi.org/10.1254/jphs.13250fp

van Calker, D., Biber, K., Domschke, K., & Serchov, T. (2019). The role of adenosine receptors in mood and anxiety disorders. *Journal of Neurochemistry, 151*(1), 11–27. https://doi.org/10.1111/jnc.14841

Waggan, I., Rissanen, E., Tuisku, J., Joutsa, J., Helin, S., Parkkola, R., ... Airas, L. (2023). Adenosine A(2A) receptor availability in patients with early- and moderate-stage Parkinson's disease. *Journal of Neurology, 270*(1), 300–310. https://doi.org/10.1007/s00415-022-11342-1

Winkler, C., Kirik, D., Björklund, A., & Cenci, M. A. (2002). L-DOPA-induced dyskinesia in the intrastriatal 6-hydroxydopamine model of parkinson's disease: Relation to motor and cellular parameters of nigrostriatal function. *Neurobiology of Disease, 10*(2), 165–186. https://doi.org/10.1006/nbdi.2002.0499

Yamada, K., Kobayashi, M., Mori, A., Jenner, P., & Kanda, T. (2013). Antidepressant-like activity of the adenosine A(2A) receptor antagonist, istradefylline (KW-6002), in the forced swim test and the tail suspension test in rodents. *Pharmacology, Biochemistry, and Behavior, 114–115*, 23–30. https://doi.org/10.1016/j.pbb.2013.10.022

Yamada, K., Kobayashi, M., Shiozaki, S., Ohta, T., Mori, A., Jenner, P., & Kanda, T. (2014). Antidepressant activity of the adenosine A2A receptor antagonist, istradefylline (KW-6002) on learned helplessness in rats. *Psychopharmacology (Berl), 231*(14), 2839–2849. https://doi.org/10.1007/s00213-014-3454-0

Yuan, X. S., Wang, L., Dong, H., Qu, W. M., Yang, S. R., Cherasse, Y., ... Huang, Z. L. (2017). Striatal adenosine A(2A) receptor neurons control active-period sleep via parvalbumin neurons in external globus pallidus. *Elife, 6*, e29055. https://doi.org/10.7554/eLife.29055

CHAPTER FIVE

# Adenosine $A_{2A}$ antagonists and Parkinson's disease

Michelle Offit[*], Brian Nagle, Gonul Ozay, Irma Zhang, Anastassia Kerasidis, Yasar Torres-Yaghi, and Fernando Pagan

MedStar Georgetown University Hospital, Neurology Department, Reservoir Rd, Washington, DC, United States
*Corresponding author. e-mail address: Michelle.Offit@medstar.net

## Contents

| | |
|---|---|
| 1. Introduction | 106 |
| 2. Receptor agonists | 108 |
|    2.1 Dopamine receptor agonists | 108 |
| 3. Inverse agonists | 112 |
|    3.1 Serotonin receptor 2A inverse agonists | 112 |
| 4. Receptor antagonists | 112 |
|    4.1 Dopamine receptor antagonists | 113 |
|    4.2 NMDA receptor antagonists | 114 |
|    4.3 Acetylcholine receptor antagonists | 115 |
|    4.4 Adenosine receptor antagonists | 115 |
| References | 117 |

## Abstract

Although there is no cure for Parkinson's disease (PD), there are several classes of medications with various mechanisms of action that can help improve the functionality of someone with PD. Dopamine derivatives are first line therapies for PD, hence dopamine receptor agonists (DAs) have been shown to improve functionality of symptoms in PD patients. The two main formulations of dopamine agonist medications in PD therapy are ergoline and non-ergoline derivatives.

Additionally, it has been shown that PD can involve irregularities in other neurotransmitters, such as acetylcholine, norepinephrine, and serotonin, hence why non-dopaminergic medications are also vital in PD management. Examples include NMDA receptor antagonists, dopamine antagonists (i.e. neuroleptics), acetylcholine receptor antagonists, serotonin receptor 2A agonists, and adenosine $A_2$ antagonists.

In general, dopaminergic medications are the most effective in improving motor involvement with PD, whereas non-dopaminergic medications tend to focus on the non-motor involvement of PD. In this chapter, we will focus on the chemistry and medication background on dopaminergic vs non-dopaminergic therapy, with a focus of adenosine $A_2$ antagonists at the end.

## 1. Introduction

Although there is currently no cure for Parkinson's disease (PD), there are several classes of medications with various mechanisms of action that can help treat symptoms and improve the functionality of someone with PD. The primary treatment is the prodrug levodopa. However, there are also other therapies that are highly utilized to help improve PD motor and non-motor symptoms. Clinical motor manifestations include resting tremor, postural instability, rigidity, and bradykinesia, whereas non-motor symptoms include sleep abnormalities, cognitive dysfunction, dysautonomia (i.e., orthostatic hypotension, constipation, sialorrhea, hyperhidrosis, urinary dysfunction, or erectile dysfunction), and neuropsychiatric symptoms such as depression, anxiety, and psychosis (Armstrong & Okun, 2020; Fox et al., 2018; Hisahara & Shimohama, 2011; Seppi et al., 2019).

The neurotransmitter dopamine plays a central role in the etiology of PD, as many symptoms stem from a decrease of dopamine-producing neurons in the substantia nigra pars compacta (SNpc) of the midbrain, which have projections to the basal ganglia (Hisahara & Shimohama, 2011; Zahoor et al., 2018). However, dysregulation of other neurotransmitters are also seen as the cause of various PD symptoms. These additional neurotransmitters include acetylcholine, norepinephrine, serotonin, and glutamate. Hence, many PD symptoms can be treated with a combination of both dopaminergic and non-dopaminergic therapies.

Normally, biochemical production of dopamine starts in the liver with the conversion of L-phenylalanine to L-tyrosine by the enzyme phenylalanine hydroxylase. It then has two pathways where it can be synthesized into dopamine. The first and foremost pathway it uses is through converting L-tyrosine to L-dopa via the action of tyrosine hydroxylase in the dopaminergic neurons located in the central nervous system. L-dopa is then converted to 3,4-dihydroxyphenethylamine, also known as dopamine, through L-3,4-dihydroxyphenylalanine decarboxylase (i.e., dopa decarboxylase). The other pathway is by the conversion of L-tyrosine to tyramine by dopa decarboxylase, followed by cytochrome P450206 (i.e., CYP206) in the substantia nigra converting tyramine to dopamine (Zahoor et al., 2018).

However, dopamine cannot be given directly because it cannot cross the blood–brain barrier. In order to increase the central levels of dopamine, the precursor form of dopamine, L-dihydroxyphenylalanine (levodopa or L-dopa) is used, as levodopa, unlike dopamine, can cross the blood–brain

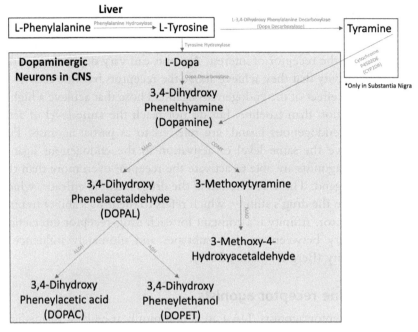

Fig. 1 Depicts the biochemistry of dopamine. Abbreviations used include MAO, monoamine oxidase; COMT, catechol-O-methyltransferase; ALDH, aldehyde dehydrogenase; CNS, central nervous system; and ADH, alcohol dehydrogenase (Zahoor et al., 2018).

barrier. Once across the blood–brain barrier, levodopa is then converted to dopamine where it interacts with its receptors. Additional therapies for treating PD symptoms include inhibitors of enzymes involved in dopamine catabolism. Monoamine oxidase-B (MAO-B) inhibitors and catechol-O-methyltransferase (COMT) inhibitors are the main two pathways for increasing dopamine by inhibiting its breakdown into inactive metabolites. See Fig. 1 for a diagram of the metabolism of dopamine (Andruşca 2018; Borea, Gessi, Merighi, Vincenzi, & Varani, 2018).

Although PD can be treated with medications that work as a prodrug or via inhibition of certain enzymes, many therapies for both motor and non-motor symptoms work via modulating receptor activity. These include agonists, antagonists, and inverse agonists. The main focus of this chapter will be on these classes of receptor-modulating medications and the receptors they influence in the treatment of both motor and non-motor features of PD (UpToDate, n.d.(b)).

## 2. Receptor agonists

Similar to endogenous ligands, receptor agonists work by increasing the activity of the receptor of interest. Agonists can vary depending on the amount of activity that they achieve above the receptors baseline activity in relation to the effect of the endogenous ligand. Those that achieve a higher level of activation than baseline, but do not reach the same level of activation as the endogenous ligand, are referred to as partial agonists. Full agonists achieve the same level of activation as the endogenous ligand, whereas superagonists are able to activate the receptor even more than the endogenous ligand. This is referred to as the drugs intrinsic efficacy, which is separate from the drug's affinity, which refers to the drug's ability to bind a specific receptor. Affinity is a constant for each drug-receptor interaction which can vary between receptor subtypes and ultimately influence a drug's selectivity (Berg & Clarke, 2018).

### 2.1 Dopamine receptor agonists

Dopamine receptor agonists (DAs) are a commonly used class of medications for treating motor symptoms of PD and can be used as a first-line treatment in early PD, as well as an adjunctive therapy (Armstrong & Okun, 2020; Fox et al., 2018; Hisahara & Shimohama, 2011; Seppi et al., 2019; Zahoor et al., 2018; UpToDate, n.d.(b)). The clinical effect of DAs stems from working immediately on pre- and post-synaptic dopamine receptors type 1 (D1) and type 2 (D2) (Brooks, 2000; Hisahara & Shimohama, 2011; Perez-Lloret & Rascol, 2010; Reichmann et al., 2006). Dopaminergic projections from the substantia nigra pars compacta (SNpc) to these receptors in the basal ganglia are key to the movement abnormalities in PD and have an effect on the balance of the direct and indirect pathways, which are central to motor circuitry (Andrųca 2018; Brooks, 2000; Hisahara & Shimohama, 2011; Perez-Lloret & Rascol, 2010; Reichmann et al., 2006).

D1 and D2 receptors are the most prominent dopamine receptors in the striatum (Brooks, 2000; Hisahara & Shimohama, 2011; Perez-Lloret & Rascol, 2010; Reichmann et al., 2006). Multiple studies have demonstrated that the depletion of dopamine correlates with an upregulation of D1 and D2 receptors. Most post-synaptic D2 receptors are located on neurons in the cortical pyramidal and striatopallidal regions (Hisahara & Shimohama, 2011). When post-synaptic D2 receptors are activated, it tends to result in initiation of movement (Andrųca, 2018). This fundamental concept is the basis for the mechanism of action of dopamine agonists. If post-synaptic dopaminergic receptors are stimulated, it can help increase the movement that was unable to be achieved due to dopamine depletion. Notably, dopamine agonists are effective because they are not dependent on pre-synaptic endogenous dopamine production. Although levodopa has greater benefits on motor symptoms in treating early PD, it also increases the risk of dyskinesias. Dopamine agonists can be used in early PD for treating motor symptoms. They can also be used adjunctively with levodopa therapy to treat breakthrough symptoms during "off periods" between levodopa doses. Side effects include orthostatic hypotension, sleep attacks, hallucinations, and impulse control disorder. Because these are more common in the elderly, dopamine agonists are reserved for patients under 60 years. There are two main classes of dopamine agonists: ergots and non-ergots (Borea et al. 2018; Brooks, 2000; Hisahara, & Shimohama, 2011; Perez-Lloret & Rascol, 2010; Reichmann et al., 2006; UpToDate, n.d.(b)).

### 2.1.1 Ergot dopamine agonists

Ergot alkaloid drugs are derived from the ergot fungus that grows on grain (Schiff, 2006). Ergotamine derivatives are an older class of dopamine agonists that predominantly act on D2 receptors, as well as D3 and D4. They have moderate-to-strong affinity to alpha-adrenergic and serotonergic receptors as well (Brooks, 2000; Perez-Lloret & Rascol, 2010; UpToDate, n.d.(b)). The main ergotamines considered useful in PD management include pergolide, bromocriptine, cabergoline, and lisuride. Bromocriptine is an oral D2 agonist and a weak D1 antagonist. It can be used as an adjunct treatment for "off periods" in Parkinson's patients experiencing motor fluctuations corresponding with the end of their levodopa dose (Brooks, 2000; Perez-Lloret & Rascol, 2010). Pergolide is an oral D1 and D2 agonist that is also used to treat motor fluctuations in Parkinson's disease patients on levodopa. Cabergoline is a highly selective D2 agonist notable for having a long half-life of between 65 and 110 h. It is an oral drug that has been useful in decreasing the "off periods" in patients taking levodopa, as well as increasing activities of daily living (ADL) scores in patients (Brooks, 2000; Perez-Lloret & Rascol, 2010; UpToDate, n.d.(b)). Its long half-life, allowing for daily dosing, also improves patient compliance (Brooks, 2000). Lastly, lisuride is a D2 agonist that also works on serotonin receptors. It comes in oral, subcutaneous, and intravenous forms. It is important to note that it has an incredibly short half-life of 2 h. It has been demonstrated to be helpful in patients who are refractory to the effects of levodopa. Similar to the other drugs listed in this class, lisuride is also shown to be beneficial in reducing motor disturbances in patients taking levodopa (Brooks, 2000; Perez-Lloret & Rascol, 2010; UpToDate, n.d.(b)).

Although rare, it is pertinent to bring up the side effects of ergots, as they are serious if they develop. Ergots have been associated with retroperitoneal, cardiovascular, and pleuropulmonary fibrosis (Reichmann et al., 2006). Additionally, these drugs may trigger Raynaud's syndrome and erythromelalgia. It has been proposed that these side effects, especially fibrosis, are due to affinity for multiple receptors. While they demonstrate similar efficacy to non-ergot dopamine agonists, ergot agonists are no longer first-line and typically avoided due to the severity of some of these side effects. Side effects of nausea, sleep attacks, orthostatic hypotension, and hallucinations are not exclusive to ergot-derivatives (Brooks, 2000; Perez-Lloret & Rascol, 2010; Reichmann et al., 2006; UpToDate, n.d.(b)).

## 2.1.2 Non-ergot dopamine agonists

Non-ergots are a newer class of dopamine agonists with higher selectivity for dopamine receptors when compared to ergots. Due to this selective and consequent absence of the side effects specific to ergot derivatives, non-ergot DAs are first-line in monotherapy for the early stages of PD and adjunctive therapy with levodopa to manage motor symptoms. They are generally well tolerated, however, the main side effects include dizziness, nausea, and orthostatic hypotension. It should be noted that all DAs may be associated with behavioral side effects, such as impulsivity, that can lead to excessive gambling or shopping, over-eating, and hypersexuality (Reichmann et al., 2006; Schiff, 2006). Patients should also be screened regularly for sleep attacks and hallucinations. Significant side effects warrant weaning the patient off of the drug. This generally requires a very gradual taper over several months to avoid DA withdrawal syndrome characterized by anxiety, excess sweating, and irritability. Additionally, dopamine agonists may also decrease prolactin levels, so breastfeeding patients should also avoid these. The four non-ergot DAs used in PD symptomatic management include ropinirole, rotigotine, apomorphine, and pramipexole (Brooks, 2000; Frampton, 2019; Hisahara & Shimohama, 2011; Kushida, 2006; Zahoor et al., 2018).

Ropinirole is a highly selective D2 and D3 receptor agonist that comes in immediate-release (IR) and extended-release (ER) oral pill forms. It was the first oral non-ergotamine DA available for treatment. Notably, ropinirole is also proven to be effective in treating restless leg syndrome and sleep disturbances, as well as improving anxiety and depression. One of the main side effects to be mindful of ropinirole is daytime drowsiness, which can lead to episodes of falling asleep, thought to be due to its inhibition of orexin neurons (Blandini & Armentero, 2014; Brooks, 2000; Hisahara & Shimohama, 2011; Kushida, 2006; Perez-Lloret & Rascol, 2010; Reichmann et al., 2006; Rektorova et al., 2008).

Rotigotine has a high affinity for D2, D3, D4, and D5 receptors, with additional agonist activity on 5-HT1A receptors and antagonist activity at α2B adrenergic receptors (Frampton, 2019). It is highly lipophilic and comes in a transdermal patch designed to deliver the drug over a 24-hour period (Frampton, 2019). Benefits of this form of delivery include continuous low-concentration release of the drug over an extended period of time. Since rotigotine comes in a transdermal patch form, it is especially beneficial as an alternative to oral DAs for patients with swallowing impairments or suboptimal kidney function. Similar to ropinirole, rotigotine has been demonstrated to treat restless leg syndrome and sleep disturbances. The patch can cause skin irritation, which can affect compliance (Frampton, 2019).

Pramipexole is a dopamine agonist with affinity to D2 receptors and selectivity for D3 receptors (Perez-Lloret & Rascol, 2010). It comes in an oral form and can be administered as an IR or ER formulation. Perhaps due to its selectivity for D3 receptors, pramipexole is particularly helpful with people with dyskinesias, mood instability, depression, and sleep dysfunction. Pramipexole undergoes minimal hepatic metabolism and is cleared renally. It should be avoided in patients with reduced kidney function (creatinine clearance <30 mL/min) (Cummings, 1999; Trenkwalder et al., 2015).

Apomorphine is given subcutaneously or sublingually and has a strong affinity for D1 and D2 receptors (Trenkwalder et al., 2015). It takes effect within 4–12 min. In the US, apomorphine is approved as on-demand therapy for treating "off periods" in between doses of levodopa. On-demand therapy comes in the form of a subcutaneous injection or sublingual film (Trenkwalder et al., 2015). In Europe, apomorphine has been administered continuously via a subcutaneous pump for over two decades, which allows doses of levodopa to be reduced significantly. The apomorphine continuous subcutaneous pump has not yet been approved by the FDA (LeWitt, 2004; Trenkwalder et al., 2015).

## 3. Inverse agonists

The effect of an inverse agonist requires that the receptor it is acting upon has a baseline level of activity in the absence of an endogenous ligand or agonist. Whereas an antagonist, as noted above, blocks the activation of a receptor, it does nothing to reduce any basal activity inherent in the receptor. Inverse agonists, on the other hand, reduce this baseline receptor. If the receptor that an inverse agonist has an affinity for does not have baseline activity in the absence of an agonist or endogenous ligand, it will behave as a competitive antagonist. So while antagonists and inverse agonists may have a similar pharmacological effect of preventing receptor activation, the inverse agonist goes a step further to reduce basal activity and further suppress the action of the receptor. Because inverse agonists bind to the same site as agonists, a receptor antagonist can block the effect of an inverse agonist (Berg & Clarke, 2018).

### 3.1 Serotonin receptor 2A inverse agonists

As discussed previously, antipsychotics tend to act on many receptors, including dopamine and serotonin. The 5-HT2A receptor is involved in hallucinations in PD through increased binding in the neocortex. Atypical antipsychotics can reduce these hallucinations through antagonism of the serotonin pathways, however, many are not selective and affect several other receptors that can result in adverse side effects or, as is in the case with dopamine blockade, worsen PD motor symptoms. Pimavanserin, a selective 5-HT2A receptor inverse agonist, is the first FDA-approved treatment for PD psychosis. By selectively targeting this receptor and not only blocking receptor activation but reducing its basal activity through inverse agonism, hallucinations can be significantly reduced compared to placebo. Moreover, the benefits do not come at the cost of worsening parkinsonism. This change remains significant regardless of age, sex, and presenting cognitive status. It is well tolerated with similar rates of adverse effects compared to placebo, though slightly more patients taking pimavanserin experienced some peripheral edema (Berg & Clarke, 2018; Cummings et al., 2014).

## 4. Receptor antagonists

Antagonists work by blocking activation of the receptor by endogenous ligand or agonists. It can do this either by competing with the ligand or agonist for the binding site (competitive antagonist) or by altering the conformation of the receptor protein, thus blocking or changing the binding site, rendering it inaccessible to the ligand or agonist (non-competitive antagonist). By blocking activation of the receptor, the receptor maintains its baseline level of activity. Similar to agonists, antagonists have a unique affinity for binding specific receptors. Again, it is a constant for each drug-receptor interaction, so one drug can have varying levels of affinity for multiple receptor subtypes, which ultimately influences its selectivity.

## 4.1 Dopamine receptor antagonists

Although the use of dopamine antagonists in patients with PD appears controversial, some of these medications can be used in the treatment of PD psychosis (UpToDate, n.d.(b), n.d.(c)). There are two main classes of medications that work via dopamine antagonism to be aware of, as using one class over the other can cause a severe worsening of PD symptoms. First-generation (typical) antipsychotics work by inhibiting post-synaptic D2 receptors cortically and in the striatal region. As mentioned previously, DAs help PD motor symptoms by stimulating the striatal D2 receptors. Hence, if a PD patient is started on a typical antipsychotic, there is a high chance the patient can worsen their PD symptoms, and in patients without PD may cause extrapyramidal symptoms (EPS), including drug-induced parkinsonism (bradykinesia, tremor, rigidity, akathisia, or dystonia), tardive dyskinesia, or even neuroleptic malignant syndrome. Typical antipsychotics to watch out for include haloperidol, pimozide, perphenazine, fluphenazine, loxapine, trifluoperazine, chlorpromazine, and thioridazine. Some of the main side effects of first-generation antipsychotics besides EPS are increased prolactin levels and sexual dysfunction (UpToDate, n.d.(a), n.d.(c)).

On the other hand, some second-generation (atypical) antipsychotics can be used over first-generation antipsychotics for PD psychosis, given lower and variable affinity for D2 receptors and affinity for other receptors. The two most common second-generation antipsychotics used in PD psychosis include quetiapine and clozapine (Connolly & Lang, 2014; Parsa & Bastani, 1998; Paz & Murer, 2021; UpToDate, n.d.(a)). Quetiapine is an antagonist against serotonin (5-HT), histamine, D1, and D2 receptors. However, what is unique about quetiapine compared to typical antagonists is that it has a more substantial effect working as an antagonist against serotonin receptors over dopamine receptors. Although theoretically this means it can be used effectively in PD psychosis without causing patients to develop severe EPS symptoms, clinical trials of its efficacy in treating PD psychosis have not been shown, and thus its use is mainly based on anecdotal evidence. At the low doses used for those with PD psychosis, it mainly has anti-histamine effects, thus treating PD psychosis mainly through sedation (Connolly & Lang, 2014; Kelly et al., 2011; Parsa & Bastani, 1998; Paz & Murer, 2021; UpToDate, n.d.(a)).

Conversely, clozapine, another atypical antipsychotic, does have clinical evidence showing efficacy in the treatment of PD psychosis. Although it antagonizes dopamine receptors, these are mainly D4 more than D2 receptors, which leads to a lower possibility of developing EPS and exacerbating already present PD symptoms. It also acts as an antagonist for serotonin receptor 2A (5-HT2A), muscarinic receptors (M1, M2, M3, and M5), histamine receptors, and α1 adrenergic receptors. Adverse side effects associated with clozapine include agranulocytosis, myocarditis, seizures, constipation, sialorrhea, and enhancement of platelet clumping, leading to pulmonary embolisms. A major drawback of this medication is the weekly blood draws to monitor for agranulocytosis (Haidary & Padhy, 2018; UpToDate, n.d.(a), n.d.(c)).

## 4.2 NMDA receptor antagonists

An alternative medication besides DAs and levodopa that can be used, especially for dyskinesias produced while on levodopa and tremor-predominant features in early PD, is the N-methyl-D-aspartate (NMDA) receptor antagonist, amantadine. NMDA receptor's endogenous ligand is glutamate, an excitatory neurotransmitter, which plays a role in the motor circuity of the basal ganglia. It has been postulated that amantadine is beneficial in PD by altering dopamine levels. For example, there has been some evidence that amantadine can help prevent dopamine from being recycled back into the metabolic pathway, causing more dopamine to be utilized. It has been shown to promote more dopamine in the body by stimulating factors to increase the release of dopamine, such as directly acting on dopamine receptors. Amantadine typically comes as an oral pill, both in IR and ER formulations. Amantadine can be used as monotherapy to treat tremor-predominant early PD. It is also commonly used adjunctively in advanced PD to reduce dyskinesias. It is important to note that amantadine is renally excreted and should be used cautiously with patients with renal failure. Typically, the dosage must be reduced to a minimum in patients with renal failure. It is noted that amantadine can cause extremity swelling and confusion. Another side effect of amantadine is its anticholinergic properties (i.e., dry eyes, constipation, hallucinations, and urinary retention), thus it should be used cautiously in the elderly (UpToDate, n.d.(d)).

## 4.3 Acetylcholine receptor antagonists

In the striatum, acetylcholine (ACh) is released by giant striatal cholinergic interneurons (SCIN) and acts upon voltage-gated nicotinic and G-protein coupled muscarinic ACh receptors to modulate output pathways. In Parkinson's disease, dopamine deficiency reduces D2 receptor activation on SCIN, reducing the inhibitory signal to SCIN, ultimately leading to increased ACh release. Increased ACh from these interneurons further reduces dopamine release in a cycle that exacerbates PD motor symptoms, especially tremors and rigidity (Haidary & Padhy, 2018; McInnis & Petursson, 1985). Trihexyphenidyl and benztropine are two anticholinergic medications that have been demonstrated to be helpful in tremor-dominant Parkinson's disease.

Trihexyphenidyl is an anticholinergic antagonist with a high affinity for the M1 (muscarinic) receptor in the parasympathetic nervous system, along with some activity at the dopamine receptors. While its mechanism of action is not well-understood, it is particularly effective at treating tremors, spasms, and decreased motor control. Benztropine acts as a dual ACh and histamine antagonist. It is approved as an adjunctive therapy to help manage motor symptoms in PD. Notably, it is also utilized in patients who developed EPS from the use of antipsychotics, especially typical antipsychotics (McInnis, & Petursson, 1985; Jilani, & Sharma, 2019).

Some side effects of anticholinergics include hallucinations, confusion, cognitive decline, constipation, urinary retention, and dry mouth. These effects are often amplified in the elderly population. Thus, anticholinergics are ideal for younger patients with tremor-dominant PD who are refractory to dopamine agonists. Similar to dopamine agonists, a slow taper is necessary for discontinuing the drug, as abrupt cessation may trigger worsening motor symptoms (McInnis & Petursson, 1985; Jilani & Sharma, 2019; Abusrair et al., 2022; Jankovic & Tan, 2020).

## 4.4 Adenosine receptor antagonists

Adenosine receptors are found in various tissues throughout the body, including the brain, heart, and kidney. There are four types of adenosine receptors: $A_1$, $A_{2A}$, $A_{2B}$, and $A_3$. All four types of adenosine receptors are G-coupled. Whereas $A_{2B}$ is a low-affinity receptor, $A_{2A}$ is a high-affinity receptor. One of the most well-known and widely used adenosine receptor antagonists is caffeine, which is non-selective. Some of the neurobehavioral effects of caffeine are due to $A_1$ and $A_{2A}$ receptors in the brain. Caffeine also works peripherally, predominantly on $A_1$ receptors, contributing to many of the familiar side effects of caffeine use, such as tachycardia, diaphoresis, and diuretic effects. Theophylline is another non-selective adenosine receptor antagonist (Borea et al., 2018; Mori et al., 2022).

Istradefylline is also an adenosine receptor antagonist, though it differs from caffeine and theophylline in its selectivity for the adenosine $A_{2A}$ receptor. The $A_{2A}$ receptor is found predominantly in the neurons that project from the striatum to the external globus pallidus in the indirect pathway. Because the indirect pathway circuitry of the basal ganglia is involved in inhibiting thalamocortical projections, an imbalance in the indirect and direct pathways favoring the indirect pathway results in decreased activity. The $A_{2A}$ receptor is also found in the nucleus accumbens and olfactory tubercle, which may prove helpful in treating neuropsychiatric symptoms. In patients on levodopa experiencing motor fluctuations, istradefylline decreased off-time by 1.8 h, with a significant difference from placebo seen as early as two weeks. Although dyskinesias were more likely in those treated with istradefylline than placebo, most dyskinesias were mild to moderate. There were no differences between the two groups regarding time spent in the on-state with troublesome dyskinesias. The istradefylline group also experienced fewer falls than the placebo group. However, occurrences were low in both groups, and the study was not powered to evaluate for a difference in the rates of falls. Further studies utilizing istradefylline or novel adenosine $A_{2A}$ receptor antagonists for specific motor and non-motor symptoms will further clarify the role of this receptor subtype in the treatment of PD (Borea et al., 2018; Mori et al., 2022; LeWitt et al., 2008; Mizuno et al., 2010).

In conclusion, adenosine $A_{2A}$ receptor antagonists, specifically istradefylline, can be particularly beneficial over the standard levodopa treatment for both motor and non-motor symptoms in PD (Fig. 2). Specifically, the majority of sleep disturbances, mood dysfunction, and cognitive impairment as a result of PD is not well managed with levodopa and can be optimized by initiating an adenosine $A_{2A}$ receptor antagonist. Istradefylline also does not cause visual or auditory hallucinations, dyskinesias, or insomnia, which can be some of the adverse effects of levodopa medications. Additionally, istradefylline can be supplemented to help with the "OFF" periods between levodopa dosages to further improve daily functionality of PD patients. Lastly, there have been some studies showing that glial and neuronal adenosine $A_{2A}$ receptors can help decrease degeneration of neurons, which may help impact the progression of the disease in the future (Borea et al., 2018; Mori et al., 2022; LeWitt et al., 2008; Mizuno et al., 2010; Pourcher & Huot, 2015).

| Dopamine Receptor Agonists | |
|---|---|
| **Ergot Dopamine Agonists** | **Non-Ergot Dopamine Agonists** |
| Pergolide | Ropinirole |
| Bromocriptin | Rotigotine |
| Cabergoline | Apomorphine |
| Lisuride | Pramipexole |
| **Dopamine Receptor Antagonists** | |
| **First Generation** | **Second Generation** |
| Haloperidol | Quetiapine |
| Pimozide | Clozapine |
| Prophenazine | |
| Fluperazine | |
| Loxapine | |
| Trifluoperazine | |
| Chlorpromazine | |
| Thioridazine | |
| **NMDA Receptor Antagonists** | |
| Amantadine | |
| **Acetylcholine Receptor Antagonists** | |
| Trihexyphenidyl | |
| Benztropine | |
| **Adenosine Receptor Antagonists** | |
| **Non-Selective** | **Selective A2A Receptor** |
| Caffeine | Istradefylline |
| Theophylline | |
| **Serontonin Receptor 2A Inverse Agonists** | |
| Pimavanserin | |

**Fig. 2** This table represents the classes of PD medications discussed in this chapter.

## References

Abusrair, A. H., Elsekaily, W., & Bohlega, S. (2022). Tremor in Parkinson's disease: From pathophysiology to advanced therapies. *Tremor and Other Hyperkinetic Movements (New York, N. Y.), 12*(1), 29.

Andryca, A.(2018). *Direct and indirect pathways of the basal ganglia*. [online] Kenhub. ⟨https://www.kenhub.com/en/library/anatomy/direct-and-indirect-pathways-of-the-basal-ganglia⟩.

Armstrong, M. J., & Okun, M. S. (2020). Diagnosis and treatment of Parkinson disease. *JAMA: The Journal of the American Medical Association, 323*(6), 548–560. https://doi.org/10.1001/jama.2019.22360

Berg, K. A., & Clarke, W. P. (2018). Making sense of pharmacology: Inverse agonism and functional selectivity. *The International Journal of Neuropsychopharmacology/Official Scientific Journal of the Collegium Internationale Neuropsychopharmacologicum (CINP), 21*(10), 962–977. https://doi.org/10.1093/ijnp/pyy071

Blandini, F., & Armentero, M. (2014). Dopamine receptor agonists for Parkinson's disease. *Expert Opinion on Investigational Drugs, 23*(3), 387–410.

Borea, P. A., Gessi, S., Merighi, S., Vincenzi, F., & Varani, K. (2018). Pharmacology of adenosine receptors: The state of the art. *Physiological Reviews, 98*(3), 1591–1625. https://doi.org/10.1152/physrev.00049.2017

Brooks, D. J. (2000). Dopamine agonists: Their role in the treatment of Parkinson's disease. *Journal of Neurology, Neurosurgery & Psychiatry, 68*(6), 685–689. https://doi.org/10.1136/jnnp.68.6.685

Connolly, B. S., & Lang, A. E. (2014). Pharmacological treatment of Parkinson disease: A review. *JAMA: The Journal of the American Medical Association, 311*(16), 1670–1683.

Cummings, J., Isaacson, S., Mills, R., et al. (2014). Pimavanserin for patients with Parkinson's disease psychosis: A randomised, placebo-controlled phase 3 trial. *Lancet*,

*383*(9916), 533–540. https://doi.org/10.1016/S0140-6736(13)62106-6 [published correction appears in Lancet. 2014 Jul 5;384(9937):28].

Cummings, J. L. (1999). *D3 receptor agonists: Combined action neurologic and neuropsychiatric agents depressive symptoms.*

Fox, S. H., Katzenschlager, R., Lim, S.-Y., Barton, B., De Bie, R. M. A., Seppi, K., & Sampaio, C. (2018). International Parkinson and movement disorder society evidence-based medicine review: Update on treatments for the motor symptoms of Parkinson's disease. *Movement Disorders, 33*(8), 1248–1266. https://doi.org/10.1002/mds.27372

Frampton, J. E. (2019). Rotigotine transdermal patch: A review in Parkinson's disease. *CNS Drugs, 33*(7), 707–718.

Haidary, H. A. and Padhy, R. K. (2018). *Clozapine.* [online]. Nih.gov. ⟨https://www.ncbi.nlm.nih.gov/books/NBK535399/⟩.

Hisahara, S., & Shimohama, S. (2011). Dopamine receptors and Parkinson's disease. *International Journal of Medicinal Chemistry, 2011,* 1–16. https://doi.org/10.1155/2011/403039

Jankovic, J., & Tan, E. K. (2020). Parkinson's disease: Etiopathogenesis and treatment. *Journal of Neurology, Neurosurgery, and Psychiatry, 91*(8), 795–808.

Jilani, T. N., & Sharma, S. (2019). *Trihexyphenidyl* [online]. Nih.gov. Available at ⟨https://www.ncbi.nlm.nih.gov/books/NBK519488/⟩.

Kelly, E. Lyons, P. & Rajesh Pahwa, M. D. (2011). The impact and management of nonmotor symptoms of Parkinson's disease (Vol. 17).

Kushida, C. A. (2006). Ropinirole for the treatment of restless legs syndrome. *Neuropsychiatric Disease and Treatment, 2*(4), 407–419.

LeWitt, P. A., Guttman, M., Tetrud, J. W., et al. (2008). Adenosine A2A receptor antagonist istradefylline (KW-6002) reduces "off" time in Parkinson's disease: A double-blind, randomized, multicenter clinical trial (6002-US-005). *Annals of Neurology, 63*(3), 295–302. https://doi.org/10.1002/ana.21315

LeWitt, P. A. (2004). Subcutaneously administered apomorphine: Pharmacokinetics and metabolism. *Neurology, 62*(6 Suppl 4), S8–S11.

McInnis, M., & Petursson, H. (1985). Withdrawal of trihexyphenidyl. *Acta Psychiatrica Scandinavica, 71*(3), 297–303.

Mizuno, Y., Hasegawa, K., Kondo, T., Kuno, S., Yamamoto, M., & Japanese Istradefylline Study Group (2010). Clinical efficacy of istradefylline (KW-6002) in Parkinson's disease: A randomized, controlled study. *Movement Disorders: Official Journal of the Movement Disorder Society, 25*(10), 1437–1443. https://doi.org/10.1002/mds.23107

Mori, A., Chen, J.-F., Uchida, S., Durlach, C., King, S. M., & Jenner, P. (2022). The pharmacological potential of adenosine A2A receptor antagonists for treating Parkinson's disease. *Molecules (Basel, Switzerland), 27*(7), 2366. https://doi.org/10.3390/molecules27072366

Parsa, M. A., & Bastani, B. (1998). Quetiapine (Seroquel) in the treatment of psychosis in patients with Parkinson's disease. *The Journal of Neuropsychiatry and Clinical Neurosciences, 10*(2), 216–219. https://doi.org/10.1176/jnp.10.2.216

Paz, R. M., & Murer, M. G. (2021). Mechanisms of antiparkinsonian anticholinergic therapy revisited. *Neuroscience, 467,* 201–217.

Perez-Lloret, S., & Rascol, O. (2010). Dopamine receptor agonists for the treatment of early or advanced Parkinson's disease. *CNS Drugs, 24*(11), 941–968.

Pourcher, E., & Huot, P. (2015). Adenosine 2A receptor antagonists for the treatment of motor symptoms in Parkinson's disease. *Movement Disorders Clinical Practice, 2*(4), 331–340. https://doi.org/10.1002/mdc3.12187

Reichmann, H., Bilsing, A., Ehret, R., Greulich, W., Schulz, J. B., Schwartz, A., & Rascol, O. (2006). Ergoline and non-ergoline derivatives in the treatment of Parkinson's disease. *Journal of Neurology, 253*(Suppl 4), IV36–IV38.

Rektorova, I., Balaz, M., Svatova, J., Zarubova, K., Honig, I., Dostal, V., & Dusek, L. (2008). Effects of ropinirole on nonmotor symptoms of Parkinson disease: A prospective multicenter study. *Clinical Neuropharmacology, 31*(5), 261–266.

Schiff, P. L. (2006). Ergot and its alkaloids. *American Journal of Pharmaceutical Education, 70*(5), 98. Available at ⟨https://www.ncbi.nlm.nih.gov/pmc/articales/PMC1637017/⟩.

Seppi, K., Ray Chaudhuri, K., Coelho, M., Fox, S. H., Katzenschlager, R., Perez Lloret, S., & Djamshidian-Tehrani, A. (2019). Update on treatments for nonmotor symptoms of Parkinson's disease—An evidence-based medicine review. *Movement Disorders, 34*(2), 180–198. https://doi.org/10.1002/mds.27602

Trenkwalder, C., Chaudhuri, K. R., García Ruiz, P. J., LeWitt, P., Katzenschlager, R., Sixel-Döring, F., & Poewe, W. (2015). Expert Consensus Group report on the use of apomorphine in the treatment of Parkinson's disease – Clinical practice recommendations. *Parkinsonism & related disorders, 21*(9), 1023–1030.

www.uptodate.com. (n.d.(a)). *UpToDate.* [online] ⟨https://www.uptodate.com/contents/first-generation-antipsychotic-medications-pharmacology-administration-and-comparative-side-effects?search=first%20generation%20antipsychotic%20drugs&source=search_result&selectedTitle=2–119&usage_type=default&display_rank=1⟩.

www.uptodate.com. (n.d.(b)). *UpToDate.* [online] ⟨https://www.uptodate.com/contents/first-generation-antipsychotic-medications-pharmacology-administration-and-comparative-side-effects?search=first%20generation%20antipsychotic%20medications&source=search_result&selectedTitle=1~150&usage_type=default&display_rank=1⟩ [Accessed 13 Apr. 2023].

www.uptodate.com. (n.d.(c)). *UpToDate.* [online] ⟨https://www.uptodate.com/contents/second-generation-antipsychotic-medications-pharmacology-administration-and-side-effects?search=D2%20receptors&source=search_result&selectedTitle=1–150&usage_type=default&display_rank=1⟩ [Accessed 13 Apr. 2023].

www.uptodate.com. (n.d.(d)). *UpToDate.* [online] ⟨https://www.uptodate.com/contents/initial-pharmacologic-treatment-of-parkinson-disease?search=amantadine%20parkinson&source=search_result&selectedTitle=1–150&usage_type=default&display_rank=1#H4045844182⟩ [Accessed 13 Apr. 2023].

Zahoor, I., Shafi, A., & Haq, E. (2018). Pharmacological treatment of Parkinson's disease. In T. B. Stoker, & J. C. Greenland (Eds.). *Parkinson's disease: Pathogenesis and clinical aspects [Internet]*Brisbane (AU): Codon Publications Chapter 7.

# CHAPTER SIX

# Effects of adenosine $A_{2A}$ receptors on cognitive function in health and disease

### Cinthia P. Garcia[a,b,c], Avital Licht-Murava[a,b,1], and Anna G. Orr[a,b,*]

[a]Appel Alzheimer's Disease Research Institute, Weill Cornell Medicine, New York, NY, United States
[b]Feil Family Brain and Mind Research Institute, Weill Cornell Medicine, New York, NY, United States
[c]Pharmacology Graduate Program, Weill Cornell Medicine, New York, NY, United States
*Corresponding author. e-mail address: ago2002@med.cornell.edu

## Contents

| | |
|---|---|
| 1. Introduction | 122 |
| 2. Effects of $A_{2A}$ receptor modulation on normal cognitive function | 124 |
|    2.1 Pharmacological manipulations | 124 |
|    2.2 Global genetic manipulations | 126 |
|    2.3 Genetic manipulations targeted to neurons | 127 |
|    2.4 Genetic manipulations targeted to astrocytes | 128 |
|    2.5 Sex-dependent effects | 129 |
| 3. Effects of $A_{2A}$ receptor modulation on cognitive impairments in aging and disease | 131 |
|    3.1 Aging and dementia | 131 |
|    3.2 Multiple sclerosis | 134 |
|    3.3 Schizophrenia | 135 |
|    3.4 Neurodevelopmental disorders | 136 |
|    3.5 Brain trauma and stroke | 137 |
| 4. Effects of $A_{2A}$ receptor modulation on cognitive impairments in stress | 137 |
| 5. Effects of $A_{2A}$ receptor modulation on drug use and addiction | 138 |
| 6. Conclusions | 139 |
| Conflicts of interests | 141 |
| Acknowledgements | 141 |
| References | 141 |

## Abstract

Adenosine $A_{2A}$ receptors have been studied extensively in the context of motor function and movement disorders such as Parkinson's disease. In addition to these roles, $A_{2A}$ receptors have also been increasingly implicated in cognitive function and cognitive impairments in diverse conditions, including Alzheimer's disease, schizophrenia, acute brain injury, and stress. We review the roles of $A_{2A}$ receptors in cog-

---

[1] Current address: Cytora, Ltd., Hayozma 6, Yokneam, Israel.

nitive processes in health and disease, focusing primarily on the effects of reducing or enhancing $A_{2A}$ expression levels or activities in animal models. Studies reveal that $A_{2A}$ receptors in neurons and astrocytes modulate multiple aspects of cognitive function, including memory and motivation. Converging evidence also indicates that $A_{2A}$ receptor levels and activities are aberrantly increased in aging, acute brain injury, and chronic disorders, and these increases contribute to neurocognitive impairments. Therapeutically targeting $A_{2A}$ receptors with selective modulators may alleviate cognitive deficits in diverse neurological and neuropsychiatric conditions. Further research on the exact neural mechanisms of these effects as well as the efficacy of selective $A_{2A}$ modulators on cognitive alterations in humans are important areas for future investigation.

## 1. Introduction

Understanding how the brain processes information, stores new knowledge, and uses experience and other inputs to alter its functions is one of the principal challenges for basic science and medicine. New research findings have provided essential insights into the cellular and molecular underpinnings of cognitive functions, including learning, memory, perception, and attention, as well as the influences of emotion and arousal on these processes. In addition, tremendous progress has been made in understanding how different environmental conditions, genetic alterations, diseases, and traumatic insults can impair these processes and cause cognitive changes. However, it remains unclear exactly how various cognitive phenomena are engaged, maintained, and regulated, and which mechanisms could be therapeutically targeted to prevent or alleviate cognitive impairments associated with diverse brain disorders.

In the central and peripheral nervous systems, ATP is a prevalent transmitter that can be co-released by neurons with other neurotransmitters, including glutamate, gamma-aminobutyric acid (GABA), acetylcholine, and dopamine (Burnstock, 2009). ATP can also be released by other neural cell types, including astrocytes (Abbracchio, Burnstock, Verkhratsky, & Zimmermann, 2009; Fields & Burnstock, 2006; Kofuji & Araque, 2021). Although the exact amounts and proportions of released ATP can vary between cell types, brain regions, and physiological conditions, substantial evidence indicates that extracellular ATP and its key breakdown product adenosine play important roles in modulating neurotransmission, neuronal plasticity, glial cell signaling, and other processes that enable and regulate cognitive function. Of note, ATP is metabolized by ATPases to generate adenosine, which can also be released directly by astrocytes (Martin et al., 2007). Indeed, the levels of adenosine can be affected by various factors and perturbations in neural function (Kovacs, Dobolyi, Kekesi, & Juhasz, 2013; Kovacs, Juhasz, Palkovits, Dobolyi, & Kekesi, 2011) and promote differential effects on cognitive processes across conditions.

Adenosine activates P1 class receptors, which include four known subtypes: $A_1$, $A_{2A}$, $A_{2B}$, and $A_3$ (Burnstock, 2020). These receptors are members of the G protein-coupled receptor (GPCR) family, the largest family of integral membrane proteins. The high-affinity $A_{2A}$ receptors are classically known to couple to $G\alpha s$ and to be expressed at high levels in dopamine receptor D2-expressing medium spiny neurons of the striatum (Bogenpohl, Ritter, Hall, & Smith, 2012; Martinez-Mir, Probst, & Palacios, 1991; Schiffmann, Fisone, Moresco, Cunha, & Ferre, 2007). However, $A_{2A}$ receptors have also been detected in other brain regions, including the hippocampus, thalamus, and neocortex (Dixon, Gubitz, Sirinathsinghji, Richardson, & Freeman, 1996; Wang et al., 2022), especially in the contexts of aging, brain injury, and disease. Indeed, various conditions associated with neurocognitive and behavioral alterations are associated with altered extracellular adenosine levels and altered $A_{2A}$ receptor expression in multiple brain regions, implicating aberrant activation of these receptors in neural impairments. In support, converging studies have revealed that $A_{2A}$ receptors modulate multiple aspects of cognitive function and behavior.

In this chapter, we review evidence that $A_{2A}$ receptors modulate cognitive processes in physiological and pathophysiological conditions. Pharmacological inhibition or genetic ablation of $A_{2A}$ receptors has been shown by multiple independent groups to enhance information processing and storage, including learning, working memory, and remote memory, whereas activation or overexpression of $A_{2A}$ receptors was sufficient to cause impairments in these cognitive domains. Among many important studies into the roles of $A_{2A}$ receptors in cognition, we highlight emerging research demonstrating that these receptors: (1) modulate goal-directed behavior and motivation, (2) can alone cause cognitive deficits upon increased activation, (3) contribute to the roles of astrocytes in memory, and (4) exhibit cognitive domain-specific and sex-dependent effects. We also discuss mounting evidence that targeting $A_{2A}$ receptors alleviates cognitive impairments and behavioral changes in diverse animal models of disease, acute brain injury, and chronic or early-life stress. Collectively, this growing body of work suggests that $A_{2A}$ receptors are important modulators of neural processes underlying learning and memory and goal-directed behavior, and that therapeutically blocking aberrant overactivation of these receptors can have beneficial effects on cognitive function and behavior in diverse contexts (Fig. 1).

## 2. Effects of $A_{2A}$ receptor modulation on normal cognitive function

### 2.1 Pharmacological manipulations

Several studies have examined the effects of selective $A_{2A}$ receptor agonists on different aspects of cognitive function. The selective $A_{2A}$ agonist CGS-21680

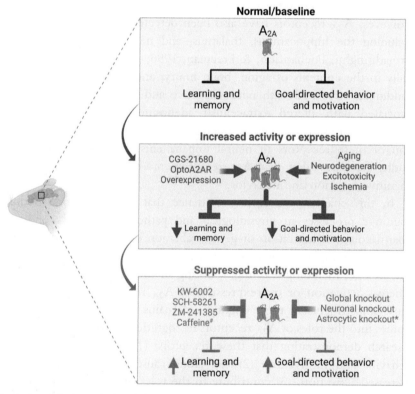

**Fig. 1** Simplified summary of the main reported effects of $A_{2A}$ receptor alterations on cognitive function. Representative disease-related conditions are listed. Additional behavioral effects, experimental details, and disorders are not included for the purpose of clarity. #nonselective targeting; *effects may not be identical among different knockouts.

can trigger memory impairments in the object recognition task, inhibitory avoidance, and modified Y maze (Pagnussat et al., 2015). In addition, direct administration of CGS-21680 into the posterior cingulate cortex is sufficient to impair memory retrieval, as assessed in the inhibitory avoidance task (Pereira et al., 2005). Motivation-related behavior was also shown to be altered by $A_{2A}$ agonists. In particular, systemic or direct administration of CGS-21680 into the nucleus accumbens caused dose-dependent suppression of lever-pressing behavior in an effort-based choice paradigm, suggesting that the agonist reduces motivation to exert effort, likely through modulation of the striatopallidal circuit. In contrast, there were no effects on lever-pressing behavior when the agonist was injected into a control brain area dorsal to the nucleus accumbens, suggesting that $A_{2A}$ receptors in different brain regions have distinct behavioral effects (Font et al., 2008; Mingote et al., 2008).

In contrast to the detrimental effects of $A_{2A}$ agonists, the nonselective antagonist caffeine enhances cognitive performance in mice and humans (Borota et al., 2014; Botton et al., 2010; Costa, Botton, Mioranzza, Souza, & Porciuncula, 2008). Moreover, treatment with the selective $A_{2A}$ receptor antagonists KW-6002 or SCH-58261 prevented memory impairments in a model of chronic unpredictable stress (Kaster et al., 2015). In addition, the $A_{2A}$ antagonist ZM241385 reversed impairments in short-term social memory in spontaneously hypertensive rats (Prediger, Fernandes, & Takahashi, 2005). The $A_{2A}$ antagonist SCH-58261 also prevented short-term memory loss caused by scopolamine, a non-selective muscarinic receptor antagonist (Pagnussat et al., 2015), and chronic treatment with the $A_{2A}$ antagonist KW-6002 reversed memory loss caused by cannabinoids (Mouro et al., 2019). In addition to alleviating memory impairments, blockade of $A_{2A}$ receptors with KW-6002 enhanced motivation-related behavior, as shown by increased goal-directed valuation (Li et al., 2018).

Of note, cognitive function and consciousness are intertwined. In addition to the effects on memory and motivation, emerging studies suggest that $A_{2A}$ agonists affect the loss of consciousness induced by anesthesia. Administration of CGS-21680 peripherally or directly into the nucleus accumbens prolonged loss of consciousness and increased the depth of sedation after propofol administration, whereas the $A_{2A}$ antagonist SCH-58261 had the opposite effects (Chen et al., 2021; Guo et al., 2022). These changes are likely related to the roles of $A_{2A}$ receptors in sleep and wakefulness (Huang, Zhang, & Qu, 2014; Li et al., 2020).

## 2.2 Global genetic manipulations

To provide further evidence that $A_{2A}$ receptors regulate cognitive function, studies have assessed the effects of global $A_{2A}$ gene knockout in animal models. Mice lacking $A_{2A}$ receptors generally had improvements in cognitive function, consistent with the reported effects of $A_{2A}$ inhibitors. For instance, Wang et al. showed that $A_{2A}$ knockout mice had improved spatial recognition memory and novelty exploration in the Y maze (Wang, Ma, & van den Buuse, 2006). Zhou et al. further showed that $A_{2A}$ knockout mice had enhanced spatial working memory in the eight-arm radial maze and repeated-trial Morris water maze (Zhou et al., 2009). Moreover, $A_{2A}$ knockout mice had improved memory following explosive blast-induced brain damage (Ning et al., 2013) or chronic intermittent hypoxia (Li et al., 2022). In contrast, Moscoso-Castro et al. have reported that $A_{2A}$ knockout mice had a range of cognitive impairments at baseline, as shown in the Y maze, passive and active avoidance tests, social interaction test, object recognition, radial arm maze, and other paradigms. These impairments were associated with changes in synaptic protein expression (Moscoso-Castro, Gracia-Rubio, Ciruela, & Valverde, 2016; Moscoso-Castro, Lopez-Cano, Gracia-Rubio, Ciruela, & Valverde, 2017). It is currently not clear if these conflicting results may be explained by differences in study design or other experimental or biological factors.

Interestingly, $A_{2A}$ receptor function has regional specificity and can modulate cognitive flexibility, instrumental behavior, and motivation. To assess these effects, Li et al. induced brain region-specific knockdown of $A_{2A}$ receptors in the nucleus accumbens using *Adora2a*-Cre mice and tested the mice in paradigms of goal-directed and habit-based behaviors (Li Y et al., 2020). These experiments revealed that $A_{2A}$ receptors in the nucleus accumbens reduce sensitivity to reward value and suppress incentive-based motivation and goal-directed behavior. Moreover, using a Cre-loxP system, Zhou et al. found that focal knockdown of $A_{2A}$ receptors in the nucleus accumbens enhanced cognitive flexibility, as demonstrated by enhanced attentional set-shifting and reversal learning (Zhou et al., 2019). Additionally, conditional knockdown of $A_{2A}$ receptors in the dorsomedial striatum enhanced effort-related motivation, but did not alter cognitive flexibility or reversal learning, whereas the lack of $A_{2A}$ receptors in the dorsolateral striatum had minimal effects on instrumental learning (Li et al., 2016; Zhou et al., 2019). Together, these compelling studies implicate $A_{2A}$ receptors in neural processes that regulate motivation, decision-making, and purposeful action.

## 2.3 Genetic manipulations targeted to neurons

$A_{2A}$ receptors can be expressed by various cell types in the brain, including different neuronal populations, microglia, astrocytes, oligodendrocytes, and endothelial cells (Borroto-Escuela & Fuxe, 2019; De Nuccio et al., 2019; Ferreira et al., 2015; Kim & Bynoe, 2015; Orr et al., 2015; Orr, Orr, Li, Gross, & Traynelis, 2009; Rebola, Lujan, Cunha, & Mulle, 2008). Several groups have characterized the effects of knocking out $A_{2A}$ selectively in neurons.

Consistent with global manipulations, forebrain or striatum-specific gene deletion of neuronal $A_{2A}$ receptors enhanced working memory and reversal learning (Jenner, Mori, & Kanda, 2020; Wei et al., 2011), suggesting that regional, neuron-targeted deletion was sufficient to improve cognitive function in certain domains. There is also evidence that deletion of neuronal $A_{2A}$ in specific brain regions causes different behavioral effects in fear conditioning and on anxiety-related behavior. Conditional deletion of $A_{2A}$ receptors in Dlx5/6-expressing striatal GABAergic neurons facilitated context and tone-related fear conditioning but did not affect anxiety, whereas deletion in CaMKII-positive excitatory forebrain neurons reduced tone-related fear conditioning and anxiety behavior (Wei et al., 2014). Also, amygdala-targeted gene deletion reduced context and tone-related fear conditioning (Simoes et al., 2016), whereas hippocampus-targeted gene deletion attenuated context-related fear conditioning (Wei et al., 2014). Collectively, these studies and others demonstrate that $A_{2A}$ receptors in different neuronal populations modulate distinct aspects of cognitive function and behavior.

The effects of $A_{2A}$ receptor overexpression targeted to neurons are largely consistent with the effects of $A_{2A}$ receptor agonists, which induce deficits in memory. Rats overexpressing human $A_{2A}$ receptors under the enolase promoter have impaired working and reference memory as measured by errors in the six-arm radial tunnel maze, novel object recognition test, and the Morris water maze. There were no detectable changes in motor or anxiety behavior (Gimenez-Llort et al., 2007). In agreement, $A_{2A}$ overexpression in mice using the neuronal CaMKII promoter was also detrimental to memory, as assessed in the Morris water maze (Temido-Ferreira et al., 2020).

In an elegant approach, Li et al. developed a chimeric rhodopsin-$A_{2A}$ receptor (opto$A_{2A}$R) that can be activated by light. The chimeric opto$A_{2A}$R includes the extracellular and transmembrane domains of rhodopsin (conferring light responsiveness and eliminating adenosine-binding pockets) and the intracellular domain of $A_{2A}$ receptors to enable $A_{2A}$-like signaling upon light stimulation. OptoA$_{2A}$R was expressed in neurons of the hippocampus or nucleus accumbens. Hippocampal optoA$_{2A}$R stimulation was sufficient to trigger CREB phosphorylation, increases in cFos levels, and deficits in memory as assessed in the Y maze. In contrast, optoA$_{2A}$R stimulation in the nucleus accumbens triggered MAPK phosphorylation and enhanced locomotor activity but did not affect memory (Li et al., 2015). Notably, optogenetic activation of neuronal optoA$_{2A}$R at the time of reward in the dorsomedial striatum suppressed goal-directed behaviors, whereas activation of optoA$_{2A}$R in the dorsolateral striatum had relatively minimal effect (Li et al., 2016). These results, together with the region-specific A$_{2A}$ gene deletion, suggest that neuronal A$_{2A}$ receptors in different brain regions trigger distinct behavioral responses due to differential neural circuit functions and/or differential A$_{2A}$ receptor functions, with the latter possibly due to biased receptor signaling and differences in downstream effectors.

## 2.4 Genetic manipulations targeted to astrocytes

Astrocytes are prevalent non-neuronal cells that influence diverse aspects of brain function, including cognitive processes, and can contribute to cognitive and behavioral deficits in pathophysiological conditions (Adamsky & Goshen, 2018; Alberini, Cruz, Descalzi, Bessieres, & Gao, 2018; Brandebura, Paumier, Onur, & Allen, 2023; Lyon & Allen, 2021; Nagai et al., 2021; Orr et al., 2015; Santello, Toni, & Volterra, 2019; Wang et al., 2022). The exact molecular mechanisms and effects of astrocytes on different cognitive processes is a topic of growing interest. Emerging studies are revealing that astrocytic GPCRs are potent modulators of memory and other cognitive processes in physiological and pathological contexts (Adamsky et al., 2018; Gao et al., 2016; Han et al., 2012; Jensen et al., 2016; Jones, Paniccia, Lebonville, Reissner, & Lysle, 2018; Kol et al., 2020; Li et al., 2022; Orr et al., 2015; Robin et al., 2018; Van Den Herrewegen et al., 2021).

Astrocytes express low levels of A$_{2A}$ receptors in normal conditions, but have increased A$_{2A}$ expression in brain injury and disease, including exposure to 1-methyl-4-phenyl-1,2,3,6-tetrahydropyridine (MPTP) or kainate, or accumulation of Aβ plaques (Matos et al., 2015; Orr et al., 2015; Yu et al., 2008; Chen et al., 2023), raising the question of whether astrocytic A$_{2A}$ receptors affect cognitive function and other processes. To address this question, two independent studies used astrocyte-specific Cre-mediated gene deletion of A$_{2A}$ receptors and examined different readouts of learning and memory, among other functional and molecular readouts. Matos et al. reported that deletion of astrocytic $A_{2A}$ receptors impaired working memory, as assessed in the Y maze and radial arm maze (Matos et al., 2015). Orr et al. reported that ablation of astrocytic $A_{2A}$ receptors improved remote memory in young and aging mice, as assessed in the Morris water maze and novel environment habituation, and increased the levels of hippocampal immediate-early gene Arc (Arg3.1) (Orr et al., 2015), which is known to be required for memory consolidation (Plath et al., 2006). In addition, Orr et al. reported that chemogenetic activation of astrocytic $G_s$-coupled signaling during learning impaired remote memory, further supporting the conclusion that astrocytic $A_{2A}$ receptors and aberrant increases in astrocytic $G_s$-coupled signaling can impair remote memory (Orr et al., 2015). The two studies provide complimentary insights into different aspects of cognitive changes upon genetic ablation of astrocytic $A_{2A}$ receptors.

## 2.5 Sex-dependent effects

Considering the manipulations in astrocytes mentioned above, it is important to also highlight that optogenetic stimulation of cAMP-dependent signaling in hippocampal astrocytes during training enhanced memory in a novel object recognition task (Zhou et al., 2021). Additionally, $G_s$-coupled adrenergic receptors in hippocampal astrocytes can facilitate memory consolidation (Gao et al., 2016). These studies suggest that astrocytic $G_s$-coupled receptors indeed enhance memory, which seemingly conflict with the findings by Orr et al. that $G_s$-coupled receptors can impair memory. One crucial difference among these studies is that most of them were conducted exclusively in male mice (Gao et al., 2016; Zhou et al., 2021), whereas Orr et al. used mixed-sex groups that included females (Orr et al., 2015). These studies did not address potential sex-dependent effects that might alter how astrocytic GPCRs, including $A_{2A}$ receptors, influence cognitive function. Interestingly, recent findings suggest that astrocytic GPCRs can promote divergent sex-dependent effects on memory (Meadows et al., 2022). Across different receptor manipulations in astrocytes, including chemogenetic stimulation of $G_s$-coupled receptors, there was a recurrent pattern of sex-specific effects on memory. In particular, chemogenetic stimulation of $G_s$-coupled receptors in hippocampal astrocytes during training in the Morris water maze enhanced memory in males but impaired memory in females. In support, astrocytes were reported to have sex differences in early development (Rurak et al., 2022) and, in adulthood, release thrombospondin-1 that induces cortical synaptogenesis in a sex-specific manner (Mazur et al., 2021). Together, these emerging studies point to sex differences in astrocytic regulation of brain function.

The effects of astrocytic $A_{2A}$ receptors in the hippocampus might also vary among males and females and cause different changes in memory-related processes and other cognitive functions across sexes. Indeed, chronic administration of the $A_{2A}$ blocker SCH-58261 caused sex-specific effects on anxiety-related behavior in young rats, as assessed in the elevated plus maze (Caetano et al., 2017). In this study, treatment with the $A_{2A}$ blocker had anxiogenic effects on baseline behavior in females but not males. Following prenatal exposure to dexamethasone to induce anxiety-like phenotypes, treatment with the same $A_{2A}$ blocker induced anxiolytic effects in males but not in females. Of note, these divergent sex-specific effects could not be accounted for by the estrous cycle or hormonal oscillations in females because no correlations were found between these factors and the observed behavioral effects (Caetano et al., 2017). A subsequent study using similar approaches and animal models reported that chronic $A_{2A}$ receptor blockade in females normalized neural synchronization in the medial prefrontal cortex-dorsal hippocampal circuit and reduced cognitive deficits but not the anxiety-related behaviors (Duarte et al., 2019). In male mice, similar beneficial effects of SCH-58261 on anxiety-related behavior have been reported in a model of chronic unpredictable stress (Kaster et al., 2015). Collectively, these findings suggest that the effects of $A_{2A}$ receptor blockade can vary across sexes and behavioral domains.

The discoveries of sex-specific effects of astrocytic $G_s$-coupled receptors and blockers of $A_{2A}$ receptors may at least partly explain the differential effects of $A_{2A}$ receptor manipulations across some studies, in addition to other potential variables. Sex differences are indeed known to affect mechanisms of learning and memory and other neural processes (Jain, Huang, & Woolley, 2019; Koss & Frick, 2017; Markowska, 1999; Velasco, Florido, Milad, & Andero, 2019; Yagi, Galea, 2019). Sex differences are also evident in the incidence, symptoms, and progression of various brain disorders (Bushnell et al., 2018; Eid, Gobinath, & Galea, 2019; Golden & Voskuhl, 2017; Guo, Zhong, Zhang, Zhang, & Cai, 2022; McEwen & Milner, 2017; Rubinow, Schmidt, 2019; Voskuhl & Itoh, 2022). However, the influence of biological sex on $A_{2A}$ receptor function and its exact roles in cognitive processes require further exploration in physiological and pathophysiological contexts, especially in conditions with altered expression of neuronal or astrocytic $A_{2A}$ receptors. The putative sex-specific effects of $A_{2A}$ receptors and their antagonists may represent novel neurobiological mechanisms and inform study design and outcome measures in potential clinical trials targeting these receptors.

## 3. Effects of $A_{2A}$ receptor modulation on cognitive impairments in aging and disease

### 3.1 Aging and dementia

Globally, the number of people over 60 is expected to triple by 2050 (United Nations. Department of Economic and Social Affairs. Population Division, 2002). By this time, the elderly will comprise 1 in every 5 persons (United Nations. Department of Economic and Social Affairs. Population Division, 2002). The elderly have high rates of aging-related memory loss and other cognitive impairments. Approximately 1 in 4 elderly has some memory loss without dementia (Barker, Jones, & Jennison, 1995; Hanninen et al., 1996; Koivisto et al., 1995; Plassman et al., 2008; Schonknecht, Pantel, Kruse, & Schroder, 2005) and 1 in 8 has memory loss due to dementia (Alzheimer's Association, 2012; Hebert, Scherr, Bienias, Bennett, & Evans, 2003).

Dementia encompasses a spectrum of multifactorial neurological disorders, including Alzheimer's disease (AD) and frontotemporal dementia (FTD), and severely impairs neurocognitive function and behavior. Neuropathological studies have shown that these disorders involve progressive proteinopathy, synaptic loss, neuroinflammation, glial alterations, and neurovascular changes. As dementia progresses, diverse neurobiological mechanisms underlying cognitive function, behavior, and sensory processing are significantly disrupted, resulting in chronic disability, increasingly limited daily functioning, and early mortality. Dementia devastates individuals, their families, and the economy, and is estimated to cost US$1 trillion annually (Elahi & Miller, 2017; Patterson, 2018). Despite the growing understanding of various molecular, cellular, and neural network mechanisms promoting brain dysfunction in aging and dementia, there continues to be a lack of effective therapeutics to alleviate or prevent cognitive and behavioral impairments in these conditions. Notably, several recent clinical trials have reported effective clearance of amyloid-β (Aβ) plaques using antibody-based approaches (Sevigny et al., 2016; van Dyck et al., 2022). Unfortunately, these treatments can cause certain adverse effects, including microhemorrhages and edema (Adhikari et al., 2022; Avgerinos, Ferrucci, & Kapogiannis, 2021), and have not yet been shown to result in extensive improvements in cognitive function or reversal of progressive functional decline.

Synaptic dysfunction and loss correlate with cognitive decline in AD and other dementias (Forner, Baglietto-Vargas, Martini, Trujillo-Estrada, & LaFerla, 2017; Hamos, DeGennaro, & Drachman, 1989; Selkoe, 2002). Therapeutically targeting synaptic and neural network alterations may represent a complementary approach that, in combination with strategies to combat proteinopathy, may alleviate or prevent cognitive decline in aging and dementia. Among various mechanisms underlying cognition, $A_{2A}$ receptors have well-established roles as modulators of synaptic function and neuronal activities in different brain regions and regulate multiple aspects of cognition and behavior, as discussed above and previously by others (Chen, 2014; Costenla et al., 2011; Cunha, 2016; Flajolet et al., 2008; Lopes, Cunha, Kull, Fredholm, & Ribeiro, 2002; Xu et al., 2022). Notably, increases in extracellular adenosine levels have been noted in aging (Cunha, Almeida, & Ribeiro, 2001) and brain disorders with altered neuronal activities, including AD and epilepsy (Alonso-Andres, Albasanz, Ferrer, & Martin, 2018; Launay et al., 2022; Tescarollo et al., 2020). In AD, adenosine levels were found to be increased in the parietal and temporal cortices, but reduced in the frontal cortex, and cortical adenosine levels were highest in early-stage AD as compared to controls and late-stage AD (Alonso-Andres, Albasanz, Ferrer, & Martin, 2018), suggesting that adenosine signaling is dysregulated early in disease. In addition to adenosine, increases in $A_{2A}$ receptor levels have been reported in aging, AD, FTD, and Parkinson's disease (PD), among other conditions (Calon et al., 2004; Canas, Duarte, Rodrigues, Kofalvi, & Cunha, 2009; Carvalho et al., 2019; Castillo et al., 2009; Coelho et al., 2014; Costenla et al., 2011; Diogenes, Assaife-Lopes, Pinto-Duarte, Ribeiro, & Sebastiao, 2007; Lopes, Cunha, & Ribeiro, 1999; Orr et al., 2015; Rebola et al., 2003; Temido-Ferreira et al., 2020).

Given these findings and other evidence implicating $A_{2A}$ receptors in neurocognitive function, studies have explored the effects of increasing or reducing $A_{2A}$ receptor levels and activities in aging and models of dementia-associated proteinopathy, as reviewed recently (Launay et al., 2022). Notably, overexpression of $A_{2A}$ receptors in excitatory forebrain neurons was sufficient to cause impairments in synaptic plasticity and learning and memory in the Morris water maze, and treatment with the $A_{2A}$ antagonists SCH-58261 or KW-6002 reversed these effects (Temido-Ferreira et al., 2020). Moreover, pharmacological blockade of $A_{2A}$ receptors reversed impairments in hippocampal plasticity and Y maze performance in aged mice as well as improved neuronal plasticity in doubly transgenic mice expressing mutant amyloid precursor protein (APP) and mutant presenilin-1 (PS1), a well-characterized model of AD-linked Aβ pathology (Temido-Ferreira et al., 2020).

These results are consistent with previous evidence in APP/PS1 mice and other models of Aβ pathology that treatment with SCH58261 or ZM241385, or genetic inactivation of $A_{2A}$ within individual CA3 neurons, restores hippocampal plasticity and reduces deficits in memory (Canas et al., 2009; Viana da Silva et al., 2016). Induction of neuronal $A_{2A}$ receptor overexpression also exacerbated spatial memory deficits in THY-Tau22 mouse line, a model of tauopathy (Carvalho et al., 2019), and global knockout of $A_{2A}$ receptors prevented memory loss in tauopathy (Laurent et al., 2016). Consistent with APP/PS1 mice, treatment of THY-Tau22 mice with the $A_{2A}$ receptor antagonist MSX-3 improved spatial memory (Laurent et al., 2016), suggesting that therapeutic targeting of $A_{2A}$ reduces cognitive impairments in multiple proteinopathies linked to AD, FTD, and other dementias. Interestingly, blockade of $A_{2A}$ receptors with SCH58261 or KW-6002 also reduced memory impairments induced by direct Aβ peptide administration but did not affect impairments induced by either scopolamine, a muscarinic receptor antagonist, or MK-801, an NMDA receptor antagonist (Cunha et al., 2008), indicating that $A_{2A}$ receptors have a context-dependent involvement in memory loss.

In addition to neuronal $A_{2A}$ receptors, astrocytic $A_{2A}$ receptors and related $G_s$-coupled signaling have also been implicated in cognitive decline linked to dementia. $A_{2A}$ receptor protein levels were reported to be increased in hippocampal astrocytes of human AD cases and mouse models with Aβ plaque pathology (Orr et al., 2015; Orr et al., 2018). Conditionally ablating astrocytic $A_{2A}$ receptors enhanced memory in mice with or without Aβ pathology, suggesting that astrocytic $A_{2A}$ receptors modulate normal memory (Orr et al., 2015). Blockade of $A_{2A}$ receptors with low-dose KW-6002 enhanced spatial memory in aging mice with Aβ pathology (Orr et al., 2018), supporting other studies highlighted above that $A_{2A}$ blockers counteract memory loss caused by proteinopathy. These findings in mouse models are also consistent with studies suggesting that increases in astrocytic $A_{2A}$ receptor levels or activities alter astrocytic functions, including clearance of extracellular glutamate and expression of inflammation-related genes (Lopes, Cunha, & Agostinho, 2021; Matos et al., 2012; Matos, Augusto, Agostinho, Cunha, & Chen, 2013; Paiva et al., 2019), which likely affect synaptic transmission and plasticity as well as pathophysiological cascades in aging and disease. However, the exact effects of astrocytic $A_{2A}$ receptor overactivation on signaling pathways and astrocytic-neuronal interactions that mediate the effects on memory remain unclear.

In addition to the selective $A_{2A}$ receptor antagonists, it has been reported that caffeine, a widely consumed nonselective $A_{2A}$ receptor antagonist, can reduce Aβ-associated impairments in cognitive function in animal models (Arendash et al., 2006; Dall'Igna et al., 2007). Intriguingly, caffeine consumption enhances or normalizes memory performance in a wide range of pathologies and contexts, which has been reviewed elsewhere (Chen, Scheltens, Groot, & Ossenkoppele, 2020; Cunha & Agostinho, 2010; Nehlig, 2010; Ribeiro & Sebastiao, 2010). However, the effects of caffeine are complex and involve multiple types of adenosine receptors and other targets.

Together, previous studies indicate that aging and dementia-associated increases in $A_{2A}$ receptor levels or activities in neurons and astrocytes can disrupt cognitive processes, especially memory. Different $A_{2A}$ receptor antagonists can alleviate or reverse these cognitive effects and hold promise as candidate therapeutics for cognitive decline in aging and dementia.

## 3.2 Multiple sclerosis

Multiple sclerosis (MS) is a disabling autoimmune disorder of the central nervous system that typically affects young adults. In MS, immune cell activation and infiltration into the brain, along with glial alterations, leads to progressive myelin damage and loss of axons. Several studies suggest that $A_{2A}$ receptors have important roles in peripheral and central inflammation and are involved in MS pathogenesis (Hasko, Linden, Cronstein, & Pacher, 2008; Ingwersen et al., 2016; Mills, Kim, Krenz, Chen, & Bynoe, 2012; Ohta & Sitkovsky, 2001; Safarzadeh, Jadidi-Niaragh, Motallebnezhad, & Yousefi, 2016). In experimental autoimmune encephalomyelitis (EAE), a widely utilized mouse model of MS, selective blockade of $A_{2A}$ receptors with the antagonist SCH58261 induced robust protection against EAE pathology (Mills et al., 2008). Chronic administration of SCH58261 also prevented deficits in spatial learning and memory in the lysolecithin-induced demyelination model in rats (Akbari, Khalili-Fomeshi, Ashrafpour, Moghadamnia, & Ghasemi-Kasman, 2018), suggesting that $A_{2A}$ receptors play key roles in MS-associated cognitive symptoms. Thus, antagonists of $A_{2A}$ receptors might have promise in ameliorating cognitive dysfunction associated with neuroinflammation and autoimmune pathologies.

## 3.3 Schizophrenia

Schizophrenia is a neuropsychiatric disorder characterized by a spectrum of cognitive and behavioral symptoms, including delusions, auditory hallucinations, disorganized speech, and cognitive impairments. Given the broad symptoms and links to neurodevelopmental changes, it has been challenging to pinpoint the exact causes and mechanisms underlying schizophrenia. Among different proposed mechanisms, adenosine dysfunction has been proposed to contribute to schizophrenia. In support, schizophrenic patients have increased catabolism of adenosine, which may diminish activation of $A_{2A}$ receptors (Boison, Singer, Shen, Feldon, & Yee, 2012). Moreover, caffeine heightens psychotic episodes in schizophrenic patients (Peng, Chiang, & Liang, 2014). It is well-established that $A_{2A}$ agonists modulate D2 dopamine receptors in striatopallidal neurons. These modulatory effects may be leveraged to treat positive symptoms of schizophrenia, including hallucinations and delusions (Ferre, 1997; Rimondini, Ferre, Ogren, & Fuxe, 1997). However, it is unclear what factors cause the putative changes in adenosine-linked signaling and whether these changes are adaptive responses to neural dysfunction or represent upstream disease-driving mechanisms in schizophrenia that might be promising targets for treatment (Domenici et al., 2019).

A recent study using in vivo PET imaging to assess $A_{2A}$ receptor availability in male patients with schizophrenia revealed no significant differences as compared to healthy controls (Marques et al., 2022). These results are contrary to *postmortem* imaging studies showing increases in $A_{2A}$ receptor levels in the hippocampus and striatum in schizophrenia cases, and to other studies demonstrating reduced striatal $A_{2A}$ receptor levels in schizophrenia (Deckert et al., 2003; Villar-Menendez et al., 2014; Zhang et al., 2012). These conflicting results highlight the need for additional research to address the exact alterations in $A_{2A}$ receptors in this disorder. Notably, a meta-analysis of six randomized controlled clinical trials that focused on the efficacy of modulators that increase adenosine levels as adjuvant therapy for schizophrenic patients found that these modulators were beneficial as compared to placebo treatment, especially the positive symptoms (Hirota & Kishi, 2013). However, the limited sample sizes in these trials indicate that additional research is needed to assess the efficacy of adenosine-targeted treatment strategies.

In animal studies, genetic deletion of astrocytic $A_{2A}$ receptors induced psychomotor and cognitive phenotypes like those seen in schizophrenia, including a decrease in working memory in an 8-baited radial arm maze and the Y maze spontaneous alternation test (Matos et al., 2015). Interestingly, deletion of neuronal $A_{2A}$ receptors enhanced working memory (Wei et al., 2011), suggesting that $A_{2A}$ gene ablation has cell type-specific effects on cognitive function. In summary, although adenosine has been implicated in schizophrenia (Rial, Lara, & Cunha, 2014; Shen et al., 2012), further research is needed to establish the precise alterations and mechanisms through which $A_{2A}$ receptors might be linked to cognitive and behavioral impairments associated with this disorder.

### 3.4 Neurodevelopmental disorders

$A_{2A}$ receptors may also be involved in neurodevelopmental disorders. In particular, fragile X syndrome (FXS) is a genetic disease that affects cognitive function, and patients can exhibit autism, behavioral hyperactivity, and seizures. In fragile X mental retardation 1 gene (Fmr1) knockout mice, the $A_{2A}$ receptor antagonist KW-6002 reduced deficits in learning by diminishing abnormal metabotropic glutamate receptor 5 (mGluR5) function and affecting other downstream targets (Ferrante et al., 2021). In addition, attention-deficit/hyperactivity disorder (ADHD) is a neurodevelopmental disorder characterized by behavioral hyperactivity, impulsiveness, and impaired attention (Sagvolden & Sergeant, 1998). Among the most affected neural pathways, dopamine signaling is known to be impaired in the frontal cortex and striatum (Arnsten, 2006). Notably, $A_{2A}$ receptors act as potent modulators of dopaminergic transmission (Ferre, von Euler, Johansson, Fredholm, & Fuxe, 1991; Fredholm, Chen, Cunha, Svenningsson, & Vaugeois, 2005; Quarta et al., 2004), implicating these receptors in ADHD and related conditions involving dopaminergic signaling.

Studies have shown that inhibitors of $A_{2A}$ receptors, either nonselective or selective antagonists, can improve alertness, attention, and memory, as discussed above and reported by others (Faivre et al., 2018; Higgins et al., 2007; Kaster et al., 2015). In animal models, deficits in object recognition were reduced by acute administration of caffeine or the $A_{2A}$ receptor blocker ZM241385 (Pires et al., 2009). A recent clinical trial evaluating the effects of L-theanine, caffeine, and their combination on sustained attention and overall cognitive function demonstrated that the combination of L-theanine and caffeine improved performance in children in the NIH Cognition Toolbox Test Battery, which involves multiple cognitive tests, including a picture sequence memory test, attention test, and oral reading recognition test. These results suggest that targeting adenosine receptors is a potential treatment strategy for ADHD-associated symptoms (Kahathuduwa et al., 2020). In contrast, in autism spectrum disorder (ASD), $A_{2A}$ agonists rather than antagonists have been reported to ameliorate social behavior impairments and improve reversal-learning performance in mouse models of ASD, suggesting that $A_{2A}$ receptors have different roles across cognitive domains and promote different effects depending on disease context and underlying neuropathological mechanisms (Amodeo, Cuevas, Dunn, Sweeney, & Ragozzino, 2018; Ansari et al., 2017).

## 3.5 Brain trauma and stroke

During ischemia and other acute insults, adenosine levels can rapidly build up in brain tissue (Chu et al., 2013; Pedata, Corsi, Melani, Bordoni, & Latini, 2001; Rudolphi, Schubert, Parkinson, & Fredholm, 1992). Thus, adenosine receptors have also been implicated in brain trauma and stroke. Studies have shown that blockade of $A_{2A}$ receptors decreases brain infarction and alleviates deficits in animals after ischemia (Melani et al., 2006; Pedata et al. 2001). However, one study indicated that the protective effects of $A_{2A}$ blockade were not observed after seven days (Melani, Dettori, Corti, Cellai, & Pedata, 2015). Administration of the $A_{2A}$ receptor agonist CGS-21680 exacerbated memory impairments in a model of chronic intermittent hypoxia (Li et al., 2022). $A_{2A}$ receptors have also been associated with other types of brain injury. Global genetic ablation of $A_{2A}$ receptors ameliorated memory deficits and lessened neuropathological changes in traumatic brain injury (Ning et al., 2013). Additionally, blockade of $A_{2A}$ receptors improved cognitive function in an animal model of epilepsy (Cognato et al., 2010).

## 4. Effects of $A_{2A}$ receptor modulation on cognitive impairments in stress

Chronic or repeated stress is known to alter neural circuits and negatively affect mood and cognitive performance. In addition to alleviating cognitive deficits induced by disease and injury, as discussed above, $A_{2A}$ receptor blockade may reverse neurocognitive deficits induced by stress. Notably, in a model of chronic unpredictable stress (CUS), treatment with KW-6002 or SCH58261, or ablation of the $A_{2A}$ gene globally or selectively in neurons, reduced depression-related behavior as well as neurochemical and electrophysiological changes (Kaster et al., 2015).

Among various measures, pharmacological or genetic targeting of $A_{2A}$ receptors prevented CUS-induced helplessness in the forced-swim test, anhedonia in the splash test, anxiety-related behavior in the elevated plus maze, low social interactions, and impaired memory in the Y maze and the Morris water maze. These important findings are consistent with prior work demonstrating that $A_{2A}$ receptor antagonists have antidepressive effects (El Yacoubi, Costentin, & Vaugeois, 2003; Minor, Rowe, Cullen, & Furst, 2008; Yamada et al., 2014). Moreover, in a maternal separation model of early-life stress, $A_{2A}$ receptor antagonist KW-6002 reduced long-term changes in anxiety-related behavior in the elevated plus maze as well as deficits in spatial learning and memory in the Morris water maze (Batalha et al., 2013). Thus, blockade of $A_{2A}$ receptors may reduce early-life stress-related changes in neurocognitive function that last into adulthood. Furthermore, in a model of anxiety associated with prenatal stress, $A_{2A}$ blockade with SCH58261 did not reduce anxiety-related behavior in the elevated plus maze but ameliorated deficits in novel object recognition (Duarte et al., 2019), suggesting potential contributions of this receptor to cognitive alterations initiated by early-life events.

## 5. Effects of $A_{2A}$ receptor modulation on drug use and addiction

Drug addiction is a major cause of disability. It is estimated that drug addiction-related crime, lost work, and related health care costs more than $740 billion annually in the United States. Current pharmacological and social treatments are inadequate for most people (Nestler & Luscher, 2019). The need for more effective treatments continues to be critical as drug use remains prevalent and overdose-related deaths among adolescents have increased substantially since 2019 (Tanz, Dinwiddie, Mattson, O'Donnell, & Davis, 2022). Different brain regions are associated with distinct aspects of addiction, such as drug reward, craving, and relapse. One of these regions involves the mesolimbic dopaminergic system located in the ventral tegmental area that extends to the nucleus accumbens and other forebrain areas.

$A_{2A}$ receptors act as potent modulators of dopaminergic transmission and are suggested to be involved in the behavioral effects of drugs of abuse. In support, pharmacological activation of $A_{2A}$ receptors with CGS-21680 abated cocaine-seeking behavior (Bachtell & Self, 2009). In agreement, a subsequent study showed that stimulation of $A_{2A}$ receptors opposes the behavioral effects of cocaine (O'Neill, LeTendre, & Bachtell, 2012). The $A_{2A}$ agonist CGS-21680 decreased cocaine self-administration in rats (Wydra et al., 2015). In a propofol self-administration rat model, a more recent study provided genetic and pharmacological evidence that $A_{2A}$ activation suppresses cue-induced relapse (Dong et al., 2021). Moreover, blockade of $A_{2A}$ receptors with MSX-3 enhanced the effects of cocaine (O'Neill et al. 2012). However, blockade of $A_{2A}$ receptors with SCH 442416 decreased susceptibility to relapse and this treatment had lasting effects (O'Neill, Hobson, Levis, & Bachtell, 2014). In addition, the $A_{2A}$ agonist CGS-21680 diminished initial extinction responses following cocaine self-administration (O'Neill et al. 2014). These contrasting findings highlight the need for additional studies to address the roles of $A_{2A}$ receptors in drug addiction. Of note, cognitive deficits such as impairments in working memory correlate with cocaine cravings and may lead to poor compliance to therapy programs (Bruijnen et al., 2019; Gobin, Shallcross, & Schwendt, 2019; Vonmoos et al., 2013). Thus, in addition to reducing relapse risk, treatments should aim to reduce drug-related effects on cognitive function (McHugh, Hearon, & Otto, 2010; Sofuoglu, DeVito, Waters, & Carroll, 2016; Verdejo-Garcia et al., 2014). In this regard, further research is needed to evaluate how the different cognitive and reward-related effects of $A_{2A}$ modulators influence drug abuse and addiction.

## 6. Conclusions

In summary, a growing number of studies suggest that the effects of $A_{2A}$ receptors on cognitive function are multifaceted, dynamic, and context-specific due in part to differential $A_{2A}$ receptor expression and activities across age, sex, brain regions, pathophysiological conditions, and environmental factors, as well as variable contributions by $A_{2A}$ receptors expressed in glial cells and neuronal subpopulations that exhibit inherent functional heterogeneity and plasticity.

Despite these neurobiological complexities, $A_{2A}$ receptor antagonists continue to represent promising therapeutic agents for alleviating cognitive impairments in diverse conditions, with the most extensive evidence reported in animal models of aging and dementia, which are conditions associated with increased brain levels of adenosine and $A_{2A}$ receptors. Based on emerging studies, $A_{2A}$ receptor antagonists may also hold potential as therapeutics for other conditions with cognitive alterations, including acute trauma, early-life or chronic stress, and substance use disorders. Blockers of $A_{2A}$ receptors may have therapeutic efficacy in a broad range of neurological and neuropsychiatric conditions at least in part due to shared neuropathogenic mechanisms of cognitive dysfunction that involve dysregulated adenosine signaling and consequent alterations in neurotransmission and plasticity.

Further studies are needed to unravel the exact molecular and neuro-circuit mechanisms of $A_{2A}$ receptor effects in memory-related processes, goal-directed behavior, and other phenomena. Additional research is also needed to determine the molecular mechanisms by which $A_{2A}$ receptors in neurons and glial cells modulate disease-associated cascades and how these effects might be similar or different across disorders. Better understanding of $A_{2A}$ receptor roles in excitation-inhibition imbalance, neuroinflammation, and circuit reorganization may inform potential treatment strategies. Importantly, clinical trials involving selective $A_{2A}$ antagonists would provide valuable insights into the potential efficacy of $A_{2A}$ receptor modulators in alleviating or preventing cognitive alterations in humans. $A_{2A}$ receptor modulators may be especially beneficial when used in combination with other therapeutic agents that target other concurrent pathogenic mechanisms. In this regard, additional studies are needed to assess whether a combinatorial treatment approach that targets multiple aspects of disease, including dysregulated $A_{2A}$ receptors, is more efficacious in ameliorating neuropathology and cognitive decline as compared to single treatments. A personalized approach may be needed given the diversity of clinical and neuropathological phenotypes that can occur for a particular neurological or neuropsychiatric disorder, as well as individual differences in treatment responses, genetic risk factors, and other variables that can affect therapeutic strategy.

Interestingly, $A_{2A}$ receptors have been proposed as potential peripheral biomarkers for neuropathology and cognitive decline. A recent study in *postmortem* AD cases reported increases in $A_{2A}$ receptor protein levels in the brain and peripheral blood-derived platelets (Merighi et al., 2021). The correlated changes in $A_{2A}$ receptor levels in central and peripheral compartments suggest that alterations in brain function might be heralded in the periphery, which could serve as an indicator of disease progression and treatment effects. In support, several studies have similarly detected increases in $A_{2A}$ receptor levels in peripheral blood mononuclear cells from patients with mild cognitive impairment or Huntington's disease (Arosio et al., 2010; Maglione et al., 2006). Noninvasive indicators of adenosine receptor function could be rapid and sensitive biomarkers of neural changes linked to cognitive dysfunction and thus warrant further investigation.

## Conflicts of interests
The authors declare no potential conflicts of interest.

## Acknowledgements
This work was supported in part by the National Institute of General Medical Sciences Initiative to Maximize Student Development Fellowship (CPG), National Institute on Aging grant K99/R00AG048222 (AGO), and the Alzheimer's Association (AGO).

## References
Alzheimer's Association (2012). Alzheimer's disease facts and figures. *Alzheimer's & Dementia: The Journal of the Alzheimer's Association, 8*(2), 131–168. https://doi.org/10.1016/j.jalz.2012.02.001

Alonso-Andres, P., Albasanz, J. L., Ferrer, I., & Martin, M. (2018). Purine-related metabolites and their converting enzymes are altered in frontal, parietal and temporal cortex at early stages of Alzheimer's disease pathology. *Brain Pathology (Zurich, Switzerland), 28*(6), 933–946. https://doi.org/10.1111/bpa.12592

Abbracchio, M. P., Burnstock, G., Verkhratsky, A., & Zimmermann, H. (2009). Purinergic signalling in the nervous system: An overview. *Trends in Neurosciences, 32*(1), 19–29. https://doi.org/10.1016/j.tins.2008.10.001

Amodeo, D. A., Cuevas, L., Dunn, J. T., Sweeney, J. A., & Ragozzino, M. E. (2018). The adenosine A(2A) receptor agonist, CGS 21680, attenuates a probabilistic reversal learning deficit and elevated grooming behavior in BTBR mice. *Autism Research, 11*(2), 223–233. https://doi.org/10.1002/aur.1901

Alberini, C. M., Cruz, E., Descalzi, G., Bessieres, B., & Gao, V. (2018). Astrocyte glycogen and lactate: New insights into learning and memory mechanisms. *Glia, 66*(6), 1244–1262. https://doi.org/10.1002/glia.23250

Adamsky, A., & Goshen, I. (2018). Astrocytes in memory function: Pioneering findings and future directions. *Neuroscience, 370*, 14–26. https://doi.org/10.1016/j.neuroscience.2017.05.033

Adamsky, A., Kol, A., Kreisel, T., Doron, A., Ozeri-Engelhard, N., Melcer, T., et al. (2018). Astrocytic activation generates de novo neuronal potentiation and memory enhancement. *Cell, 174*(1), 59–71.e14. https://doi.org/10.1016/j.cell.2018.05.002

Adhıkarı, U. K., Khan, R., Mikhael, M., Balez, R., David, M. A., Mahns, D., et al. (2022). Therapeutic anti-amyloid beta antibodies cause neuronal disturbances. *Alzheimer's & Dementia: The Journal of the Alzheimer's Association.* https://doi.org/10.1002/alz.12833

Akbari, A., Khalili-Fomeshi, M., Ashrafpour, M., Moghadamnia, A. A., & Ghasemi-Kasman, M. (2018). Adenosine A(2A) receptor blockade attenuates spatial memory deficit and extent of demyelination areas in lyolecithin-induced demyelination model. *Life Sciences, 205*, 63–72. https://doi.org/10.1016/j.lfs.2018.05.007

Ansari, M. A., Nadeem, A., Attia, S. M., Bakheet, S. A., Raish, M., & Ahmad, S. F. (2017). Adenosine A2A receptor modulates neuroimmune function through Th17/retinoid-related orphan receptor gamma t (RORgammat) signaling in a BTBR T(+) Itpr3(tf)/J mouse model of autism. *Cellular Signalling, 36*, 14–24. https://doi.org/10.1016/j.cellsig.2017.04.014

Arendash, G. W., Schleif, W., Rezai-Zadeh, K., Jackson, E. K., Zacharia, L. C., Cracchiolo, J. R., et al. (2006). Caffeine protects Alzheimer's mice against cognitive impairment and reduces brain beta-amyloid production. *Neuroscience, 142*(4), 941–952. https://doi.org/10.1016/j.neuroscience.2006.07.021

Arnsten, A. F. (2006). Fundamentals of attention-deficit/hyperactivity disorder: circuits and pathways. *The Journal of Clinical Psychiatry, 67*(Suppl 8), 7–12.

Arosio, B., Viazzoli, C., Mastronardi, L., Bilotta, C., Vergani, C., & Bergamaschini, L. (2010). Adenosine A2A receptor expression in peripheral blood mononuclear cells of patients with mild cognitive impairment. *Journal of Alzheimer's Disease: JAD, 20*(4), 991–996. https://doi.org/10.3233/JAD-2010-090814

Avgerinos, K. I., Ferrucci, L., & Kapogiannis, D. (2021). Effects of monoclonal antibodies against amyloid-beta on clinical and biomarker outcomes and adverse event risks: A systematic review and meta-analysis of phase III RCTs in Alzheimer's disease. *Ageing Research Reviews, 68*, 101339. https://doi.org/10.1016/j.arr.2021.101339

Bachtell, R. K., & Self, D. W. (2009). Effects of adenosine A2A receptor stimulation on cocaine-seeking behavior in rats. *Psychopharmacology (Berl), 206*(3), 469–478. https://doi.org/10.1007/s00213-009-1624-2

Barker, A., Jones, R., & Jennison, C. (1995). A prevalence study of age-associated memory impairment. *The British Journal of Psychiatry: The Journal of Mental Science, 167*(5), 642–648.

Batalha, V. L., Pego, J. M., Fontinha, B. M., Costenla, A. R., Valadas, J. S., Baqi, Y., et al. (2013). Adenosine A(2A) receptor blockade reverts hippocampal stress-induced deficits and restores corticosterone circadian oscillation. *Molecular Psychiatry, 18*(3), 320–331. https://doi.org/10.1038/mp.2012.8

Bogenpohl, J. W., Ritter, S. L., Hall, R. A., & Smith, Y. (2012). Adenosine A2A receptor in the monkey basal ganglia: Ultrastructural localization and colocalization with the metabotropic glutamate receptor 5 in the striatum. *The Journal of Comparative Neurology, 520*(3), 570–589. https://doi.org/10.1002/cne.22751

Boison, D., Singer, P., Shen, H. Y., Feldon, J., & Yee, B. K. (2012). Adenosine hypothesis of schizophrenia—opportunities for pharmacotherapy. *Neuropharmacology, 62*(3), 1527–1543. https://doi.org/10.1016/j.neuropharm.2011.01.048

Borota, D., Murray, E., Keceli, G., Chang, A., Watabe, J. M., Ly, M., et al. (2014). Post-study caffeine administration enhances memory consolidation in humans. *Nature Neuroscience, 17*(2), 201–203. https://doi.org/10.1038/nn.3623

Borroto-Escuela, D. O., & Fuxe, K. (2019). Adenosine heteroreceptor complexes in the basal ganglia are implicated in Parkinson's disease and its treatment. *The Journal of Neural Transmission (Vienna), 126*(4), 455–471. https://doi.org/10.1007/s00702-019-01969-2

Botton, P. H., Costa, M. S., Ardais, A. P., Mioranzza, S., Souza, D. O., da Rocha, J. B., et al. (2010). Caffeine prevents disruption of memory consolidation in the inhibitory avoidance and novel object recognition tasks by scopolamine in adult mice. *Behavioural Brain Research, 214*(2), 254–259. https://doi.org/10.1016/j.bbr.2010.05.034

Brandebura, A. N., Paumier, A., Onur, T. S., & Allen, N. J. (2023). Astrocyte contribution to dysfunction, risk and progression in neurodegenerative disorders. *Nature Reviews. Neuroscience, 24*(1), 23–39. https://doi.org/10.1038/s41583-022-00641-1

Bruijnen, C., Dijkstra, B. A. G., Walvoort, S. J. W., Markus, W., VanDerNagel, J. E. L., Kessels, R. P. C., et al. (2019). Prevalence of cognitive impairment in patients with substance use disorder. *Drug and Alcohol Review, 38*(4), 435–442. https://doi.org/10.1111/dar.12922

Burnstock, G. (2009). Purinergic cotransmission. *Experimental Physiology, 94*(1), 20–24. https://doi.org/10.1113/expphysiol.2008.043620

Burnstock, G. (2020). Introduction to purinergic signaling. *Methods in Molecular Biology, 2041*, 1–15. https://doi.org/10.1007/978-1-4939-9717-6_1

Bushnell, C. D., Chaturvedi, S., Gage, K. R., Herson, P. S., Hurn, P. D., Jimenez, M. C., et al. (2018). Sex differences in stroke: Challenges and opportunities. *Journal of Cerebral Blood Flow and Metabolism: Official Journal of the International Society of Cerebral Blood Flow and Metabolism, 38*(12), 2179–2191. https://doi.org/10.1177/0271678X18793324

Caetano, L., Pinheiro, H., Patricio, P., Mateus-Pinheiro, A., Alves, N. D., Coimbra, B., et al. (2017). Adenosine A(2A) receptor regulation of microglia morphological remodeling-gender bias in physiology and in a model of chronic anxiety. *Molecular Psychiatry, 22*(7), 1035–1043. https://doi.org/10.1038/mp.2016.173

Calon, F., Dridi, M., Hornykiewicz, O., Bedard, P. J., Rajput, A. H., & Di Paolo, T. (2004). Increased adenosine A2A receptors in the brain of Parkinson's disease patients with dyskinesias. *Brain, 127*(Pt 5), 1075–1084. https://doi.org/10.1093/brain/awh128

Canas, P. M., Duarte, J. M., Rodrigues, R. J., Kofalvi, A., & Cunha, R. A. (2009). Modification upon aging of the density of presynaptic modulation systems in the hippocampus. *Neurobiology of Aging, 30*(11), 1877–1884. https://doi.org/10.1016/j.neurobiolaging.2008.01.003

Canas, P. M., Porciuncula, L. O., Cunha, G. M., Silva, C. G., Machado, N. J., Oliveira, J. M., et al. (2009). Adenosine A2A receptor blockade prevents synaptotoxicity and memory dysfunction caused by beta-amyloid peptides via p38 mitogen-activated protein kinase pathway. *The Journal of Neuroscience, 29*(47), 14741–14751. https://doi.org/10.1523/JNEUROSCI.3728-09.2009

Carvalho, K., Faivre, E., Pietrowski, M. J., Marques, X., Gomez-Murcia, V., Deleau, A., et al. (2019). Exacerbation of C1q dysregulation, synaptic loss and memory deficits in tau pathology linked to neuronal adenosine A2A receptor. *Brain, 142*(11), 3636–3654. https://doi.org/10.1093/brain/awz288

Castillo, C. A., Albasanz, J. L., Leon, D., Jordan, J., Pallas, M., Camins, A., et al. (2009). Age-related expression of adenosine receptors in brain from the senescence-accelerated mouse. *Experimental Gerontology, 44*(6–7), 453–461. https://doi.org/10.1016/j.exger.2009.04.006

Chen, Z. P., Wang, S., & Zhao, X. (2023). Lipid-accumulated reactive astrocytes promote disease progression in epilepsy. *Nat Neurosci, 26*, 542–554.

Chen, J. F. (2014). Adenosine receptor control of cognition in normal and disease. *International Review of Neurobiology, 119*, 257–307. https://doi.org/10.1016/B978-0-12-801022-8.00012-X

Chen, J. Q. A., Scheltens, P., Groot, C., & Ossenkoppele, R. (2020). Associations between caffeine consumption, cognitive decline, and dementia: A systematic review. *Journal of Alzheimer's Disease: JAD, 78*(4), 1519–1546. https://doi.org/10.3233/JAD-201069

Chen, L., Li, S., Zhou, Y., Liu, T., Cai, A., Zhang, Z., et al. (2021). Neuronal mechanisms of adenosine A(2A) receptors in the loss of consciousness induced by propofol general anesthesia with functional magnetic resonance imaging. *Journal of Neurochemistry, 156*(6), 1020–1032. https://doi.org/10.1111/jnc.15146

Chu, S., Xiong, W., Zhang, D., Soylu, H., Sun, C., Albensi, B. C., et al. (2013). Regulation of adenosine levels during cerebral ischemia. *Acta Pharmacologica Sinica, 34*(1), 60–66. https://doi.org/10.1038/aps.2012.127

Coelho, J. E., Alves, P., Canas, P. M., Valadas, J. S., Shmidt, T., Batalha, V. L., et al. (2014). Overexpression of adenosine A2A receptors in rats: Effects on depression, locomotion, and anxiety. *Frontiers in Psychiatry, 5*, 67. https://doi.org/10.3389/fpsyt.2014.00067

Cognato, G. P., Agostinho, P. M., Hockemeyer, J., Muller, C. E., Souza, D. O., & Cunha, R. A. (2010). Caffeine and an adenosine A(2A) receptor antagonist prevent memory impairment and synaptotoxicity in adult rats triggered by a convulsive episode in early life. *Journal of Neurochemistry, 112*(2), 453–462. https://doi.org/10.1111/j.1471-4159.2009.06465.x

Costa, M. S., Botton, P. H., Mioranzza, S., Souza, D. O., & Porciuncula, L. O. (2008). Caffeine prevents age-associated recognition memory decline and changes brain-derived neurotrophic factor and tirosine kinase receptor (TrkB) content in mice. *Neuroscience, 153*(4), 1071–1078. https://doi.org/10.1016/j.neuroscience.2008.03.038

Costenla, A. R., Diogenes, M. J., Canas, P. M., Rodrigues, R. J., Nogueira, C., Maroco, J., et al. (2011). Enhanced role of adenosine A(2A) receptors in the modulation of LTP in the rat hippocampus upon ageing. *The European Journal of Neuroscience, 34*(1), 12–21. https://doi.org/10.1111/j.1460-9568.2011.07719.x

Cunha, G. M., Canas, P. M., Melo, C. S., Hockemeyer, J., Muller, C. E., Oliveira, C. R., et al. (2008). Adenosine A2A receptor blockade prevents memory dysfunction caused by beta-amyloid peptides but not by scopolamine or MK-801. *Experimental Neurology, 210*(2), 776–781. https://doi.org/10.1016/j.expneurol.2007.11.013

Cunha, R. A. (2016). How does adenosine control neuronal dysfunction and neurodegeneration? *Journal of Neurochemistry, 139*(6), 1019–1055. https://doi.org/10.1111/jnc.13724

Cunha, R. A., & Agostinho, P. (2010). Chronic caffeine consumption prevents memory disturbance in different animal models of memory decline. *Journal of Alzheimer's Disease: JAD, 20*(Suppl 1), S95–S116. https://doi.org/10.3233/JAD-2010-1408

Cunha, R. A., Almeida, T., & Ribeiro, J. A. (2001). Parallel modification of adenosine extracellular metabolism and modulatory action in the hippocampus of aged rats. *Journal of Neurochemistry, 76*(2), 372–382. https://doi.org/10.1046/j.1471-4159.2001.00095.x

Dall'Igna, O. P., Fett, P., Gomes, M. W., Souza, D. O., Cunha, R. A., & Lara, D. R. (2007). Caffeine and adenosine A(2a) receptor antagonists prevent beta-amyloid (25-35)-induced cognitive deficits in mice. *Experimental Neurology, 203*(1), 241–245. https://doi.org/10.1016/j.expneurol.2006.08.008

De Nuccio, C., Bernardo, A., Ferrante, A., Pepponi, R., Martire, A., Falchi, M., et al. (2019). Adenosine A(2A) receptor stimulation restores cell functions and differentiation in Niemann-Pick type C-like oligodendrocytes. *Sci Rep, 9*(1), 9782. https://doi.org/10.1038/s41598-019-46268-8

Deckert, J., Brenner, M., Durany, N., Zochling, R., Paulus, W., Ransmayr, G., et al. (2003). Up-regulation of striatal adenosine A(2A) receptors in schizophrenia. *Neuroreport, 14*(3), 313–316. https://doi.org/10.1097/00001756-200303030-00003

Diogenes, M. J., Assaife-Lopes, N., Pinto-Duarte, A., Ribeiro, J. A., & Sebastiao, A. M. (2007). Influence of age on BDNF modulation of hippocampal synaptic transmission: interplay with adenosine A2A receptors. *Hippocampus, 17*(7), 577–585. https://doi.org/10.1002/hipo.20294

Dixon, A. K., Gubitz, A. K., Sirinathsinghji, D. J., Richardson, P. J., & Freeman, T. C. (1996). Tissue distribution of adenosine receptor mRNAs in the rat. *British Journal of Pharmacology, 118*(6), 1461–1468. https://doi.org/10.1111/j.1476-5381.1996.tb15561.x

Domenici, M. R., Ferrante, A., Martire, A., Chiodi, V., Pepponi, R., Tebano, M. T., et al. (2019). Adenosine A(2A) receptor as potential therapeutic target in neuropsychiatric disorders. *Pharmacological Research: The Official Journal of the Italian Pharmacological Society, 147*, 104338. https://doi.org/10.1016/j.phrs.2019.104338

Dong, Z., Huang, B., Jiang, C., Chen, J., Lin, H., Lian, Q., et al. (2021). The adenosine A2A receptor activation in nucleus accumbens suppress cue-Induced reinstatement of propofol self-administration in rats. *Neurochemical Research, 46*(5), 1081–1091. https://doi.org/10.1007/s11064-021-03238-9

Duarte, J. M., Gaspar, R., Caetano, L., Patricio, P., Soares-Cunha, C., Mateus-Pinheiro, A., et al. (2019). Region-specific control of microglia by adenosine A(2A) receptors: Uncoupling anxiety and associated cognitive deficits in female rats. *Glia, 67*(1), 182–192. https://doi.org/10.1002/glia.23476

Eid, R. S., Gobinath, A. R., & Galea, L. A. M. (2019). Sex differences in depression: Insights from clinical and preclinical studies. *Progress in Neurobiology, 176*, 86–102. https://doi.org/10.1016/j.pneurobio.2019.01.006

El Yacoubi, M., Costentin, J., & Vaugeois, J. M. (2003). Adenosine A2A receptors and depression. *Neurology, 61*(11 Suppl 6), S82–S87. https://doi.org/10.1212/01.wnl.0000095220.87550.f6

Elahi, F. M., & Miller, B. L. (2017). A clinicopathological approach to the diagnosis of dementia. *Nature Reviews Neurology, 13*(8), 457–476. https://doi.org/10.1038/nrneurol.2017.96

Faivre, E., Coelho, J. E., Zornbach, K., Malik, E., Baqi, Y., Schneider, M., et al. (2018). Beneficial effect of a selective adenosine A(2A) receptor antagonist in the APPswe/PS1dE9 mouse model of Alzheimer's disease. *Frontiers in Molecular Neuroscience, 11*, 235. https://doi.org/10.3389/fnmol.2018.00235

Ferrante, A., Boussadia, Z., Borreca, A., Mallozzi, C., Pedini, G., Pacini, L., et al. (2021). Adenosine A(2A) receptor inhibition reduces synaptic and cognitive hippocampal alterations in Fmr1 KO mice. *Translational Psychiatry, 11*(1), 112. https://doi.org/10.1038/s41398-021-01238-5

Ferre, S. (1997). Adenosine-dopamine interactions in the ventral striatum. Implications for the treatment of schizophrenia. *Psychopharmacology (Berl), 133*(2), 107–120. https://doi.org/10.1007/s002130050380

Ferre, S., von Euler, G., Johansson, B., Fredholm, B. B., & Fuxe, K. (1991). Stimulation of high-affinity adenosine A2 receptors decreases the affinity of dopamine D2 receptors in rat striatal membranes. *Proceeding of the National Academy of Sciences of the United States of America, 88*(16), 7238–7241. https://doi.org/10.1073/pnas.88.16.7238

Ferreira, S. G., Goncalves, F. Q., Marques, J. M., Tome, A. R., Rodrigues, R. J., Nunes-Correia, I., et al. (2015). Presynaptic adenosine A2A receptors dampen cannabinoid CB1 receptor-mediated inhibition of corticostriatal glutamatergic transmission. *British Journal of Pharmacology, 172*(4), 1074–1086. https://doi.org/10.1111/bph.12970

Fields, R. D., & Burnstock, G. (2006). Purinergic signalling in neuron-glia interactions. *Nature Reviews. Neuroscience, 7*(6), 423–436. https://doi.org/10.1038/nrn1928

Flajolet, M., Wang, Z., Futter, M., Shen, W., Nuangchamnong, N., Bendor, J., et al. (2008). FGF acts as a co-transmitter through adenosine A(2A) receptor to regulate synaptic plasticity. *Nature Neuroscience, 11*(12), 1402–1409. https://doi.org/10.1038/nn.2216

Font, L., Mingote, S., Farrar, A. M., Pereira, M., Worden, L., Stopper, C., et al. (2008). Intra-accumbens injections of the adenosine A2A agonist CGS 21680 affect effort-related choice behavior in rats. *Psychopharmacology (Berl), 199*(4), 515–526. https://doi.org/10.1007/s00213-008-1174-z

Forner, S., Baglietto-Vargas, D., Martini, A. C., Trujillo-Estrada, L., & LaFerla, F. M. (2017). Synaptic impairment in Alzheimer's disease: A dysregulated symphony. *Trends in Neurosciences, 40*(6), 347–357. https://doi.org/10.1016/j.tins.2017.04.002

Fredholm, B. B., Chen, J. F., Cunha, R. A., Svenningsson, P., & Vaugeois, J. M. (2005). Adenosine and brain function. *International Review of Neurobiology, 63*, 191–270. https://doi.org/10.1016/S0074-7742(05)63007-3

Gao, V., Suzuki, A., Magistretti, P. J., Lengacher, S., Pollonini, G., Steinman, M. Q., et al. (2016). Astrocytic beta2-adrenergic receptors mediate hippocampal long-term memory consolidation. *Proceeding of the National Academy of Sciences of the United States of America, 113*(30), 8526–8531. https://doi.org/10.1073/pnas.1605063113

Gimenez-Llort, L., Schiffmann, S. N., Shmidt, T., Canela, L., Camon, L., Wassholm, M., et al. (2007). Working memory deficits in transgenic rats overexpressing human adenosine A2A receptors in the brain. *Neurobiology of Learning and Memory, 87*(1), 42–56. https://doi.org/10.1016/j.nlm.2006.05.004

Gobin, C., Shallcross, J., & Schwendt, M. (2019). Neurobiological substrates of persistent working memory deficits and cocaine-seeking in the prelimbic cortex of rats with a history of extended access to cocaine self-administration. *Neurobiology of Learning and Memory, 161*, 92–105. https://doi.org/10.1016/j.nlm.2019.03.007

Golden, L. C., & Voskuhl, R. (2017). The importance of studying sex differences in disease: The example of multiple sclerosis. *Journal of Neuroscience Research, 95*(1–2), 633–643. https://doi.org/10.1002/jnr.23955

Guo, L., Zhong, M. B., Zhang, L., Zhang, B., & Cai, D. (2022). Sex differences in Alzheimer's disease: Insights from the multiomics landscape. *Biological Psychiatry, 91*(1), 61–71. https://doi.org/10.1016/j.biopsych.2021.02.968

Guo, M., Wang, J., Yuan, Y., Chen, L., He, J., Wei, W., et al. (2022). Role of adenosine A(2A) receptors in the loss of consciousness induced by propofol anesthesia. *Journal of Neurochemistry*. https://doi.org/10.1111/jnc.15734

Hamos, J. E., DeGennaro, L. J., & Drachman, D. A. (1989). Synaptic loss in Alzheimer's disease and other dementias. *Neurology, 39*(3), 355–361. https://doi.org/10.1212/wnl.39.3.355

Han, J., Kesner, P., Metna-Laurent, M., Duan, T., Xu, L., Georges, F., et al. (2012). Acute cannabinoids impair working memory through astroglial CB1 receptor modulation of hippocampal LTD. *Cell, 148*(5), 1039–1050. https://doi.org/10.1016/j.cell.2012.01.037

Hanninen, T., Koivisto, K., Reinikainen, K. J., Helkala, E. L., Soininen, H., Mykkanen, L., et al. (1996). Prevalence of ageing-associated cognitive decline in an elderly population. *Age and Ageing, 25*(3), 201–205.

Hasko, G., Linden, J., Cronstein, B., & Pacher, P. (2008). Adenosine receptors: Therapeutic aspects for inflammatory and immune diseases. *Nature Reviews. Drug Discovery, 7*(9), 759–770. https://doi.org/10.1038/nrd2638

Hebert, L., Scherr, P., Bienias, J., Bennett, D., & Evans, D. (2003). Alzheimer disease in the US population: Prevalence estimates using the 2000 census. *Archives of Neurology, 60*, 1119–1122.

Higgins, G. A., Grzelak, M. E., Pond, A. J., Cohen-Williams, M. E., Hodgson, R. A., & Varty, G. B. (2007). The effect of caffeine to increase reaction time in the rat during a test of attention is mediated through antagonism of adenosine A2A receptors. *Behavioural Brain Research, 185*(1), 32–42. https://doi.org/10.1016/j.bbr.2007.07.013

Hirota, T., & Kishi, T. (2013). Adenosine hypothesis in schizophrenia and bipolar disorder: A systematic review and meta-analysis of randomized controlled trial of adjuvant purinergic modulators. *Schizophrenia Research, 149*(1–3), 88–95. https://doi.org/10.1016/j.schres.2013.06.038

Huang, Z. L., Zhang, Z., & Qu, W. M. (2014). Roles of adenosine and its receptors in sleep-wake regulation. *International Review of Neurobiology, 119*, 349–371. https://doi.org/10.1016/B978-0-12-801022-8.00014-3

Ingwersen, J., Wingerath, B., Graf, J., Lepka, K., Hofrichter, M., Schroter, F., et al. (2016). Dual roles of the adenosine A2a receptor in autoimmune neuroinflammation. *Journal of Neuroinflammation, 13*, 48. https://doi.org/10.1186/s12974-016-0512-z

Jain, A., Huang, G. Z., & Woolley, C. S. (2019). Latent sex differences in molecular signaling that underlies excitatory synaptic potentiation in the hippocampus. *The Journal of Neuroscience, 39*(9), 1552–1565. https://doi.org/10.1523/JNEUROSCI.1897-18.2018

Jenner, P., Mori, A., & Kanda, T. (2020). Can adenosine A(2A) receptor antagonists be used to treat cognitive impairment, depression or excessive sleepiness in Parkinson's disease? *Parkinsonism & Related Disorders, 80*(Suppl 1), S28–S36. https://doi.org/10.1016/j.parkreldis.2020.09.022

Jensen, C. J., Demol, F., Bauwens, R., Kooijman, R., Massie, A., Villers, A., et al. (2016). Astrocytic beta2 adrenergic receptor gene deletion affects memory in aged mice. *PLoS One, 11*(10), e0164721. https://doi.org/10.1371/journal.pone.0164721

Jones, M. E., Paniccia, J. E., Lebonville, C. L., Reissner, K. J., & Lysle, D. T. (2018). Chemogenetic manipulation of dorsal hippocampal astrocytes protects against the development of stress-enhanced fear learning. *Neuroscience, 388*, 45–56. https://doi.org/10.1016/j.neuroscience.2018.07.015

Kahathuduwa, C. N., Wakefield, S., West, B. D., Blume, J., Dassanayake, T. L., Weerasinghe, V. S., et al. (2020). Effects of L-theanine-caffeine combination on sustained attention and inhibitory control among children with ADHD: A proof-of-concept neuroimaging RCT. *Scientific Reports, 10*(1), 13072. https://doi.org/10.1038/s41598-020-70037-7

Kaster, M. P., Machado, N. J., Silva, H. B., Nunes, A., Ardais, A. P., Santana, M., et al. (2015). Caffeine acts through neuronal adenosine A2A receptors to prevent mood and memory dysfunction triggered by chronic stress. *Proceeding of the National Academy of Sciences of the United States of America, 112*(25), 7833–7838. https://doi.org/10.1073/pnas.1423088112

Kim, D. G., & Bynoe, M. S. (2015). A2A adenosine receptor regulates the human blood-brain barrier permeability. *Molecular Neurobiology, 52*(1), 664–678. https://doi.org/10.1007/s12035-014-8879-2

Kofuji, P., & Araque, A. (2021). G-Protein-coupled receptors in astrocyte-neuron communication. *Neuroscience, 456*, 71–84. https://doi.org/10.1016/j.neuroscience.2020.03.025

Koivisto, K., Reinikainen, K. J., Hanninen, T., Vanhanen, M., Helkala, E. L., Mykkanen, L., et al. (1995). Prevalence of age-associated memory impairment in a randomly selected population from eastern Finland. *Neurology, 45*(4), 741–747.

Kol, A., Adamsky, A., Groysman, M., Kreisel, T., London, M., & Goshen, I. (2020). Astrocytes contribute to remote memory formation by modulating hippocampal-cortical communication during learning. *Nature Neuroscience, 23*(10), 1229–1239. https://doi.org/10.1038/s41593-020-0679-6

Koss, W. A., & Frick, K. M. (2017). Sex differences in hippocampal function. *Journal of Neuroscience Research, 95*(1–2), 539–562. https://doi.org/10.1002/jnr.23864

Kovacs, Z., Dobolyi, A., Kekesi, K. A., & Juhasz, G. (2013). 5'-Nucleotidases, nucleosides and their distribution in the brain: Pathological and therapeutic implications. *Current Medicinal Chemistry, 20*(34), 4217–4240. https://doi.org/10.2174/09298673113203400003

Kovacs, Z., Juhasz, G., Palkovits, M., Dobolyi, A., & Kekesi, K. A. (2011). Area, age and gender dependence of the nucleoside system in the brain: A review of current literature. *Current Topics in Medicinal Chemistry, 11*(8), 1012–1033. https://doi.org/10.2174/156802611795347636

Launay, A., Nebie, O., Vijayashankara, J., Lebouvier, T., Buee, L., Faivre, E., et al. (2022). The role of adenosine A(2A) receptors in Alzheimer's disease and tauopathies. *Neuropharmacology,* 109379. https://doi.org/10.1016/j.neuropharm.2022.109379

Laurent, C., Burnouf, S., Ferry, B., Batalha, V. L., Coelho, J. E., Baqi, Y., et al. (2016). A2A adenosine receptor deletion is protective in a mouse model of Tauopathy. *Molecular Psychiatry, 21*(1), 97–107. https://doi.org/10.1038/mp.2014.151

Li, P., Rial, D., Canas, P. M., Yoo, J. H., Li, W., Zhou, X., et al. (2015). Optogenetic activation of intracellular adenosine A2A receptor signaling in the hippocampus is sufficient to trigger CREB phosphorylation and impair memory. *Molecular Psychiatry, 20*(11), 1339–1349. https://doi.org/10.1038/mp.2014.182

Li, Y., Ruan, Y., He, Y., Cai, Q., Pan, X., Zhang, Y., et al. (2020). Striatopallidal adenosine A(2A) receptors in the nucleus accumbens confer motivational control of goal-directed behavior. *Neuropharmacology, 168*, 108010. https://doi.org/10.1016/j.neuropharm.2020.108010

Li, W. P., Su, X. H., Hu, N. Y., Hu, J., Li, X. W., Yang, J. M., et al. (2022). Astrocytes mediate cholinergic regulation of adult hippocampal neurogenesis and memory through M1 muscarinic receptor. *Biological Psychiatry, 92*(12), 984–998. https://doi.org/10.1016/j.biopsych.2022.04.019

Li, X. C., Hong, F. F., Tu, Y. J., Li, Y. A., Ma, C. Y., Yu, C. Y., et al. (2022). Blockade of adenosine A(2A) receptor alleviates cognitive dysfunction after chronic exposure to intermittent hypoxia in mice. *Experimental Neurology, 350*, 113929. https://doi.org/10.1016/j.expneurol.2021.113929

Li, Y., He, Y., Chen, M., Pu, Z., Chen, L., Li, P., et al. (2016). Optogenetic activation of adenosine A2A receptor signaling in the dorsomedial striatopallidal neurons suppresses goal-directed behavior. *Neuropsychopharmacology: Official Publication of the American College of Neuropsychopharmacology, 41*(4), 1003–1013. https://doi.org/10.1038/npp.2015.227

Li, Y., Pan, X., He, Y., Ruan, Y., Huang, L., Zhou, Y., et al. (2018). Pharmacological blockade of adenosine A(2A) but not A(1) receptors enhances goal-directed valuation in satiety-based instrumental behavior. *Frontiers in Pharmacology, 9*, 393. https://doi.org/10.3389/fphar.2018.00393

Lopes, C. R., Cunha, R. A., & Agostinho, P. (2021). Astrocytes and adenosine A(2A) receptors: Active players in Alzheimer's disease. *Frontiers in Neuroscience, 15*, 666710. https://doi.org/10.3389/fnins.2021.666710

Lopes, L. V., Cunha, R. A., Kull, B., Fredholm, B. B., & Ribeiro, J. A. (2002). Adenosine A(2A) receptor facilitation of hippocampal synaptic transmission is dependent on tonic A(1) receptor inhibition. *Neuroscience, 112*(2), 319–329. https://doi.org/10.1016/s0306-4522(02)00080-5

Lopes, L. V., Cunha, R. A., & Ribeiro, J. A. (1999). Increase in the number, G protein coupling, and efficiency of facilitatory adenosine A2A receptors in the limbic cortex, but not striatum, of aged rats. *Journal of Neurochemistry, 73*(4), 1733–1738. https://doi.org/10.1046/j.1471-4159.1999.731733.x

Lyon, K. A., & Allen, N. J. (2021). From synapses to circuits, astrocytes regulate behavior. *Frontiers in Neural Circuits, 15*, 786293. https://doi.org/10.3389/fncir.2021.786293

Maglione, V., Cannella, M., Martino, T., De Blasi, A., Frati, L., & Squitieri, F. (2006). The platelet maximum number of A2A-receptor binding sites (Bmax) linearly correlates with age at onset and CAG repeat expansion in Huntington's disease patients with predominant chorea. *Neuroscience Letters, 393*(1), 27–30. https://doi.org/10.1016/j.neulet.2005.09.037

Markowska, A. L. (1999). Sex dimorphisms in the rate of age-related decline in spatial memory: Relevance to alterations in the estrous cycle. *The Journal of Neuroscience, 19*(18), 8122–8133.

Marques, T. R., Natesan, S., Rabiner, E. A., Searle, G. E., Gunn, R., Howes, O. D., et al. (2022). Adenosine A(2A) receptor in schizophrenia: An in vivo brain PET imaging study. *Psychopharmacology (Berl), 239*(11), 3439–3445. https://doi.org/10.1007/s00213-021-05900-0

Martin, E. D., Fernandez, M., Perea, G., Pascual, O., Haydon, P. G., Araque, A., et al. (2007). Adenosine released by astrocytes contributes to hypoxia-induced modulation of synaptic transmission. *Glia, 55*(1), 36–45. https://doi.org/10.1002/glia.20431

Martinez-Mir, M. I., Probst, A., & Palacios, J. M. (1991). Adenosine A2 receptors: selective localization in the human basal ganglia and alterations with disease. *Neuroscience, 42*(3), 697–706. https://doi.org/10.1016/0306-4522(91)90038-p

Matos, M., Augusto, E., Agostinho, P., Cunha, R. A., & Chen, J. F. (2013). Antagonistic interaction between adenosine A2A receptors and Na+/K+-ATPase-alpha2 controlling glutamate uptake in astrocytes. *The Journal of Neuroscience, 33*(47), 18492–18502. https://doi.org/10.1523/JNEUROSCI.1828-13.2013

Matos, M., Augusto, E., Santos-Rodrigues, A. D., Schwarzschild, M. A., Chen, J. F., Cunha, R. A., et al. (2012). Adenosine A2A receptors modulate glutamate uptake in cultured astrocytes and gliosomes. *Glia, 60*(5), 702–716. https://doi.org/10.1002/glia.22290

Matos, M., Shen, H. Y., Augusto, E., Wang, Y., Wei, C. J., Wang, Y. T., et al. (2015). Deletion of adenosine A2A receptors from astrocytes disrupts glutamate homeostasis leading to psychomotor and cognitive impairment: relevance to schizophrenia. *Biological Psychiatry, 78*(11), 763–774. https://doi.org/10.1016/j.biopsych.2015.02.026

Mazur, A., Bills, E. H., DeSchepper, K. M., Williamson, J. C., Henderson, B. J., & Risher, W. C. (2021). Astrocyte-derived thrombospondin induces cortical synaptogenesis in a sex-specific manner. *eNeuro, 8*(4), ENEURO.0014-21.2021. https://doi.org/10.1523/ENEURO.0014-21.2021

McEwen, B. S., & Milner, T. A. (2017). Understanding the broad influence of sex hormones and sex differences in the brain. *Journal of Neuroscience Research, 95*(1–2), 24–39. https://doi.org/10.1002/jnr.23809

McHugh, R. K., Hearon, B. A., & Otto, M. W. (2010). Cognitive behavioral therapy for substance use disorders. *The Psychiatric Clinics of North America, 33*(3), 511–525. https://doi.org/10.1016/j.psc.2010.04.012

Meadows, S. M., Palaguachi, F., Licht-Murava, A., Barnett, D., Zimmer, T. S., Zhou, C., et al. (2022). Astrocytes regulate spatial memory in a sex-specific manner. *bioRxiv, 2022*(2011), 2003–511881. https://doi.org/10.1101/2022.11.03.511881

Melani, A., Dettori, I., Corti, F., Cellai, L., & Pedata, F. (2015). Time-course of protection by the selective A2A receptor antagonist SCH58261 after transient focal cerebral ischemia. *Neurological Sciences: Official Journal of the Italian Neurological Society and of the Italian Society of Clinical Neurophysiology, 36*(8), 1441–1448. https://doi.org/10.1007/s10072-015-2160-y

Melani, A., Gianfriddo, M., Vannucchi, M. G., Cipriani, S., Baraldi, P. G., Giovannini, M. G., et al. (2006). The selective A2A receptor antagonist SCH 58261 protects from neurological deficit, brain damage and activation of p38 MAPK in rat focal cerebral ischemia. *Brain Research, 1073*(1074), 470–480. https://doi.org/10.1016/j.brainres.2005.12.010

Merighi, S., Battistello, E., Casetta, I., Gragnaniello, D., Poloni, T. E., Medici, V., et al. (2021). Upregulation of cortical A2A adenosine receptors is reflected in platelets of patients with Alzheimer's disease. *Journal of Alzheimer's Disease: JAD, 80*(3), 1105–1117. https://doi.org/10.3233/JAD-201437

Mills, J. H., Kim, D. G., Krenz, A., Chen, J. F., & Bynoe, M. S. (2012). A2A adenosine receptor signaling in lymphocytes and the central nervous system regulates inflammation during experimental autoimmune encephalomyelitis. *Journal of Immunology, 188*(11), 5713–5722. https://doi.org/10.4049/jimmunol.1200545

Mills, J. H., Thompson, L. F., Mueller, C., Waickman, A. T., Jalkanen, S., Niemela, J., et al. (2008). CD73 is required for efficient entry of lymphocytes into the central nervous system during experimental autoimmune encephalomyelitis. *Proceeding of the National Academy of Sciences of the United States of America, 105*(27), 9325–9330. https://doi.org/10.1073/pnas.0711175105

Mingote, S., Font, L., Farrar, A. M., Vontell, R., Worden, L. T., Stopper, C. M., et al. (2008). Nucleus accumbens adenosine A2A receptors regulate exertion of effort by acting on the ventral striatopallidal pathway. *The Journal of Neuroscience, 28*(36), 9037–9046. https://doi.org/10.1523/JNEUROSCI.1525-08.2008

Minor, T. R., Rowe, M., Cullen, P. K., & Furst, S. (2008). Enhancing brain adenosine signaling with the nucleoside transport blocker NBTI (S-(4-nitrobenzyl)-6-theoinosine) mimics the effects of inescapable shock on later shuttle-escape performance in rats. *Behavioral Neuroscience, 122*(6), 1236–1247. https://doi.org/10.1037/a0013143

Moscoso-Castro, M., Gracia-Rubio, I., Ciruela, F., & Valverde, O. (2016). Genetic blockade of adenosine A2A receptors induces cognitive impairments and anatomical changes related to psychotic symptoms in mice. *European Neuropsychopharmacology: The Journal of the European College of Neuropsychopharmacology, 26*(7), 1227–1240. https://doi.org/10.1016/j.euroneuro.2016.04.003

Moscoso-Castro, M., Lopez-Cano, M., Gracia-Rubio, I., Ciruela, F., & Valverde, O. (2017). Cognitive impairments associated with alterations in synaptic proteins induced by the genetic loss of adenosine A(2A) receptors in mice. *Neuropharmacology, 126*, 48–57. https://doi.org/10.1016/j.neuropharm.2017.08.027

Mouro, F. M., Kofalvi, A., Andre, L. A., Baqi, Y., Muller, C. E., Ribeiro, J. A., et al. (2019). Memory deficits induced by chronic cannabinoid exposure are prevented by adenosine A(2A)R receptor antagonism. *Neuropharmacology, 155*, 10–21. https://doi.org/10.1016/j.neuropharm.2019.05.003

Nagai, J., Yu, X., Papouin, T., Cheong, E., Freeman, M. R., Monk, K. R., et al. (2021). Behaviorally consequential astrocytic regulation of neural circuits. *Neuron, 109*(4), 576–596. https://doi.org/10.1016/j.neuron.2020.12.008

Nehlig, A. (2010). Is caffeine a cognitive enhancer? *Journal of Alzheimer's Disease: JAD*, *20*(Suppl 1), S85–S94. https://doi.org/10.3233/JAD-2010-091315

Nestler, E. J., & Luscher, C. (2019). The molecular basis of drug addiction: Linking epigenetic to synaptic and circuit mechanisms. *Neuron*, *102*(1), 48–59. https://doi.org/10.1016/j.neuron.2019.01.016

Ning, Y. L., Yang, N., Chen, X., Xiong, R. P., Zhang, X. Z., Li, P., et al. (2013). Adenosine A2A receptor deficiency alleviates blast-induced cognitive dysfunction. *Journal of Cerebral Blood Flow and Metabolism: Official Journal of the International Society of Cerebral Blood Flow and Metabolism*, *33*(11), 1789–1798. https://doi.org/10.1038/jcbfm.2013.127

O'Neill, C. E., Hobson, B. D., Levis, S. C., & Bachtell, R. K. (2014). Persistent reduction of cocaine seeking by pharmacological manipulation of adenosine A1 and A2A receptors during extinction training in rats. *Psychopharmacology (Berl)*, *231*(16), 3179–3188. https://doi.org/10.1007/s00213-014-3489-2

O'Neill, C. E., LeTendre, M. L., & Bachtell, R. K. (2012). Adenosine A2A receptors in the nucleus accumbens bi-directionally alter cocaine seeking in rats. *Neuropsychopharmacology. Official Publication of the American College of Neuropsychopharmacology*, *37*(5), 1245–1256. https://doi.org/10.1038/npp.2011.312

Ohta, A., & Sitkovsky, M. (2001). Role of G-protein-coupled adenosine receptors in downregulation of inflammation and protection from tissue damage. *Nature*, *414*(6866), 916–920. https://doi.org/10.1038/414916a

Orr, A. G., Hsiao, E. C., Wang, M. M., Ho, K., Kim, D. H., Wang, X., et al. (2015). Astrocytic adenosine receptor A2A and Gs-coupled signaling regulate memory. *Nature Neuroscience*, *18*(3), 423–434. https://doi.org/10.1038/nn.3930

Orr, A. G., Lo, I., Schumacher, H., Ho, K., Gill, M., Guo, W., et al. (2018). Istradefylline reduces memory deficits in aging mice with amyloid pathology. *Neurobiology of Disease*, *110*, 29–36. https://doi.org/10.1016/j.nbd.2017.10.014

Orr, A. G., Orr, A. L., Li, X. J., Gross, R. E., & Traynelis, S. F. (2009). Adenosine A(2A) receptor mediates microglial process retraction. *Nature Neuroscience*, *12*(7), 872–878. https://doi.org/10.1038/nn.2341

Pagnussat, N., Almeida, A. S., Marques, D. M., Nunes, F., Chenet, G. C., Botton, P. H., et al. (2015). Adenosine A(2A) receptors are necessary and sufficient to trigger memory impairment in adult mice. *British Journal of Pharmacology*, *172*(15), 3831–3845. https://doi.org/10.1111/bph.13180

Paiva, I., Carvalho, K., Santos, P., Cellai, L., Pavlou, M. A. S., Jain, G., et al. (2019). A(2A) R-induced transcriptional deregulation in astrocytes: An in vitro study. *Glia*, *67*(12), 2329–2342. https://doi.org/10.1002/glia.23688

Patterson, C. (2018). World alzheimer report 2018.

Pedata, F., Corsi, C., Melani, A., Bordoni, F., & Latini, S. (2001). Adenosine extracellular brain concentrations and role of A2A receptors in ischemia. *Annals of the New York Academy of Sciences*, *939*, 74–84. https://doi.org/10.1111/j.1749-6632.2001.tb03614.x

Peng, P. J., Chiang, K. T., & Liang, C. S. (2014). Low-dose caffeine may exacerbate psychotic symptoms in people with schizophrenia. *The Journal of Neuropsychiatry and Clinical Neurosciences*, *26*(2), E41. https://doi.org/10.1176/appi.neuropsych.13040098

Pereira, G. S., Rossato, J. I., Sarkis, J. J., Cammarota, M., Bonan, C. D., & Izquierdo, I. (2005). Activation of adenosine receptors in the posterior cingulate cortex impairs memory retrieval in the rat. *Neurobiology of Learning and Memory*, *83*(3), 217–223. https://doi.org/10.1016/j.nlm.2004.12.002

Pires, V. A., Pamplona, F. A., Pandolfo, P., Fernandes, D., Prediger, R. D., & Takahashi, R. N. (2009). Adenosine receptor antagonists improve short-term object-recognition ability of spontaneously hypertensive rats: a rodent model of attention-deficit hyperactivity disorder. *Behavioural Pharmacology*, *20*(2), 134–145. https://doi.org/10.1097/FBP.0b013e32832a80bf

Plassman, B. L., Langa, K. M., Fisher, G. G., Heeringa, S. G., Weir, D. R., Ofstedal, M. B., et al. (2008). Prevalence of cognitive impairment without dementia in the United States. *Annals of Internal Medicine, 148*(6), 427–434 doi:148/6/427 [pii].

Plath, N., Ohana, O., Dammermann, B., Errington, M. L., Schmitz, D., Gross, C., et al. (2006). Arc/Arg3.1 is essential for the consolidation of synaptic plasticity and memories. *Neuron, 52*(3), 437–444. https://doi.org/10.1016/j.neuron.2006.08.024

Prediger, R. D., Fernandes, D., & Takahashi, R. N. (2005). Blockade of adenosine A2A receptors reverses short-term social memory impairments in spontaneously hypertensive rats. *Behavioural Brain Research, 159*(2), 197–205. https://doi.org/10.1016/j.bbr.2004.10.017

Quarta, D., Borycz, J., Solinas, M., Patkar, K., Hockemeyer, J., Ciruela, F., et al. (2004). Adenosine receptor-mediated modulation of dopamine release in the nucleus accumbens depends on glutamate neurotransmission and N-methyl-D-aspartate receptor stimulation. *Journal of Neurochemistry, 91*(4), 873–880. https://doi.org/10.1111/j.1471-4159.2004.02761.x

Rebola, N., Lujan, R., Cunha, R. A., & Mulle, C. (2008). Adenosine A2A receptors are essential for long-term potentiation of NMDA-EPSCs at hippocampal mossy fiber synapses. *Neuron, 57*(1), 121–134. https://doi.org/10.1016/j.neuron.2007.11.023

Rebola, N., Sebastiao, A. M., de Mendonca, A., Oliveira, C. R., Ribeiro, J. A., & Cunha, R. A. (2003). Enhanced adenosine A2A receptor facilitation of synaptic transmission in the hippocampus of aged rats. *Journal of Neurophysiology, 90*(2), 1295–1303. https://doi.org/10.1152/jn.00896.2002

Rial, D., Lara, D. R., & Cunha, R. A. (2014). The adenosine neuromodulation system in schizophrenia. *International Review of Neurobiology, 119*, 395–449. https://doi.org/10.1016/B978-0-12-801022-8.00016-7

Ribeiro, J. A., & Sebastiao, A. M. (2010). Caffeine and adenosine. *Journal of Alzheimer's Disease: JAD, 20*(Suppl 1), S3–S15. https://doi.org/10.3233/JAD-2010-1379

Rimondini, R., Ferre, S., Ogren, S. O., & Fuxe, K. (1997). Adenosine A2A agonists: A potential new type of atypical antipsychotic. *Neuropsychopharmacology: Official Publication of the American College of Neuropsychopharmacology, 17*(2), 82–91. https://doi.org/10.1016/S0893-133X(97)00033-X

Robin, L. M., Oliveira da Cruz, J. F., Langlais, V. C., Martin-Fernandez, M., Metna-Laurent, M., Busquets-Garcia, A., et al. (2018). Astroglial CB1 receptors determine synaptic D-serine availability to enable recognition memory. *Neuron, 98*(5), 935–944.e935. https://doi.org/10.1016/j.neuron.2018.04.034

Rubinow, D. R., & Schmidt, P. J. (2019). Sex differences and the neurobiology of affective disorders. *Neuropsychopharmacology: Official Publication of the American College of Neuropsychopharmacology, 44*(1), 111–128. https://doi.org/10.1038/s41386-018-0148-z

Rudolphi, K. A., Schubert, P., Parkinson, F. E., & Fredholm, B. B. (1992). Neuroprotective role of adenosine in cerebral ischaemia. *Trends in Pharmacological Sciences, 13*(12), 439–445. https://doi.org/10.1016/0165-6147(92)90141-r

Rurak, G. M., Simard, S., Freitas-Andrade, M., Lacoste, B., Charih, F., Van Geel, A., et al. (2022). Sex differences in developmental patterns of neocortical astroglia: A mouse translatome database. *Cell Rep, 38*(5), 110310. https://doi.org/10.1016/j.celrep.2022.110310

Safarzadeh, E., Jadidi-Niaragh, F., Motallebnezhad, M., & Yousefi, M. (2016). The role of adenosine and adenosine receptors in the immunopathogenesis of multiple sclerosis. *Inflammation Research: Official Journal of the European Histamine Research Society, 65*(7), 511–520. https://doi.org/10.1007/s00011-016-0936-z

Sagvolden, T., & Sergeant, J. A. (1998). Attention deficit/hyperactivity disorder—From brain dysfunctions to behaviour. *Behavioural Brain Research, 94*(1), 1–10.

Santello, M., Toni, N., & Volterra, A. (2019). Astrocyte function from information processing to cognition and cognitive impairment. *Nature Neuroscience, 22*(2), 154–166. https://doi.org/10.1038/s41593-018-0325-8

Schiffmann, S. N., Fisone, G., Moresco, R., Cunha, R. A., & Ferre, S. (2007). Adenosine A2A receptors and basal ganglia physiology. *Progress in Neurobiology, 83*(5), 277–292. https://doi.org/10.1016/j.pneurobio.2007.05.001

Schonknecht, P., Pantel, J., Kruse, A., & Schroder, J. (2005). Prevalence and natural course of aging-associated cognitive decline in a population-based sample of young-old subjects. *The American Journal of Psychiatry, 162*(11), 2071–2077. https://doi.org/10.1176/appi.ajp.162.11.2071

Selkoe, D. J. (2002). Alzheimer's disease is a synaptic failure. *Science (New York, N. Y.), 298*(5594), 789–791. https://doi.org/10.1126/science.1074069

Sevigny, J., Chiao, P., Bussiere, T., Weinreb, P. H., Williams, L., Maier, M., et al. (2016). The antibody aducanumab reduces Abeta plaques in Alzheimer's disease. *Nature, 537*(7618), 50–56. https://doi.org/10.1038/nature19323

Shen, H. Y., Singer, P., Lytle, N., Wei, C. J., Lan, J. Q., Williams-Karnesky, R. L., et al. (2012). Adenosine augmentation ameliorates psychotic and cognitive endophenotypes of schizophrenia. *The Journal of Clinical Investigation, 122*(7), 2567–2577. https://doi.org/10.1172/JCI62378

Simoes, A. P., Machado, N. J., Goncalves, N., Kaster, M. P., Simoes, A. T., Nunes, A., et al. (2016). Adenosine A(2A) receptors in the amygdala control synaptic plasticity and contextual fear memory. *Neuropsychopharmacology: Official Publication of the American College of Neuropsychopharmacology, 41*(12), 2862–2871. https://doi.org/10.1038/npp.2016.98

Sofuoglu, M., DeVito, E. E., Waters, A. J., & Carroll, K. M. (2016). Cognitive function as a transdiagnostic treatment target in stimulant use disorders. *Journal of Dual Diagnosis, 12*(1), 90–106. https://doi.org/10.1080/15504263.2016.1146383

Tanz, L. J., Dinwiddie, A. T., Mattson, C. L., O'Donnell, J., & Davis, N. L. (2022). Drug overdose deaths among persons aged 10–19 years—United States, July 2019–December 2021. *MMWR. Morbidity and Mortality Weekly Report, 71*, 1576–1582. https://doi.org/10.15585/mmwr.mm7150a2

Temido-Ferreira, M., Ferreira, D. G., Batalha, V. L., Marques-Morgado, I., Coelho, J. E., Pereira, P., et al. (2020). Age-related shift in LTD is dependent on neuronal adenosine A(2A) receptors interplay with mGluR5 and NMDA receptors. *Molecular Psychiatry, 25*(8), 1876–1900. https://doi.org/10.1038/s41380-018-0110-9

Tescarollo, F. C., Rombo, D. M., DeLiberto, L. K., Fedele, D. E., Alharfoush, E., Tome, A. R., et al. (2020). Role of adenosine in epilepsy and seizures. *Journal of Caffeine and Adenosine Research, 10*(2), 45–60. https://doi.org/10.1089/caff.2019.0022

United Nations. Department of Economic and Social Affairs. Population Division. (2002). *World population ageing, 1950–2050*. New York: United Nations.

Van Den Herrewegen, Y., Sanderson, T. M., Sahu, S., De Bundel, D., Bortolotto, Z. A., & Smolders, I. (2021). Side-by-side comparison of the effects of Gq- and Gi-DREADD-mediated astrocyte modulation on intracellular calcium dynamics and synaptic plasticity in the hippocampal CA1. *Molecular Brain, 14*, 144. https://doi.org/10.1186/s13041-021-00856-w

van Dyck, C. H., Swanson, C. J., Aisen, P., Bateman, R. J., Chen, C., Gee, M., et al. (2022). Lecanemab in early Alzheimer's disease. *The New England Journal of Medicine, 388*(1), 9–21. https://doi.org/10.1056/NEJMoa2212948

Velasco, E. R., Florido, A., Milad, M. R., & Andero, R. (2019). Sex differences in fear extinction. *Neuroscience and Biobehavioral Reviews, 103*, 81–108. https://doi.org/10.1016/j.neubiorev.2019.05.020

Verdejo-Garcia, A., Albein-Urios, N., Martinez-Gonzalez, J. M., Civit, E., de la Torre, R., & Lozano, O. (2014). Decision-making impairment predicts 3-month hair-indexed cocaine relapse. *Psychopharmacology (Berl), 231*(21), 4179–4187. https://doi.org/10.1007/s00213-014-3563-9

Viana da Silva, S., Haberl, M. G., Zhang, P., Bethge, P., Lemos, C., Goncalves, N., et al. (2016). Early synaptic deficits in the APP/PS1 mouse model of Alzheimer's disease involve neuronal adenosine A2A receptors. *Nature Communications, 7*, 11915. https://doi.org/10.1038/ncomms11915

Villar-Menendez, I., Diaz-Sanchez, S., Blanch, M., Albasanz, J. L., Pereira-Veiga, T., Monje, A., et al. (2014). Reduced striatal adenosine A2A receptor levels define a molecular subgroup in schizophrenia. *Journal of Psychiatric Research, 51*, 49–59. https://doi.org/10.1016/j.jpsychires.2013.12.013

Vonmoos, M., Hulka, L. M., Preller, K. H., Jenni, D., Baumgartner, M. R., Stohler, R., et al. (2013). Cognitive dysfunctions in recreational and dependent cocaine users: Role of attention-deficit hyperactivity disorder, craving and early age at onset. *The British Journal of Psychiatry: The Journal of Mental Science, 203*(1), 35–43. https://doi.org/10.1192/bjp.bp.112.118091

Voskuhl, R., & Itoh, Y. (2022). The X factor in neurodegeneration. *The Journal of Experimental Medicine, 219*(12), e20211488. https://doi.org/10.1084/jem.20211488

Wang, J. H., Ma, Y. Y., & van den Buuse, M. (2006). Improved spatial recognition memory in mice lacking adenosine A2A receptors. *Experimental Neurology, 199*(2), 438–445. https://doi.org/10.1016/j.expneurol.2006.01.005

Wang, M., Li, Z., Song, Y., Sun, Q., Deng, L., Lin, Z., et al. (2022). Genetic tagging of the adenosine A2A receptor reveals its heterogeneous expression in brain regions. *Frontiers in Neuroanatomy, 16*, 978641. https://doi.org/10.3389/fnana.2022.978641

Wei, C. J., Augusto, E., Gomes, C. A., Singer, P., Wang, Y., Boison, D., et al. (2014). Regulation of fear responses by striatal and extrastriatal adenosine A2A receptors in forebrain. *Biological Psychiatry, 75*(11), 855–863. https://doi.org/10.1016/j.biopsych.2013.05.003

Wei, C. J., Singer, P., Coelho, J., Boison, D., Feldon, J., Yee, B. K., et al. (2011). Selective inactivation of adenosine A(2A) receptors in striatal neurons enhances working memory and reversal learning. *Learning & Memory (Cold Spring Harbor, N. Y.), 18*(7), 459–474. https://doi.org/10.1101/lm.2136011

Wydra, K., Golembiowska, K., Suder, A., Kaminska, K., Fuxe, K., & Filip, M. (2015). On the role of adenosine (A)(2)A receptors in cocaine-induced reward: A pharmacological and neurochemical analysis in rats. *Psychopharmacology (Berl), 232*(2), 421–435. https://doi.org/10.1007/s00213-014-3675-2

Xu, X., Beleza, R. O., Goncalves, F. Q., Valbuena, S., Alcada-Morais, S., Goncalves, N., et al. (2022). Adenosine A(2A) receptors control synaptic remodeling in the adult brain. *Scientific Reports, 12*(1), 14690. https://doi.org/10.1038/s41598-022-18884-4

Yagi, S., & Galea, L. A. M. (2019). Sex differences in hippocampal cognition and neurogenesis. *Neuropsychopharmacology: Official Publication of the American College of Neuropsychopharmacology, 44*(1), 200–213. https://doi.org/10.1038/s41386-018-0208-4

Yamada, K., Kobayashi, M., Shiozaki, S., Ohta, T., Mori, A., Jenner, P., et al. (2014). Antidepressant activity of the adenosine A2A receptor antagonist, istradefylline (KW-6002) on learned helplessness in rats. *Psychopharmacology (Berl), 231*(14), 2839–2849. https://doi.org/10.1007/s00213-014-3454-0

Yu, L., Shen, H. Y., Coelho, J. E., Araujo, I. M., Huang, Q. Y., Day, Y. J., et al. (2008). Adenosine A2A receptor antagonists exert motor and neuroprotective effects by distinct cellular mechanisms. *Annals of Neurology, 63*(3), 338–346. https://doi.org/10.1002/ana.21313

Zhang, J., Abdallah, C. G., Wang, J., Wan, X., Liang, C., Jiang, L., et al. (2012). Upregulation of adenosine A2A receptors induced by atypical antipsychotics and its correlation with sensory gating in schizophrenia patients. *Psychiatry Research, 200*(2–3), 126–132. https://doi.org/10.1016/j.psychres.2012.04.021

Zhou, J., Wu, B., Lin, X., Dai, Y., Li, T., Zheng, W., et al. (2019). Accumbal adenosine A(2A) receptors enhance cognitive flexibility by facilitating strategy shifting. *Frontiers in Cellular Neuroscience, 13*, 130. https://doi.org/10.3389/fncel.2019.00130

Zhou, S. J., Zhu, M. E., Shu, D., Du, X. P., Song, X. H., Wang, X. T., et al. (2009). Preferential enhancement of working memory in mice lacking adenosine A(2A) receptors. *Brain Research, 1303*, 74–83. https://doi.org/10.1016/j.brainres.2009.09.082

Zhou, Z., Okamoto, K., Onodera, J., Hiragi, T., Andoh, M., Ikawa, M., et al. (2021). Astrocytic cAMP modulates memory via synaptic plasticity. *Proceeding of the National Academy of Sciences of the United States of America, 118*(3), e2016584118. https://doi.org/10.1073/pnas.2016584118

CHAPTER SEVEN

# Adenosine $A_{2A}$ receptors and sleep

**Mustafa Korkutata[a,*] and Michael Lazarus[b,*]**
[a]Department of Neurology, Division of Sleep Medicine, Beth Israel Deaconess Medical Center and Harvard Medical School, Boston, MA, USA
[b]International Institute for Integrative Sleep Medicine (WPI-IIIS) and Institute of Medicine, University of Tsukuba, Tsukuba, Japan
*Corresponding authors. e-mail address: mkorkuta@bidmc.harvard.edu; Lazarus.michael.ka@u.tsukuba.ac.jp

## Contents

1. Introduction  155
2. Adenosine and its receptors  156
   2.1 Adenosine  156
   2.2 Purinergic P1 receptors  157
   2.3 Adenosine $A_{2A}$ receptors  158
3. The role of adenosine and $A_{2A}$Rs in sleep-wake regulation  160
   3.1 Adenosine and sleep  160
   3.2 $A_{2A}$R agonists and sleep  162
   3.3 Adenosine $A_{2A}$R antagonists and sleep  163
   3.4 Natural $A_{2A}$R agonists and sleep  164
   3.5 Allosteric modulation of adenosine $A_{2A}$Rs and sleep  165
4. Conclusion  169
References  169

## Abstract

Adenosine, a known endogenous somnogen, induces sleep via $A_1$ and $A_{2A}$ receptors. In this chapter, we review the current knowledge regarding the role of the adenosine $A_{2A}$ receptor and its agonists, antagonists, and allosteric modulators in sleep-wake regulation. Although many adenosine $A_{2A}$ receptor agonists, antagonists, and allosteric modulators have been identified, only a few have been tested to see if they can promote sleep or wakefulness. In addition, the growing popularity of natural sleep aids has led to an investigation of natural compounds that may improve sleep by activating the adenosine $A_{2A}$ receptor. Finally, we discuss the potential therapeutic advantage of allosteric modulators of adenosine $A_{2A}$ receptors over classic agonists and antagonists for treating sleep and neurologic disorders.

## 1. Introduction

The need for sleep, also known as sleep drive, has remained mysterious at the cellular and neurological levels. Classical studies suggest that one or more endogenous somnogens (sleep-inducing substances, also referred to as "hypnotoxins") gradually accumulate during waking hours and dissipate during sleep as a reason for sleep homeostatic pressure (Inoué, Honda, & Komoda, 1995; Kubota, 1989), which was initially demonstrated by the presence of hypnogenic substances in the cerebrospinal fluid of sleep-deprived dogs (Ishimori, 1909; Legendre & Piéron, 1912). Following these early studies, several putative hypnogenic substances involved in the sleep homeostatic process have been identified (Urade & Hayaishi, 2011), including prostaglandin $D_2$ (Cherasse et al., 2018; Qu et al., 2006; Ueno, Ishikawa, Nakayama, & Hayaishi, 1982), anandamide (García-García, Acosta-Peña, Venebra-Muñoz, & Murillo-Rodríguez, 2009), urotensin II peptide (Huitron-Resendiz et al., 2005), cytokines (Krueger, Walter, Dinarello, Wolff, & Chedid, 1984), and adenosine (Feldberg & Sherwood, 1954; Porkka-Heiskanen et al., 1997). A growing body of evidence also suggests that sleep regulation has links with components of the immune system, such as pro-inflammatory cytokines (Krueger & Majde, 2003; Krueger, Obál, Fang, Kubota, & Taishi, 2001; Mullington et al., 2000; Mullington, Hinze-Selch, & Pollmächer, 2001) and prostaglandins (Lazarus et al., 2007; Oishi et al., 2015; Urade & Lazarus, 2013; Ushikubi et al., 1998). Several excellent reviews describe how sleep and wakefulness are regulated at the neuronal level (Fuller, Yamanaka, & Lazarus, 2015; Saper, Fuller, Pedersen, Lu, & Scammell, 2010; Saper, Scammell, & Lu, 2005; Shinnosuke Yasugaki, Yu Hayashi, & Lazarus, 2023). In this chapter, we focus on the roles of adenosine and adenosine $A_{2A}$ receptors ($A_{2A}Rs$) in sleep regulation.

## 2. Adenosine and its receptors
### 2.1 Adenosine

Adenosine is a naturally occurring endogenous purine nucleoside made up of adenine and D-ribose and is formed through hydrolysis of S-adenosylhomocysteine or adenosine monophosphate (Fredholm, 2007; Schrader, 1983). The intracellular activity of the enzyme S-adenosylhomocysteine hydrolase is required for the synthesis of adenosine from S-adenosylhomocysteine (Ballarín, Fredholm, Ambrosio, & Mahy, 1991). Both intracellular and extracellular levels of adenosine are generated from adenosine monophosphate (AMP) through the activity of 5′-nucleotidases (Zimmermann, 2000). The ecto-5′-nucleotidase, also known as CD73, is the exclusive enzyme in the brain that converts extracellular AMP to adenosine (Lovatt et al., 2012; Resta, Yamashita, & Thompson, 1998; Wall & Dale, 2013). Adenosine can also be produced independently of CD73 by releasing adenosine in the cytosol of neurons due to metabolic exhaustion (Lovatt et al., 2012). This may represent an autonomic feedback system that inhibits excitatory transmission during prolonged activity under pathological and physiological conditions, such as cortical seizures and perhaps those related to the control of sleep and wakefulness. High levels of adenosine are lowered by adenosine deaminase (ADA) activity or by uptake into cells and rapid phosphorylation to AMP by adenosine kinase (AdK), which efficiently regulates intracellular adenosine levels (Fredholm, Chen, Cunha, Svenningsson, & Vaugeois, 2005; Oishi, Huang, Fredholm, Urade, & Hayaishi, 2008; Parkinson et al., 2011). AdK expression in the adult central nervous system is found primarily in glia, and consequently, adenosine levels are coupled to the glial metabolism (Studer et al., 2006; Zhou et al., 2019).

Adenosine regulates various physiologic functions, including vasodilation, blood vessel formation, inflammation and wound healing, cardiac contraction, learning, memory, sleep, and arousal (Adair, 2005; Chen, 2014; Feoktistov, Biaggioni, & Cronstein, 2009; Headrick, Ashton, Rose'meyer, & Peart, 2013; Hein, Wang, Zoghi, Muthuchamy, & Kuo, 2001; Lazarus, Chen, Huang, Urade, & Fredholm, 2019; Ma et al., 2023; Ohta & Sitkovsky, 2001). Adenosine also has a direct function in synaptic processes and the regulation of various neurotransmitters in the central nervous system (CNS), although it does not act at synapses and is not stored in synaptic vesicles. The release and reuptake of adenosine are mediated by nucleoside transporters across a concentration gradient between the intracellular and extracellular compartments. As a result, adenosine was proposed to be a modulator affecting neurotransmitter release, neuronal hyper- or depolarization, and glial cell regulation (Boison, Singer, Shen, Feldon, & Yee, 2012). Adenosine also has neurotransmitter properties due to the presence of adenosine-producing enzymes in synapses (Cacciari et al., 2005).

## 2.2 Purinergic P1 receptors

Adenosine and ATP, short for adenosine triphosphate, are purine molecules that naturally bind to purinergic receptors, first identified in 1978 (Burnstock, Cocks, Crowe, & Kasakov, 1978). A different pharmacologic profile of these receptors led to identifying P1 and P2 as two distinct subtypes of purinergic receptors (Matsumoto, Tostes, & Webb, 2012). P1 receptors recognize adenosine as the major natural ligand and are therefore also referred to as adenosine receptors. The $A_1$, $A_{2A}$, $A_{2B}$, or $A_3$ subtypes of adenosine receptors each have their own distinct pharmacological profile and are members of the G protein-coupled receptors (GPCR) superfamily (Fredholm, IJzerman, Jacobson, Linden, & Müller, 2011; Göblyös & Ijzerman, 2009). The Gs-coupled $A_{2A}Rs$ and adenosine $A_{2B}$ receptors ($A_{2B}Rs$) increase the activity of adenylyl cyclase, which initiates the synthesis of cyclic AMP (cAMP). In contrast, activation of Gi/q-coupled adenosine $A_1$ and $A_3$ receptors ($A_1Rs$ and $A_3Rs$) by adenosine or agonist molecules suppresses adenylyl cyclase activity and inhibits cAMP synthesis in cells. (Cunha, 2001; Fredholm et al., 2011; Paes-De-Carvalho, 2002).

$A_{2B}R$ activation has been reported to require a high concentration of adenosine, whereas physiological levels of adenosine are sufficient to activate $A_1Rs$, $A_{2A}Rs$, and $A_3Rs$ with roughly comparable efficacy (Fredholm et al., 2011). However, the number of receptors present on cells determines the pharmacological potency of an endogenous ligand or agonist at its receptor. When only a few receptors are present, larger amounts of adenosine are required to produce an effect. Extracellular adenosine interacts with neurons primarily through $A_1Rs$ and $A_{2A}Rs$, which are thought to be more abundant in the brain compared to the other two adenosine receptors (Fredholm et al., 2005; Lazarus et al., 2019).

## 2.3 Adenosine $A_{2A}$ receptors

Libert and colleagues first discovered $A_{2A}Rs$ after cloning several orphan GPCRs from the thyroid of dogs (Maenhaut et al., 1990). $A_{2A}Rs$ have subsequently been cloned from a variety of other species, including guinea pigs, mice, rats, and humans (Chern, King, Lai, & Lai, 1992; Furlong, Pierce, Selbie, & Shine, 1992; Ledent et al., 1997; Meng et al., 1994). The classical secondary messenger pathways are induced by $A_{2A}Rs$, much like the other GPCRs. The $A_{2A}R$ signaling pathway may differ depending on the cell and tissue type in which the receptors are found. For example, Gs is the major G-protein associated with $A_{2A}Rs$ in the peripheral system. However, in the striatum, where they are extensively expressed, $A_{2A}Rs$ exert their effects in rats primarily through activating Golf. Active Gs proteins and Golf proteins activate adenylyl cyclase, increasing cellular cAMP levels and activating protein kinase A (PKA), which phosphorylates and promotes the cAMP-responsive element binding protein 1 (de Lera Ruiz, Lim, & Zheng, 2014; Kull, Svenningsson, & Fredholm, 2000). Extracellular signal-regulated kinases (ERK) and several other members of the mitogen-activated protein kinase (MAPK) family are also triggered by the activation of $A_{2A}Rs$, which in turn elicits several cellular responses (Schulte & Fredholm, 2000). $A_{2A}Rs$ and other GPCRs combine to form heterodimer structures [e.g., metabotropic glutamate type 5 receptor/$A_{2A}R$, cannabinoid receptor type 1/$A_{2A}R$, dopamine $D_2$ receptor ($D_2R$)/$A_{2A}R$, dopamine $D_3$ receptor/$A_{2A}R$], and even cannabinoid receptor type 1/$A_{2A}R$/$D_2R$ heterotrimers (Ferré et al., 2002; Ferré, Goldberg, Lluis, & Franco, 2009; Fuxe et al., 2005; Navarro et al., 2008; Torvinen et al., 2005).

$A_{2A}Rs$ are found primarily on GABAergic medium-sized spiny neurons in the striatum, the nucleus accumbens (NAc) core and shell, the olfactory tubercle, and dopamine-rich regions of the brain (de Lera Ruiz et al., 2014) (Fig. 1). The indirect pathways of the basal ganglia in the brain are significantly regulated by $A_{2A}Rs$ (Schulte & Fredholm, 2000). The basal ganglia play an important role in learning habits, goal-directed behaviors, and locomotion that has been preserved throughout evolution (Grillner & Robertson, 2016). In contrast to neurons of the indirect pathway, which express inhibitory $D_2Rs$ and excitatory $A_{2A}Rs$, direct pathway neurons exhibit both inhibitory $D_1Rs$ and excitatory $A_1Rs$ (Oishi & Lazarus, 2017). Studies in mice have shown that both direct and indirect pathway medium spiny neurons are active during mouse movement but passive during inactive phases (Cui et al., 2013). Chemogenetic stimulation of direct and indirect pathway neurons increases and decreases locomotor activity, respectively (Zhu, Ottenheimer, & DiLeone, 2016).

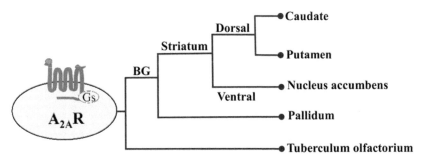

**Fig. 1** Expression profile of $A_{2A}Rs$ in the central nervous system. The basal ganglia (BG), including the dorsal pallidum, the nucleus accumbens in the ventral striatum, the dorsal striatum consisting of the caudate and the putamen, and the olfactory tubercle are the main sites of $A_{2A}R$ expression.

## 3. The role of adenosine and $A_{2A}$Rs in sleep-wake regulation

### 3.1 Adenosine and sleep

In 1954, Feldberg and colleagues discovered that adenosine had hypnotic effects in the cat brain, more than 60 years after it was first isolated from heart tissue extracts (Drury & Szent-Györgyi, 1929; Feldberg & Sherwood, 1954). The somnogenic effects of adenosine were then observed in various species, including dogs, birds, rats, and mice (Dunwiddie & Worth, 1982; Haulică, Ababei, Brănișteanu, & Topoliceanu, 1973; Marley & Nistico, 1972; Radulovacki, Virus, Djuricic-Nedelson, & Green, 1984; Radulovacki, Virus, Rapoza, & Crane, 1985; Ticho & Radulovacki, 1991). However, the primary function of adenosine in controlling sleep/wake behavior, the types of brain cells affected by the sleep-promoting effects of adenosine (Fig. 2), and the respective contributions of $A_1R$ and $A_{2A}R$ to sleep/wake regulation, on the other hand, are all still controversial. Adenosine is a sign of relative energy deficiency. Early theories of sleep/wake regulation assumed that the urge to sleep is due, at least in part, to the periodic need of the brain to replace depleted energy resources (Pull & McIlwain, 1972; Tobler & Scherschlicht, 1990; Van Wylen, Park, Rubio, & Berne, 1986). Indeed, in-vivo microdialysis studies of extracellular adenosine levels in the

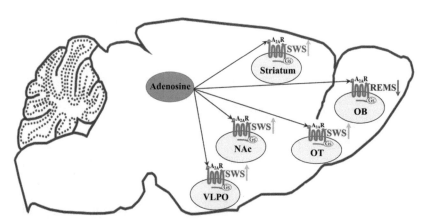

**Fig. 2** Effects of $A_{2A}$Rs in sleep-wake regulation. Adenosine activates excitatory $A_{2A}$Rs in the rostral, centromedial, and centrolateral striatum, olfactory tubercle (OT), the nucleus accumbens (NAc), and the ventrolateral preoptic region (VLPO) to induce slow-wave sleep (SWS) or inhibit REM sleep in the olfactory bulb (OB).

hippocampus and neostriatum of freely behaving rats have shown that adenosine levels are higher during the inactive period than the active period, providing the first evidence for this assumption (Huston et al., 1996).

In addition, ATP depletion is associated with increased extracellular adenosine levels and sleep (Kalinchuk et al., 2003; Porkka-Heiskanen et al., 1997). Adenosine levels were higher during sleep than waking in all brain regions studied when samples were collected by in-vivo microdialysis from different brain regions of the cat during spontaneous sleep-wake cycles (Porkka-Heiskanen et al., 1997; Porkka-Heiskanen, Strecker, & McCarley, 2000). In-vivo microdialysis experiments in the cat brain also showed a twofold increase in basal forebrain (BF) adenosine levels over a continuous 6-h arousal period compared with adenosine levels at the onset of sleep deprivation (Porkka-Heiskanen et al., 1997, 2000). However, these observations are difficult to reconcile with recent work that measured adenosine in real-time and showed rapid adenosine dynamics dependent on neuronal activity, although this work did not track adenosine changes during and after sleep deprivation (Peng et al., 2020). The energy theory of sleep and its relationship to extracellular adenosine levels appears to be even more complicated than first thought (Benington, Kodali, & Heller, 1995; Petit, Burlet-Godinot, Magistretti, & Allaman, 2015; Scharf, Naidoo, Zimmerman, & Pack, 2008). Glucose, the primary energy source in most species, has been shown to block orexin neurons that increase wakefulness and modulate food intake. However, glucose also activates sleep-active neurons in the ventrolateral preoptic region (VLPO), a known sleep center in the brain (Burdakov, Luckman, & Verkhratsky, 2005; Gallopin et al., 2000, 2005; Moore et al., 2012; Saper et al., 2005) because microinjections of glucose into the VLPO promote both sleep and the expression of c-fos, a known neuronal activity marker (Varin et al., 2015).

Cre-dependent adeno-associated virus encoding humanized Renilla green fluorescent protein was used as a tracer for long axonal projections in mice expressing Cre recombinase under the control of the $A_{2A}R$ promoter to map the projections of NAc $A_{2A}R$ neurons. The results show that NAc $A_{2A}R$ neurons might regulate sleep by suppressing wakefulness since they project to the lateral hypothalamus, the ventral tegmental area (VTA), the dorsal raphe, and the TMN, all of which include wake-promoting neurons (Zhang et al., 2013). Indeed, chemogenetic or optogenetic stimulation of NAc core $A_{2A}R$ neurons projecting to the ventral pallidum in the BF significantly promotes slow-wave sleep (SWS). In contrast, chemogenetic inhibition of these neurons inhibits sleep induction but does not affect the homeostatic sleep rebound (Oishi et al., 2017). Interestingly, motivational stimuli reduce the activity of NAc $A_{2A}R$ neurons, which project to the ventral pallidum and prevent sleep.

NAc medium spiny neurons can be divided into two groups that respond differently to activation by dopamine or adenosine. In contrast to indirect pathway neurons, which express inhibitory $D_2Rs$ and excitatory $A_{2A}R$, direct pathway neurons express excitatory dopamine $D_1$ receptors ($D_1Rs$) and inhibitory $A_1Rs$. Previous studies based on in-vivo electrophysiologic recordings show that sleep-active neurons coexist with wake-active neurons in the rat NAc area (Callaway & Henriksen, 1992; Osaka & Matsumura, 1995; Tellez, Perez, Simon, & Gutierrez, 2012). Indeed, chemogenetic stimulation prolongs arousal, while optogenetic activation of NAc $D_1R$-expressing neurons that project to the midbrain and lateral hypothalamus causes a rapid transition from SWS to wakefulness (Luo et al., 2018). The ability NAc indirect pathway to control sleep may explain why humans tend to fall asleep in the absence of motivating stimuli, i.e., when bored. In cats and rats, striatal ablation reduces sleep (Mena-Segovia, Cintra, Prospéro-García, & Giordano, 2002; Villablanca, 1972), and chemogenetic activation of $A_{2A}R$-expressing neurons in the rostral, centromedial, and centrolateral striatum, increases sleep, whereas $A_{2A}R$-expressing neurons in the caudal striatum are unable to do so (Yuan et al., 2017). In addition, abundant $A_{2A}Rs$ are also expressed in the olfactory tubercle, which increase or decrease SWS when chemogenetically activated or inhibited, respectively (Li et al., 2020).

## 3.2 $A_{2A}R$ agonists and sleep

The highly selective $A_{2A}R$ agonist CGS 21680 significantly increases SWS and REM sleep in mice following injection into the lateral ventricle or subarachnoid space under the ventral surface area of the rostral BF (Satoh et al. 1999; Satoh, Matsumura, & Hayaishi, 1998; Satoh, Matsumura, Suzuki, & Hayaishi, 1996; Urade et al., 2003). Infusions of CGS 21680 into the BF decrease histamine release in the frontal cortex and medial preoptic region in a dose-dependent manner while increasing the release of γ-aminobutyric acid (GABA) in the hypothalamic tuberomammillary nucleus (TMN) but not in the frontal cortex (Hong et al., 2005). The blockade of histamine release induced by CGS 21680 is abolished when the TMN is perfused with the GABA antagonist picrotoxin, suggesting that the $A_{2A}R$ agonist causes sleep by suppressing the histaminergic system via an increased GABA release in the TMN. The activation of sleep neurons could promote sleep in the VLPO and reciprocal suppression of histaminergic wake neurons in the TMN by GABAergic and galaninergic inhibitory projections (Sherin, Elmquist, Torrealba, & Saper, 1998; Sherin, Shiromani, McCarley, & Saper, 1996).

Whole-cell recordings of VLPO neurons in rat brain slices revealed the existence of two unique types of VLPO neurons in terms of their responses to monoamines, acetylcholine, and adenosine receptor agonists (Gallopin et al., 2005). The VLPO neurons are inhibited by noradrenaline, acetylcholine, and an $A_1R$ agonist, whereas serotonin inhibits type-1 neurons but excites type-2 neurons. However, an $A_{2A}R$ agonist postsynaptically excites type-2, but not type-1, neurons. These results suggest that type-2 neurons are involved in initiating sleep, whereas type-1 neurons may contribute to sleep consolidation because they are only activated without inhibitory effects from wake-inducing systems. However, administration of CGS 21680 into the rostral BF results in c-fos expression not only in the VLPO but also in the NAc shell and the medial part of the olfactory tubercle (Satoh et al., 1999; Scammell et al., 2001). Notably, the $A_{2A}R$ agonist promotes SWS and REM sleep when infused directly into the NAc, which is approximately three-quarters of the amount of sleep reported when the $A_{2A}R$ agonist is infused into the subarachnoid space (Satoh et al., 1999). Although $A_{2A}Rs$ appear to be primarily responsible for the control of SWS, some studies suggest that $A_{2A}R$ may also be involved in regulating REM sleep. Infusion of CGS 21680 into the region of the medial pontine reticular formation increases REM sleep (Marks, Shaffery, Speciale, & Birabil, 2003).

## 3.3 Adenosine $A_{2A}R$ antagonists and sleep

Caffeine acts as an antagonist of the $A_1R$ and $A_{2A}R$ and blocks adenosine to increase wakefulness (Fredholm, Bättig, Holmén, Nehlig, & Zvartau, 1999). Experiments with $A_1R$ and $A_{2A}R$ knockout mice have shown that $A_{2A}Rs$, but not $A_1Rs$, mediate the arousal-inducing effect of caffeine (Huang et al., 2005). The specific function of $A_{2A}R$ in the basal ganglia has been well-studied by site-specific gene modification techniques, such as conditional $A_{2A}R$ knockout mice based on Cre/lox technology or local infection with an adeno-associated virus carrying an $A_{2A}R$ short-hairpin RNA to suppress $A_{2A}R$ expression. Caffeine-induced wakefulness is prevented by selective $A_{2A}R$ ablation in the NAc shell (Lazarus et al., 2011). The ability of caffeine to act as an $A_{2A}R$ antagonist depends on adenosine's ability to tonically activate excitatory $A_{2A}R$ inside the NAc shell (Rosin, Robeva, Woodard, Guyenet, & Linden, 1998; Svenningsson et al., 1999).

Extracellular glucose increases extracellular adenosine levels in the VLPO to induce sleep, which can be inhibited by the glial toxin fluoroacetate or the $A_{2A}R$ antagonist ZM 241385 (Scharbarg et al., 2016). These findings highlight the critical function of glial cells in coordinating neuronal firing with the local blood supply via astrocyte-derived adenosine and $A_{2A}R$ activation in neurons. Also, a recent study demonstrated that inhibiting $A_{2A}R$ or $A_{2A}R$-expressing neurons in the olfactory bulb of mice increased REM sleep, suggesting that the olfactory bulb is a crucial site for controlling REM sleep by the $A_{2A}R$ system (Wang et al., 2012). REM sleep is likely related to the perception of odors in the olfactory bulb as an $A_{2A}R$ antagonist, such as caffeine or ZM 241385, can improve olfactory dysfunction (Prediger, Batista, & Takahashi, 2005). Notably, people with REM sleep behavior disorders have a lower sense of smell (Stiasny-Kolster, Clever, Möller, Oertel, & Mayer, 2007).

New data support the theory that $A_{2A}R$ antagonists can alter the sleep-wake cycle, which is consistent with the idea that the arousal effect of caffeine in mice is mediated exclusively by this receptor. For example, the recently developed dual $A_{2A}R/A_1R$ antagonist JNJ-40255293 increases wakefulness in rats in a dose-dependent manner (Atack et al., 2014). A report on four patients revealed that evening use of the $A_{2A}R$ antagonist istradefylline (also known as KW-6002), which has been approved for the management of motor fluctuations in Japan (treatment of 'wearing-off' phenomenon) and the USA (treatment of 'OFF' episodes) (Jenner, Mori, Aradi, & Hauser, 2021), shortened nighttime sleep duration and made them sleepier during the day (Matsuura & Tomimoto, 2015). Therefore, selective $A_{2A}R$ antagonists may have significant potential as eugeroics (drugs that promote wakefulness) while avoiding some of the severe side effects of coffee (e.g., anxiety) or adverse effects of other psychostimulants (e.g., dependence).

## 3.4 Natural $A_{2A}R$ agonists and sleep

While there are currently no approved drugs that target adenosine receptors for treating sleep disorders, a range of natural substances that activate adenosine receptors induce sleep. Supplementation with Japanese sake yeast improves sleep quality in humans and mice (Monoi et al., 2016; Nakamura et al., 2016; Nishimon, Sakai, & Nishino, 2021). At the same time, pretreatment with the $A_{2A}R$ antagonist ZM 241385 abolishes sake yeast-induced SWS in mice, providing evidence for the involvement of $A_{2A}Rs$ in promoting sleep. Since sake yeast contains a significant amount of S-adenosyl-L-methionine and its metabolite methylthioadenosine, but other *Saccharomyces cerevisiae* yeasts (such as baker's and brewer's yeast) do not, the ability of sake yeast to induce sleep is most likely due to the activation of $A_{2A}R$ by these substances (Nakamura et al., 2016). N6-(4-hydroxybenzyl) adenine riboside isolated from *Gastrodia elata* also exhibits hypnotic effects in mice (Zhang et al., 2012) and may enhance SWS through processes involving $A_1R$ and $A_{2A}R$ in a dose-dependent manner. Cordycepin (3-deoxyadenosine), an adenosine analog derived from *Cordyceps* fungus, improves SWS in rats, although it is unclear whether this is due to the activation of adenosine receptors (Hu et al., 2013).

## 3.5 Allosteric modulation of adenosine $A_{2A}Rs$ and sleep

The most common method of activating receptors in biochemistry and pharmacology involves targeting orthosteric regions with endogenous ligands, agonists, or antagonists. Research also shows that small compounds that bind to an allosteric site different from where the endogenous ligand, agonist, or antagonist would bind can alter receptor activity (Ritter et al., 2020). Unlike endogenous ligands, agonists, or antagonists, an allosteric modulator does not activate or inactivate receptors but modifies the receptor's response to orthosteric substrates (Korkutata et al., 2017, 2019; Korkutata, Agrawal, & Lazarus, 2022; Lin et al., 2023).

The oldest known allosterically regulated GPCRs are adenosine receptors. Early studies showed that amiloride and its analogs are allosteric $A_{2A}R$ inhibitors (Gao, Kim, Ijzerman, & Jacobson, 2005; Göblyös & Ijzerman, 2009; Jacobson & Gao, 2006). Subsequently, studies have discovered numerous allosteric modulators of the $A_{2A}R$ (Korkutata et al., 2022). Although $A_{2A}R$ agonists have potent sleep-inducing effects (Methippara, Kumar, Alam, Szymusiak, & McGinty, 2005; Satoh et al., 1999; Scammell et al., 2001; Urade et al., 2003), they also have negative cardiovascular consequences and cannot be utilized in clinical sleep medicine. In addition, the development of adenosine analogs for treating CNS disorders, such as insomnia, is impeded by poor drug transport across the blood-brain barrier. In a recent study in mice, the monocarboxylate [3,4-difluoro-2-((2-fluoro-4-iodophenyl)amino)-benzoic acid] ($A_{2A}RPAM-1$), which is permeable to the blood-brain barrier, was found to enhance $A_{2A}R$ signaling in the brain, thereby inducing sleep (Korkutata et al., 2017, 2019; Lin et al., 2023). Surprisingly, $A_{2A}RPAM-1$ did not exhibit the typical adverse cardiovascular and body temperature effects of $A_{2A}R$ agonists. $A_{2A}RPAM-1$

Table 1 Pharmacologic and natural substances that promote sleep or wakefulness via $A_{2A}$Rs.

| Name | Type | Pharmacology | Chemical structure | Physiological effects |
|---|---|---|---|---|
| Adenosine | Natural ligand | Activates $A_{2A}$R signaling pathway | | Induces SWS (Porkka-Heiskanen et al., 1997). |
| CGS 21680 | Selective agonist | Activates $A_{2A}$R signaling pathway | | Induces SWS and REM sleep in mice following injection into the lateral ventricle or subarachnoid space under the ventral surface area of the rostral BF (Satoh et al., 1996, 1998, 1999; Urade et al., 2003). |
| Caffeine | Antagonist/ natural substance | Blocks activation of $A_{2A}$R signaling pathway. | | Increases wakefulness (Fredholm et al., 1999; Huang et al., 2005; Lazarus et al., 2011). |

| Name | Type | Mechanism | Structure | Effect |
|---|---|---|---|---|
| ZM 241385 | Selective antagonist | Blocks activation of $A_{2A}R$ signaling pathway. | | Inhibits SWS (Scharbarg et al., 2016). |
| JNJ-40255293 | Antagonist | Blocks activation of $A_{2A}R$ signaling pathway. | | Induces wakefulness in rats in a dose-dependent manner (Atack et al., 2014). |
| Istradefylline (KW-6002) | Antagonist | Blocks activation of $A_{2A}R$ signaling pathway. | | Reduces the total amount of SWS time in the dark period (Matsuura & Tomimoto, 2015). |
| 3,4-difluoro-2-((2-fluoro-4-iodophenyl)amino) benzoic acid ($A_{2A}$RPAM-1) | Allosteric enhancer/ modulator | Enhances adenosine signaling at $A_{2A}R$ | | Induces SWS in male mice without impacting body temperature or cardiovascular function (Korkutata et al., 2017, 2019; Korkutata, Agrawal, & Lazarus, 2022; Lin et al., 2023). |

*(continued)*

Table 1 Pharmacologic and natural substances that promote sleep or wakefulness via $A_{2A}$Rs. (cont'd)

| Name | Type | Pharmacology | Chemical structure | Physiological effects |
|---|---|---|---|---|
| Sake yeast | Agonist/natural substance | Activates $A_{2A}$R signaling pathway | n/a | Improves sleep quality in humans. The ability of sake yeast to induce sleep is most likely due to the activation of $A_{2A}$R (Monoi et al., 2016; Nakamura et al., 2016; Nishimon, Sakai, & Nishino, 2021). |
| N6-(4-hydroxybenzyl) adenine riboside | Agonist/natural substance | Activates $A_{2A}$R signaling pathway |  | Active ingredient of *Gastrodia elata* rhizomes exhibits hypnotic effects in mice and may enhance SWS through processes involving $A_1$R and $A_{2A}$R (Zhang et al., 2012). |
| Cordycepin (3-deoxyadenosine) | Adenosine analog | Unclear |  | Enhances SWS in rats, however it is unclear whether this is due to adenosine receptor activation (Hu et al., 2013). |

selectively increased $A_{2A}R$ signaling in $A_{2A}R$-expressing Chinese hamster ovary (CHO) cells but not in $A_{2A}R$-deficient CHO cells or the absence of adenosine. The effect of the $A_{2A}R$ agonist CGS 21680's was not affected by the $A_{2A}RPAM$-1 (Mustafa, 2019). $A_{2A}RPAM$-1 was administered intracerebroventricularly and intraperitoneally to induce extended SWS, but not rapid eye movement sleep in wild-type mice but not in $A_{2A}R$ knockout animals. In the study, $A_2ARPAM$-1 was found not to affect body temperature, cardiac function, or blood pressure, unlike $A_{2A}R$ agonists (Korkutata et al., 2019), suggesting that adenosine or $A_{2A}R$ expression levels in the cardiovascular system are not sufficient to trigger an $A_{2A}RPAM$-1 response under normal physiologic conditions. Consequently, agents that allosterically increase $A_{2A}R$ signaling may help people with insomnia to fall asleep more quickly.

## 4. Conclusion

Here, we discussed recent developments regarding the role of adenosine, $A_{2A}R$, and $A_{2A}R$ agonists, antagonists, and allosteric modulators in sleep-wake control. Although numerous $A_{2A}R$ agonists, antagonists, and allosteric modulators have been reported (Table 1), few of them have been tested in sleep-wake regulation. Moreover, some existing $A_{2A}R$ agonists are known to have potent sleep-promoting effects, but their clinical use has been ruled out because of their cardiovascular side effects and poor permeability of the brain-blood barrier. On the other hand, allosteric modulators have a potential therapeutic advantage over conventional agonists or antagonists because they act only where and when the orthosteric ligand is released. Therefore, small compounds that allosterically modulate $A_{2A}Rs$ could help people with sleep problems to fall asleep and thus represent a potential therapy for neurologic disorders.

## References

Adair, T. H. (2005). Growth regulation of the vascular system: An emerging role for adenosine. *American Journal of, 289*(2), R283–R296. https://doi.org/10.1152/ajpregu.00840.2004

Atack, J. R., Shook, B. C., Rassnick, S., Jackson, P. F., Rhodes, K., Drinkenburg, W. H., ... Megens, A. A. H. P. (2014). JNJ-40255293, a novel adenosine A2A/A1 antagonist with efficacy in preclinical models of Parkinson's disease. *ACS Chemical Neuroscience, 5*(10), 1005–1019. https://doi.org/10.1021/cn5001606

Ballarín, M., Fredholm, B. B., Ambrosio, S., & Mahy, N. (1991). Extracellular levels of adenosine and its metabolites in the striatum of awake rats: Inhibition of uptake and metabolism. *Acta Physiologica Scandinavica, 142*(1), 97–103. https://doi.org/10.1111/j.1748-1716.1991.tb09133.x

Benington, J. H., Kodali, S. K., & Heller, H. C. (1995). Stimulation of A1 adenosine receptors mimics the electroencephalographic effects of sleep deprivation. *Brain Research, 692*(1–2), 79–85.

Boison, D., Singer, P., Shen, H.-Y., Feldon, J., & Yee, B. K. (2012). Adenosine hypothesis of schizophrenia—Opportunities for pharmacotherapy. *Neuropharmacology, 62*(3), 1527–1543. https://doi.org/10.1016/j.neuropharm.2011.01.048

Burdakov, D., Luckman, S. M., & Verkhratsky, A. (2005). Glucose-sensing neurons of the hypothalamus. *Philosophical Transactions of the Royal Society of London. Series B, Biological Sciences, 360*(1464), 2227–2235. https://doi.org/10.1098/rstb.2005.1763

Burnstock, G., Cocks, T., Crowe, R., & Kasakov, L. (1978). Purinergic innervation of the guinea-pig urinary bladder. *British Journal of Pharmacology, 63*(1), 125–138. https://doi.org/10.1111/j.1476-5381.1978.tb07782.x

Cacciari, B., Pastorin, G., Bolcato, C., Spalluto, G., Bacilieri, M., & Moro, S. (2005). A2B adenosine receptor antagonists: Recent developments. *Mini Reviews in Medicinal Chemistry, 5*(12), 1053–1060. https://doi.org/10.2174/138955705774933374

Callaway, C. W., & Henriksen, S. J. (1992). Neuronal firing in the nucleus accumbens is associated with the level of cortical arousal. *Neuroscience, 51*(3), 547–553. https://doi.org/10.1016/0306-4522(92)90294-C

Chen, J.-F. (2014). Adenosine receptor control of cognition in normal and disease. *International Review of Neurobiology, 119*, 257–307. https://doi.org/10.1016/B978-0-12-801022-8.00012-X

Cherasse, Y., Aritake, K., Oishi, Y., Kaushik, M. K., Korkutata, M., & Urade, Y. (2018). The leptomeninges produce prostaglandin D2 involved in sleep regulation in mice. *Frontiers in Cellular Neuroscience, 12*, 357. ⟨https://www.frontiersin.org/articles/10.3389/fncel.2018.00357⟩.

Chern, Y., King, K., Lai, H. L., & Lai, H. T. (1992). Molecular cloning of a novel adenosine receptor gene from rat brain. *Biochemical and Biophysical Research Communications, 185*(1), 304–309. https://doi.org/10.1016/s0006-291x(05)90000-4

Cui, G., Jun, S. B., Jin, X., Pham, M. D., Vogel, S. S., Lovinger, D. M., & Costa, R. M. (2013). Concurrent activation of striatal direct and indirect pathways during action initiation. *Nature, 494*(7436), 238–242. https://doi.org/10.1038/nature11846

Cunha, R. A. (2001). Adenosine as a neuromodulator and as a homeostatic regulator in the nervous system: Different roles, different sources and different receptors. *Neurochemistry International, 38*(2), 107–125. https://doi.org/10.1016/s0197-0186(00)00034-6

de Lera Ruiz, M., Lim, Y.-H., & Zheng, J. (2014). Adenosine A2A receptor as a drug discovery target. *Journal of Medicinal Chemistry, 57*(9), 3623–3650. https://doi.org/10.1021/jm4011669

Drury, A. N., & Szent-Györgyi, A. (1929). The physiological activity of adenine compounds with especial reference to their action upon the mammalian heart. *The Journal of Physiology, 68*(3), 213–237.

Dunwiddie, T. V., & Worth, T. (1982). Sedative and anticonvulsant effects of adenosine analogs in mouse and rat. *The Journal of Pharmacology and Experimental Therapeutics, 220*(1), 70–76.

Feldberg, W., & Sherwood, S. L. (1954). Injections of drugs into the lateral ventricle of the cat. *The Journal of Physiology, 123*(1), 148–167.

Feoktistov, I., Biaggioni, I., & Cronstein, B. N. (2009). Adenosine receptors in wound healing, fibrosis and angiogenesis. *Handbook of Experimental Pharmacology, 193*, 383–397. https://doi.org/10.1007/978-3-540-89615-9_13

Ferré, S., Goldberg, S. R., Lluis, C., & Franco, R. (2009). Looking for the role of cannabinoid receptor heteromers in striatal function. *Neuropharmacology, 56*(Suppl 1), 226–234. https://doi.org/10.1016/j.neuropharm.2008.06.076

Ferré, S., Karcz-Kubicha, M., Hope, B. T., Popoli, P., Burgueño, J., Gutiérrez, M. A., ... Ciruela, F. (2002). Synergistic interaction between adenosine A2A and glutamate mGlu5 receptors: Implications for striatal neuronal function. *Proceedings of the National Academy of Sciences of the United States of America, 99*(18), 11940–11945. https://doi.org/10.1073/pnas.172393799

Fredholm, B. B. (2007). Adenosine, an endogenous distress signal, modulates tissue damage and repair. *Cell Death and Differentiation, 14*(7), 1315–1323. https://doi.org/10.1038/sj.cdd.4402132

Fredholm, B. B., Bättig, K., Holmén, J., Nehlig, A., & Zvartau, E. E. (1999). Actions of caffeine in the brain with special reference to factors that contribute to its widespread use. *Pharmacological Reviews, 51*(1), 83–133.

Fredholm, B. B., Chen, J.-F., Cunha, R. A., Svenningsson, P., & Vaugeois, J.-M. (2005). Adenosine and brain function. *International Review of Neurobiology, 63*, 191–270. https://doi.org/10.1016/S0074-7742(05)63007-3

Fredholm, B. B., IJzerman, A. P., Jacobson, K. A., Linden, J., & Müller, C. E. (2011). International Union of Basic and Clinical Pharmacology. LXXXI. Nomenclature and classification of adenosine receptors—An update. *Pharmacological Reviews, 63*(1), 1–34. https://doi.org/10.1124/pr.110.003285

Fuller, P. M., Yamanaka, A., & Lazarus, M. (2015). How genetically engineered systems are helping to define, and in some cases redefine, the neurobiological basis of sleep and wake. *Temperature, 2*(3), 406–417. https://doi.org/10.1080/23328940.2015.1075095

Furlong, T. J., Pierce, K. D., Selbie, L. A., & Shine, J. (1992). Molecular characterization of a human brain adenosine A2 receptor. *Brain, 15*(1–2), 62–66. https://doi.org/10.1016/0169-328x(92)90152-2

Fuxe, K., Ferré, S., Canals, M., Torvinen, M., Terasmaa, A., Marcellino, D., ... Franco, R. (2005). Adenosine A2A and dopamine D2 heteromeric receptor complexes and their function. *Journal of Molecular Neuroscience: MN, 26*(2–3), 209–220. https://doi.org/10.1385/JMN:26:2-3:209

Gallopin, T., Fort, P., Eggermann, E., Cauli, B., Luppi, P. H., Rossier, J., ... Serafin, M. (2000). Identification of sleep-promoting neurons in vitro. *Nature, 404*(6781), 992–995. https://doi.org/10.1038/35010109

Gallopin, T., Luppi, P.-H., Cauli, B., Urade, Y., Rossier, J., Hayaishi, O., ... Fort, P. (2005). The endogenous somnogen adenosine excites a subset of sleep-promoting neurons via A2A receptors in the ventrolateral preoptic nucleus. *Neuroscience, 134*(4), 1377–1390. https://doi.org/10.1016/j.neuroscience.2005.05.045

Gao, Z.-G., Kim, S.-K., Ijzerman, A. P., & Jacobson, K. A. (2005). Allosteric modulation of the adenosine family of receptors. *Mini Reviews in Medicinal Chemistry, 5*(6), 545–553. https://doi.org/10.2174/1389557054023242

García-García, F., Acosta-Peña, E., Venebra-Muñoz, A., & Murillo-Rodríguez, E. (2009). Sleep-inducing factors. *CNS & Neurological Disorders Drug Targets, 8*(4), 235–244. https://doi.org/10.2174/187152709788921672

Göblyös, A., & Ijzerman, A. P. (2009). Allosteric modulation of adenosine receptors. *Purinergic Signalling, 5*(1), 51–61. https://doi.org/10.1007/s11302-008-9105-3

Grillner, S., & Robertson, B. (2016). The basal ganglia over 500 million years. *Current Biology: CB, 26*(20), R1088–R1100. https://doi.org/10.1016/j.cub.2016.06.041

Haulică, I., Ababei, L., Brănişteanu, D., & Topoliceanu, F. (1973). Letter: Preliminary data on the possible hypnogenic role of adenosine. *Journal of Neurochemistry, 21*(4), 1019–1020. https://doi.org/10.1111/j.1471-4159.1973.tb07549.x

Headrick, J. P., Ashton, K. J., Rose'meyer, R. B., & Peart, J. N. (2013). Cardiovascular adenosine receptors: Expression, actions and interactions. *Pharmacology & Therapeutics, 140*(1), 92–111. https://doi.org/10.1016/j.pharmthera.2013.06.002

Hein, T. W., Wang, W., Zoghi, B., Muthuchamy, M., & Kuo, L. (2001). Functional and molecular characterization of receptor subtypes mediating coronary microvascular dilation to adenosine. *Journal of Molecular and Cellular Cardiology, 33*(2), 271–282. https://doi.org/10.1006/jmcc.2000.1298

Hong, Z.-Y., Huang, Z.-L., Qu, W.-M., Eguchi, N., Urade, Y., & Hayaishi, O. (2005). An adenosine A receptor agonist induces sleep by increasing GABA release in the tuberomammillary nucleus to inhibit histaminergic systems in rats. *Journal of Neurochemistry, 92*(6), 1542–1549. https://doi.org/10.1111/j.1471-4159.2004.02991.x

Hu, Z., Lee, C.-I., Shah, V. K., Oh, E.-H., Han, J.-Y., Bae, J.-R., ... Oh, K.-W. (2013). Cordycepin increases nonrapid eye movement sleep via adenosine receptors in rats. *Evidence-Based Complementary and Alternative Medicine: ECAM, 2013*, 840134. https://doi.org/10.1155/2013/840134

Huang, Z.-L., Qu, W.-M., Eguchi, N., Chen, J.-F., Schwarzschild, M. A., Fredholm, B. B., ... Hayaishi, O. (2005). Adenosine A2A, but not A1, receptors mediate the arousal effect of caffeine. *Nature Neuroscience, 8*(7), 858–859. https://doi.org/10.1038/nn1491

Huitron-Resendiz, S., Kristensen, M. P., Sánchez-Alavez, M., Clark, S. D., Grupke, S. L., Tyler, C., ... de Lecea, L. (2005). Urotensin II modulates rapid eye movement sleep through activation of brainstem cholinergic neurons. *The Journal of Neuroscience: The Official Journal of the Society for Neuroscience, 25*(23), 5465–5474. https://doi.org/10.1523/JNEUROSCI.4501-04.2005

Huston, J. P., Haas, H. L., Boix, F., Pfister, M., Decking, U., Schrader, J., & Schwarting, R. K. (1996). Extracellular adenosine levels in neostriatum and hippocampus during rest and activity periods of rats. *Neuroscience, 73*(1), 99–107. https://doi.org/10.1016/0306-4522(96)00021-8

Inoué, S., Honda, K., & Komoda, Y. (1995). Sleep as neuronal detoxification and restitution. *Behavioural Brain Research, 69*(1–2), 91–96. https://doi.org/10.1016/0166-4328(95)00014-k

Ishimori, K. (1909). True cause of sleep: A hypnogenic substance as evidenced in the brain of sleep-deprived animals. *Tokyo Igakkai Zasshi, 23*, 429–457.

Jacobson, K. A., & Gao, Z.-G. (2006). Adenosine receptors as therapeutic targets. *Nature Reviews. Drug Discovery, 5*(3), 247–264. https://doi.org/10.1038/nrd1983

Jenner, P., Mori, A., Aradi, S. D., & Hauser, R. A. (2021). Istradefylline—A first generation adenosine A2A antagonist for the treatment of Parkinson's disease. *Expert Review of Neurotherapeutics, 21*(3), 317–333. https://doi.org/10.1080/14737175.2021.1880896

Kalinchuk, A. V., Urrila, A.-S., Alanko, L., Heiskanen, S., Wigren, H.-K., Suomela, M., ... Porkka-Heiskanen, T. (2003). Local energy depletion in the basal forebrain increases sleep. *The European Journal of Neuroscience, 17*(4), 863–869.

Korkutata, M., Agrawal, L., & Lazarus, M. (2022). Allosteric modulation of adenosine A2A receptors as a new therapeutic avenue. *International Journal of Molecular Sciences, 23*(4), 2101. https://doi.org/10.3390/ijms23042101

Korkutata, M., Saitoh, T., Feng, D., Murakoshi, N., Sugiyama, F., Cherasse, Y., ... Lazarus, M. (2017). Allosteric modulation of adenosine A2A receptors in mice induces slow-wave sleep without cardiovascular effects. *Sleep Medicine, 40*, e181. https://doi.org/10.1016/j.sleep.2017.11.530

Korkutata, M., Saitoh, T., Cherasse, Y., Ioka, S., Duo, F., Qin, R., ... Lazarus, M. (2019). Enhancing endogenous adenosine A2A receptor signaling induces slow-wave sleep without affecting body temperature and cardiovascular function. *Neuropharmacology, 144*, 122–132. https://doi.org/10.1016/j.neuropharm.2018.10.022

Krueger, J. M., & Majde, J. A. (2003). Humoral links between sleep and the immune system: Research issues. *Annals of the New York Academy of Sciences, 992*, 9–20. https://doi.org/10.1111/j.1749-6632.2003.tb03133.x

Krueger, J. M., Walter, J., Dinarello, C. A., Wolff, S. M., & Chedid, L. (1984). Sleep-promoting effects of endogenous pyrogen (interleukin-1). *The American Journal of Physiology, 246*(6 Pt 2), R994–R999. https://doi.org/10.1152/ajpregu.1984.246.6.R994

Krueger, J. M., Obál, F. J., Fang, J., Kubota, T., & Taishi, P. (2001). The role of cytokines in physiological sleep regulation. *Annals of the New York Academy of Sciences, 933*, 211–221. https://doi.org/10.1111/j.1749-6632.2001.tb05826.x

Kubota, K. (1989). Kuniomi Ishimori and the first discovery of sleep-inducing substances in the brain. *Neuroscience Research, 6*(6), 497–518. https://doi.org/10.1016/0168-0102(89)90041-2

Kull, B., Svenningsson, P., & Fredholm, B. B. (2000). Adenosine A2A receptors are colocalized with and activate golf in rat striatum. *Molecular Pharmacology, 58*(4), 771–777. https://doi.org/10.1124/mol.58.4.771

Lazarus, M., Chen, J.-F., Huang, Z.-L., Urade, Y., & Fredholm, B. B. (2019). Adenosine and sleep. *Handbook of Experimental Pharmacology, 253*, 359–381. https://doi.org/10.1007/164_2017_36

Lazarus, M., Yoshida, K., Coppari, R., Bass, C. E., Mochizuki, T., Lowell, B. B., & Saper, C. B. (2007). EP3 prostaglandin receptors in the median preoptic nucleus are critical for fever responses. *Nature Neuroscience, 10*(9), 1131–1133. https://doi.org/10.1038/nn1949

Lazarus, M., Shen, H.-Y., Cherasse, Y., Qu, W.-M., Huang, Z.-L., Bass, C. E., ... Chen, J.-F. (2011). Arousal effect of caffeine depends on adenosine A2A receptors in the shell of the nucleus accumbens. *The Journal of Neuroscience: The Official Journal of the Society for Neuroscience, 31*(27), 10067–10075. https://doi.org/10.1523/JNEUROSCI.6730-10.2011

Ledent, C., Vaugeois, J. M., Schiffmann, S. N., Pedrazzini, T., El Yacoubi, M., Vanderhaeghen, J. J., ... Parmentier, M. (1997). Aggressiveness, hypoalgesia and high blood pressure in mice lacking the adenosine A2a receptor. *Nature, 388*(6643), 674–678. https://doi.org/10.1038/41771

Legendre, R., & Piéron, H. (1912). *Recherches sur le besoin de sommeil consécutif à une veille prolongée.* Verlag von Gustav Fischer.

Li, R., Wang, Y.-Q., Liu, W.-Y., Zhang, M.-Q., Li, L., Cherasse, Y., ... Huang, Z.-L. (2020). Activation of adenosine A2A receptors in the olfactory tubercle promotes sleep in rodents. *Neuropharmacology, 168*, 107923. https://doi.org/10.1016/j.neuropharm.2019.107923

Lin, Y., Roy, K., Ioka, S., Otani, R., Amezawa, M., Ishikawa, Y., ... Lazarus, M. (2023). Positive allosteric adenosine A2A receptor modulation suppresses insomnia associated with mania and schizophrenia like behaviors in mice. *Front. Pharmacol. 14*, 1138666. https://doi.org/10.3389/fphar.2023.1138666

Lovatt, D., Xu, Q., Liu, W., Takano, T., Smith, N. A., Schnermann, J., ... Nedergaard, M. (2012). Neuronal adenosine release, and not astrocytic ATP release, mediates feedback inhibition of excitatory activity. *Proceedings of the National Academy of Sciences of the United States of America, 109*(16), 6265–6270. https://doi.org/10.1073/pnas.1120997109

Luo, Y.-J., Li, Y.-D., Wang, L., Yang, S.-R., Yuan, X.-S., Wang, J., ... Huang, Z.-L. (2018). Nucleus accumbens controls wakefulness by a subpopulation of neurons expressing dopamine D1 receptors. *Nature Communications, 9*(1), 1576. https://doi.org/10.1038/s41467-018-03889-3

Ma, W., Yuan, P., Zhang, H., Kong, L., Lazarus, M., Qu, W., ... Huang, Z. (2023). Adenosine and P1 receptors: key targets in the regulation of sleep, torpor, and hibernation. *Frontiers in Pharmacology, 14*, 2023. https://doi.org/10.3389/fphar.2023.1098976

Maenhaut, C., Van Sande, J., Libert, F., Abramowicz, M., Parmentier, M., Vanderhaegen, J. J., ... Schiffmann, S. (1990). RDC8 codes for an adenosine A2 receptor with physiological constitutive activity. *Biochemical and Biophysical Research Communications, 173*(3), 1169–1178. https://doi.org/10.1016/s0006-291x(05)80909-x

Marks, G. A., Shaffery, J. P., Speciale, S. G., & Birabil, C. G. (2003). Enhancement of rapid eye movement sleep in the rat by actions at A1 and A2a adenosine receptor subtypes with a differential sensitivity to atropine. *Neuroscience, 116*(3), 913–920. https://doi.org/10.1016/s0306-4522(02)00561-4

Marley, E., & Nistico, G. (1972). Effects of catecholamines and adenosine derivatives given into the brain of fowls. *British Journal of Pharmacology, 46*(4), 619–636. https://doi.org/10.1111/j.1476-5381.1972.tb06888.x

Matsumoto, T., Tostes, R. C., & Webb, R. C. (2012). Alterations in vasoconstrictor responses to the endothelium-derived contracting factor uridine adenosine tetraphosphate are region specific in DOCA-salt hypertensive rats. *Pharmacological Research, 65*(1), 81–90. https://doi.org/10.1016/j.phrs.2011.09.005

Matsuura, K., & Tomimoto, H. (2015). Istradefylline is recommended for morning use: A report of 4 cases. *Internal Medicine (Tokyo, Japan), 54*(5), 509–511. https://doi.org/10.2169/internalmedicine.54.3522

Mena-Segovia, J., Cintra, L., Próspero-García, O., & Giordano, M. (2002). Changes in sleep-waking cycle after striatal excitotoxic lesions. *Behavioural Brain Research, 136*(2), 475–481. https://doi.org/10.1016/s0166-4328(02)00201-2

Meng, F., Xie, G. X., Chalmers, D., Morgan, C., Watson, S. J., & Akil, H. (1994). Cloning and expression of the A2a adenosine receptor from guinea pig brain. *Neurochemical Research, 19*(5), 613–621. https://doi.org/10.1007/BF00971338

Methippara, M. M., Kumar, S., Alam, M. N., Szymusiak, R., & McGinty, D. (2005). Effects on sleep of microdialysis of adenosine A1 and A2a receptor analogs into the lateral preoptic area of rats. *American Journal of, 289*(6), R1715–R1723. https://doi.org/10.1152/ajpregu.00247.2005

Monoi, N., Matsuno, A., Nagamori, Y., Kimura, E., Nakamura, Y., Oka, K., ... Urade, Y. (2016). Japanese sake yeast supplementation improves the quality of sleep: A double-blind randomised controlled clinical trial. *Journal of Sleep Research, 25*(1), 116–123. https://doi.org/10.1111/jsr.12336

Moore, J. T., Chen, J., Han, B., Meng, Q. C., Veasey, S. C., Beck, S. G., & Kelz, M. B. (2012). Direct activation of sleep-promoting VLPO neurons by volatile anesthetics contributes to anesthetic hypnosis. *Current Biology: CB, 22*(21), 2008–2016. https://doi.org/10.1016/j.cub.2012.08.042

Mullington, J., Korth, C., Hermann, D. M., Orth, A., Galanos, C., Holsboer, F., & Pollmächer, T. (2000). Dose-dependent effects of endotoxin on human sleep. *American Journal of Physiology. Regulatory, Integrative and Comparative Physiology, 278*(4), R947–R955. https://doi.org/10.1152/ajpregu.2000.278.4.R947

Mullington, J. M., Hinze-Selch, D., & Pollmächer, T. (2001). Mediators of inflammation and their interaction with sleep: Relevance for chronic fatigue syndrome and related conditions. *Annals of the New York Academy of Sciences, 933*, 201–210. https://doi.org/10.1111/j.1749-6632.2001.tb05825.x

Mustafa, K. (2019). *A potential treatment for insomnia by positive allosteric modulation of adenosine A2A receptors* (Ph.D. thesis, 筑波大学 University of Tsukuba). ⟨https://ci.nii.ac.jp/naid/500001352635/⟩.

Nakamura, Y., Midorikawa, T., Monoi, N., Kimura, E., Murata-Matsuno, A., Sano, T., ... Urade, Y. (2016). Oral administration of Japanese sake yeast (*Saccharomyces cerevisiae* sake) promotes non-rapid eye movement sleep in mice via adenosine A2A receptors. *Journal of Sleep Research, 25*(6), 746–753. https://doi.org/10.1111/jsr.12434

Navarro, G., Carriba, P., Gandía, J., Ciruela, F., Casadó, V., Cortés, A., ... Franco, R. (2008). Detection of heteromers formed by cannabinoid CB1, dopamine D2, and adenosine A2A G-protein-coupled receptors by combining bimolecular fluorescence complementation and bioluminescence energy transfer. *The Scientific World Journal, 8*, 1088–1097. https://doi.org/10.1100/tsw.2008.136

Nishimon, S., Sakai, N., & Nishino, S. (2021). Sake yeast induces the sleep-promoting effects under the stress-induced acute insomnia in mice. *Sci Rep, 11*(1), 20816. https://doi.org/10.1038/s41598-021-00271-0. PMID: 34675261; PMCID: PMC8531297.

Ohta, A., & Sitkovsky, M. (2001). Role of G-protein-coupled adenosine receptors in downregulation of inflammation and protection from tissue damage. Article 6866 *Nature, 414*(6866), https://doi.org/10.1038/414916a

Oishi, Y., & Lazarus, M. (2017). The control of sleep and wakefulness by mesolimbic dopamine systems. *Neuroscience Research, 118*, 66–73. https://doi.org/10.1016/j.neures.2017.04.008

Oishi, Y., Huang, Z.-L., Fredholm, B. B., Urade, Y., & Hayaishi, O. (2008). Adenosine in the tuberomammillary nucleus inhibits the histaminergic system via A1 receptors and promotes non-rapid eye movement sleep. *Proceedings of the National Academy of Sciences of the United States of America, 105*(50), 19992–19997. https://doi.org/10.1073/pnas.0810926105

Oishi, Y., Yoshida, K., Scammell, T. E., Urade, Y., Lazarus, M., & Saper, C. B. (2015). The roles of prostaglandin E2 and D2 in lipopolysaccharide-mediated changes in sleep. *Brain, Behavior, and Immunity, 47*, 172–177. https://doi.org/10.1016/j.bbi.2014.11.019

Oishi, Y., Xu, Q., Wang, L., Zhang, B.-J., Takahashi, K., Takata, Y., ... Lazarus, M. (2017). Slow-wave sleep is controlled by a subset of nucleus accumbens core neurons in mice. *Nature Communications, 8*(1), 734. https://doi.org/10.1038/s41467-017-00781-4

Osaka, T., & Matsumura, H. (1995). Noradrenaline inhibits preoptic sleep-active neurons through alpha 2-receptors in the rat. *Neuroscience Research, 21*(4), 323–330. https://doi.org/10.1016/0168-0102(94)00871-c

Paes-De-Carvalho, R. (2002). Adenosine as a signaling molecule in the retina: Biochemical and developmental aspects. *Anais da Academia Brasileira de Ciencias, 74*(3), 437–451. https://doi.org/10.1590/s0001-37652002000300007

Parkinson, F. E., Damaraju, V. L., Graham, K., Yao, S. Y. M., Baldwin, S. A., Cass, C. E., & Young, J. D. (2011). Molecular biology of nucleoside transporters and their distributions and functions in the brain. *Current Topics in Medicinal Chemistry, 11*(8), 948–972. https://doi.org/10.2174/156802611795347582

Peng, W., Wu, Z., Song, K., Zhang, S., Li, Y., & Xu, M. (2020). Regulation of sleep homeostasis mediator adenosine by basal forebrain glutamatergic neurons. *Science (New York, N. Y.), 369*(6508), eabb0556. https://doi.org/10.1126/science.abb0556

Petit, J.-M., Burlet-Godinot, S., Magistretti, P. J., & Allaman, I. (2015). Glycogen metabolism and the homeostatic regulation of sleep. *Metabolic Brain Disease, 30*(1), 263–279. https://doi.org/10.1007/s11011-014-9629-x

Porkka-Heiskanen, T., Strecker, R. E., & McCarley, R. W. (2000). Brain site-specificity of extracellular adenosine concentration changes during sleep deprivation and spontaneous sleep: An in vivo microdialysis study. *Neuroscience, 99*(3), 507–517.

Porkka-Heiskanen, T., Strecker, R. E., Thakkar, M., Bjorkum, A. A., Greene, R. W., & McCarley, R. W. (1997). Adenosine: A mediator of the sleep-inducing effects of prolonged wakefulness. *Science (New York, N. Y.), 276*(5316), 1265–1268. https://doi.org/10.1126/science.276.5316.1265

Prediger, R. D. S., Batista, L. C., & Takahashi, R. N. (2005). Caffeine reverses age-related deficits in olfactory discrimination and social recognition memory in rats. Involvement of adenosine A1 and A2A receptor. *Neurobiology of Aging, 26*(6), 957–964. https://doi.org/10.1016/j.neurobiolaging.2004.08.012

Pull, I., & McIlwain, H. (1972). Metabolism of [14C]adenine and derivatives by cerebral tissues, superfused and electrically stimulated. *Biochemical Journal, 126*(4), 965–973.

Qu, W.-M., Huang, Z.-L., Xu, X.-H., Aritake, K., Eguchi, N., Nambu, F., ... Hayaishi, O. (2006). Lipocalin-type prostaglandin D synthase produces prostaglandin D2 involved in regulation of physiological sleep. *Proceedings of the National Academy of Sciences of the United States of America, 103*(47), 17949–17954. https://doi.org/10.1073/pnas.0608581103

Radulovacki, M., Virus, R. M., Djuricic-Nedelson, M., & Green, R. D. (1984). Adenosine analogs and sleep in rats. *The Journal of Pharmacology and Experimental Therapeutics, 228*(2), 268–274.

Radulovacki, M., Virus, R. M., Rapoza, D., & Crane, R. A. (1985). A comparison of the dose response effects of pyrimidine ribonucleosides and adenosine on sleep in rats. *Psychopharmacology, 87*(2), 136–140. https://doi.org/10.1007/BF00431796

Resta, R., Yamashita, Y., & Thompson, L. F. (1998). Ecto-enzyme and signaling functions of lymphocyte CD73. *Immunological Reviews, 161*, 95–109.

Rosin, D. L., Robeva, A., Woodard, R. L., Guyenet, P. G., & Linden, J. (1998). Immunohistochemical localization of adenosine A2A receptors in the rat central nervous system. *The Journal of Comparative Neurology, 401*(2), 163–186.

Saper, C. B., Scammell, T. E., & Lu, J. (2005). Hypothalamic regulation of sleep and circadian rhythms. *Nature, 437*(7063), 1257–1263. https://doi.org/10.1038/nature04284

Saper, C. B., Fuller, P. M., Pedersen, N. P., Lu, J., & Scammell, T. E. (2010). Sleep state switching. *Neuron, 68*(6), 1023–1042. https://doi.org/10.1016/j.neuron.2010.11.032

Satoh, S., Matsumura, H., & Hayaishi, O. (1998). Involvement of adenosine A2A receptor in sleep promotion. *European Journal of Pharmacology, 351*(2), 155–162.

Satoh, S., Matsumura, H., Suzuki, F., & Hayaishi, O. (1996). Promotion of sleep mediated by the A2a-adenosine receptor and possible involvement of this receptor in the sleep induced by prostaglandin D2 in rats. *Proceedings of the National Academy of Sciences of the United States of America, 93*(12), 5980–5984.

Satoh, S., Matsumura, H., Koike, N., Tokunaga, Y., Maeda, T., & Hayaishi, O. (1999). Region-dependent difference in the sleep-promoting potency of an adenosine A2A receptor agonist. *The European Journal of Neuroscience, 11*(5), 1587–1597.

Scammell, T. E., Gerashchenko, D. Y., Mochizuki, T., McCarthy, M. T., Estabrooke, I. V., Sears, C. A., ... Hayaishi, O. (2001). An adenosine A2a agonist increases sleep and induces Fos in ventrolateral preoptic neurons. *Neuroscience, 107*(4), 653–663.

Scharbarg, E., Daenens, M., Lemaître, F., Geoffroy, H., Guille-Collignon, M., Gallopin, T., & Rancillac, A. (2016). Astrocyte-derived adenosine is central to the hypnogenic effect of glucose. *Scientific Reports, 6*(1), 19107. https://doi.org/10.1038/srep19107

Scharf, M. T., Naidoo, N., Zimmerman, J. E., & Pack, A. I. (2008). The energy hypothesis of sleep revisited. *Progress in Neurobiology, 86*(3), 264–280. https://doi.org/10.1016/j.pneurobio.2008.08.003

Schrader, J. (1983). *Metabolism of adenosine and sites of production in the heart. Regulatory function of adenosine.* Boston, MA: Springer, 133–156. 〈https://doi.org/10.1007/978-1-4613-3909-0_9〉.

Schulte, G., & Fredholm, B. B. (2000). Human adenosine A(1), A(2A), A(2B), and A(3) receptors expressed in Chinese hamster ovary cells all mediate the phosphorylation of extracellular-regulated kinase 1/2. *Molecular Pharmacology, 58*(3), 477–482.

Sherin, J. E., Shiromani, P. J., McCarley, R. W., & Saper, C. B. (1996). Activation of ventrolateral preoptic neurons during sleep. *Science (New York, N. Y.), 271*(5246), 216–219.

Sherin, J. E., Elmquist, J. K., Torrealba, F., & Saper, C. B. (1998). Innervation of histaminergic tuberomammillary neurons by GABAergic and galaninergic neurons in the ventrolateral preoptic nucleus of the rat. *The Journal of Neuroscience: The Official Journal of the Society for Neuroscience, 18*(12), 4705–4721.

Yasugaki, S., Hayashi, Y., & Lazarus, M. (2023). NREM-REM sleep regulation. *Clete A. Kushida, Encyclopedia of Sleep and Circadian Rhythms* (pp. 128–136). Academic Press. https://doi.org/10.1016/B978-0-12-822963-7.00229-2. Second Edition.

Stiasny-Kolster, K., Clever, S.-C., Möller, J. C., Oertel, W. H., & Mayer, G. (2007). Olfactory dysfunction in patients with narcolepsy with and without REM sleep behaviour disorder. *Brain: A Journal of Neurology, 130*(Pt 2), 442–449. https://doi.org/10.1093/brain/awl343

Studer, F. E., Fedele, D. E., Marowsky, A., Schwerdel, C., Wernli, K., Vogt, K., ... Boison, D. (2006). Shift of adenosine kinase expression from neurons to astrocytes during postnatal development suggests dual functionality of the enzyme. *Neuroscience, 142*(1), 125–137. https://doi.org/10.1016/j.neuroscience.2006.06.016

Svenningsson, P., Fourreau, L., Bloch, B., Fredholm, B. B., Gonon, F., & Le, C. M. (1999). Opposite tonic modulation of dopamine and adenosine on c-fos gene expression in striatopallidal neurons. *Neuroscience, 89*(3), 827–837. https://doi.org/10.1016/S0306-4522(98)00403-5

Tellez, L. A., Perez, I. O., Simon, S. A., & Gutierrez, R. (2012). Transitions between sleep and feeding states in rat ventral striatum neurons. *Journal of Neurophysiology, 108*(6), 1739–1751. https://doi.org/10.1152/jn.00394.2012

Ticho, S. R., & Radulovacki, M. (1991). Role of adenosine in sleep and temperature regulation in the preoptic area of rats. *Pharmacology, Biochemistry, and Behavior, 40*(1), 33–40. https://doi.org/10.1016/0091-3057(91)90317-U

Tobler, I., & Scherschlicht, R. (1990). Sleep and EEG slow-wave activity in the domestic cat: Effect of sleep deprivation. *Behavioural Brain Research, 37*(2), 109–118. https://doi.org/10.1016/0166-4328(90)90086-t

Torvinen, M., Marcellino, D., Canals, M., Agnati, L. F., Lluis, C., Franco, R., & Fuxe, K. (2005). Adenosine A2A receptor and dopamine D3 receptor interactions: Evidence of functional A2A/D3 heteromeric complexes. *Molecular Pharmacology, 67*(2), 400–407. https://doi.org/10.1124/mol.104.003376

Ueno, R., Ishikawa, Y., Nakayama, T., & Hayaishi, O. (1982). Prostaglandin D2 induces sleep when microinjected into the preoptic area of conscious rats. *Biochemical and Biophysical Research Communications, 109*(2), 576–582.

Urade, Y., & Hayaishi, O. (2011). Prostaglandin D2 and sleep/wake regulation. *Sleep Medicine Reviews, 15*(6), 411–418. https://doi.org/10.1016/j.smrv.2011.08.003

Urade, Y., & Lazarus, M. (2013). Prostaglandin D2in the regulation of sleep. In M. Tafti, M. J. Thorpy, & P. Shaw (Eds.). *The genetic basis of sleep and sleep disorders* (pp. 73–83). Cambridge University Press. https://doi.org/10.1017/CBO9781139649469.010

Urade, Y., Eguchi, N., Qu, W.-M., Sakata, M., Huang, Z.-L., Chen, J.-F., ... Hayaishi, O. (2003). Sleep regulation in adenosine A2A receptor-deficient mice. *Neurology, 61*(11 Suppl 6), S94–S96. https://doi.org/10.1212/01.wnl.0000095222.41066.5e

Ushikubi, F., Segi, E., Sugimoto, Y., Murata, T., Matsuoka, T., Kobayashi, T., ... Narumiya, S. (1998). Impaired febrile response in mice lacking the prostaglandin E receptor subtype EP3. *Nature, 395*(6699), 281–284. https://doi.org/10.1038/26233

Van Wylen, D. G., Park, T. S., Rubio, R., & Berne, R. M. (1986). Increases in cerebral interstitial fluid adenosine concentration during hypoxia, local potassium infusion, and ischemia. *Journal of Cerebral Blood Flow and Metabolism: Official Journal of the International Society of Cerebral Blood Flow and Metabolism, 6*(5), 522–528. https://doi.org/10.1038/jcbfm.1986.97

Varin, C., Rancillac, A., Geoffroy, H., Arthaud, S., Fort, P., & Gallopin, T. (2015). Glucose induces slow-wave sleep by exciting the sleep-promoting neurons in the ventrolateral preoptic nucleus: A new link between sleep and metabolism. *Journal of Neuroscience, 35*(27), 9900–9911. https://doi.org/10.1523/JNEUROSCI.0609-15.2015

Villablanca, J. (1972). Permanent reduction in sleep after removal of cerebral cortex and striatum in cats. *Brain Research, 36*(2), 463–468. https://doi.org/10.1016/0006-8993(72)90756-1

Wall, M. J., & Dale, N. (2013). Neuronal transporter and astrocytic ATP exocytosis underlie activity-dependent adenosine release in the hippocampus. *The Journal of Physiology, 591*(16), 3853–3871. https://doi.org/10.1113/jphysiol.2013.253450

Wang, Y.-Q., Tu, Z.-C., Xu, X.-Y., Li, R., Qu, W.-M., Urade, Y., & Huang, Z.-L. (2012). Acute administration of fluoxetine normalizes rapid eye movement sleep abnormality, but not depressive behaviors in olfactory bulbectomized rats. *Journal of Neurochemistry, 120*(2), 314–324. https://doi.org/10.1111/j.1471-4159.2011.07558.x

Yuan, X.-S., Wang, L., Dong, H., Qu, W.-M., Yang, S.-R., Cherasse, Y., ... Huang, Z.-L. (2017). Striatal adenosine A2A receptor neurons control active-period sleep via parvalbumin neurons in external globus pallidus. *ELife, 6*, e29055. https://doi.org/10.7554/eLife.29055

Zhang, J., Xu, Q., Yuan, X., Chérasse, Y., Schiffmann, S., De Kerchove D'ExaErde, A., ... Li, R. (2013). Projections of nucleus accumbens adenosine A2A receptor neurons in the mouse brain and their implications in mediating sleep-wake regulation. *Frontiers in Neuroanatomy, 7,* ⟨https://www.frontiersin.org/articles/10.3389/fnana.2013.00043⟩.

Zhang, Y., Li, M., Kang, R.-X., Shi, J.-G., Liu, G.-T., & Zhang, J.-J. (2012). NHBA isolated from *Gastrodia elata* exerts sedative and hypnotic effects in sodium pentobarbital-treated mice. *Pharmacology, Biochemistry, and Behavior, 102*(3), 450–457. https://doi.org/10.1016/j.pbb.2012.06.002

Zhou, X., Oishi, Y., Cherasse, Y., Korkutata, M., Fujii, S., Lee, C.-Y., & Lazarus, M. (2019). Extracellular adenosine and slow-wave sleep are increased after ablation of nucleus accumbens core astrocytes and neurons in mice. *Neurochemistry International, 124*, 256–263. https://doi.org/10.1016/j.neuint.2019.01.020

Zhu, X., Ottenheimer, D., & DiLeone, R. J. (2016). Activity of D1/2 receptor expressing neurons in the nucleus accumbens regulates running, locomotion, and food intake. *Frontiers in Behavioral Neuroscience, 10,* 66. https://doi.org/10.3389/fnbeh.2016.00066

Zimmermann, H. (2000). Extracellular metabolism of ATP and other nucleotides. *Naunyn-Schmiedeberg's Archives of Pharmacology, 362*(4–5), 299–309. https://doi.org/10.1007/s002100000309

CHAPTER EIGHT

# Adenosine $A_{2A}$ signals and dystonia

## Makio Takahashi*
Department of Neurodegenerative disorders, Kansai Medical University, Hirakata, Osaka, Japan
*Corresponding author. e-mail address: ta@kuhp.kyoto-u.ac.jp

## Contents

| | |
|---|---|
| Body | 179 |
| References | 183 |

## Abstract

Dystonia is a movement disorder characterized by sustained or intermittent involuntary muscle contractions, which is also seen in an advanced stage of Parkinson's disease (PD) as camptocormia, torticollis, and Pisa syndrome. Istradefylline, an adenosine $A_{2A}$ receptor antagonist, can be used for the treatment of PD to reduce 'off'-time period, and several clinical studies demonstrated the improvement of camptocormia, which have many similar features to dopa-responsive/non-responsive dystonia. Many animal models of dystonia showed that adenosine $A_{2A}$ receptor colocalized with dopamine D2 positive spiny projection neurons in indirect pathway of basal ganglia circuit, and also in the cholinergic interneurons that affects the balance of indirect and direct pathway of basal ganglia. In this chapter, the potential effect of adenosine $A_{2A}$ antagonism on dystonia was discussed in view of clinical studies of PD with postural abnormalities and the findings of dystonia mouse models.

## Body

Dystonia is a movement disorder characterized by involuntary, sustained, patterned, and often repetitive muscle contractions of opposing muscles, and frequently causing twisting movements and/or abnormal postures (Albanese, Di Giovanni, & Lalli, 2019). The pathophysiology of different types of idiopathic dystonia is not well understood, but there are several evidences that basal ganglia dysfunctions can cause dystonia (Bhatia & Marsden, 1994, Scarduzio, M, 2022). The existence of variation in phenotypic and genotypic subtypes indicates that the pathogenesis of idiopathic dystonia is heterogeneous (Frucht, 2013).

Anti-cholinergic medications have proved useful in patients with generalized and focal isolated dystonia syndromes, whereas levodopa is effective in some patients with dopa-responsive dystonia/parkinsonism. Other agents that have been used in dystonia are baclofen, carbamazepine and benzodiazepines, either alone or in combination, but the effectiveness is still controversial in most cases. Botulinum neurotoxin is the first-line treatment for patients with blepharospasm and cervical dystonia (spasmodic torticollis). It is also effective in laryngeal and limb dystonia. Evidence of the efficacy of bilateral DBS of the GPi on DYT1 dystonia is confirmed by several studies. GPi-DBS has also proved efficacious in acquired dystonia syndromes. Further, a meta-analysis of several studies also showed that STN/GPi-DBS is helpful for camptocormia in Parkinson's disease (PD) cases (Schulz-Schaeffer et al., 2015).

Accumulation of clinical evidences shows that adenosine $A_{2A}$ antagonists exert anti-parkinsonian action (Kanda, Tashiro, Kuwana, & Jenner, 1998), while adenosine receptor agonists have been suggested as new therapeutic agents for hyperkinetic basal ganglia disorders (Ferré et al., 1997). Therefore, the adenosine–dopamine receptor–receptor interactions might provide new therapeutic approaches for basal ganglia disorders. A combination of adenosine antagonist and dopamine agonist might be useful in the treatment of Parkinson's disease (Takahashi, Fujita, Asai, Saki, & Mori, 2018; Takahashi, Ito, Tsuji, & Horiguchi, 2022).

On the contrary, it is plausible that tardive dyskinesia, which usually occurs in response to the chronic blockade of striatal $D_2$ receptors, might be treated with adenosine $A_{2A}$ agonists. Further, treatment with adenosine $A_{2A}$ agonists might be helpful in the early stages of Huntington's chorea with hyperkinesia caused by abnormal striate-pallidal neurons. Similarly, a dopamine $D_1$ antagonist, the effect of which could be potentiated by an adenosine $A_1$ agonist, might reduce the activity of the dominant direct striatal GABAergic efferent pathway, reducing the hyperkinesia (Ferré et al., 1997).

Istradefylline, an adenosine $A_{2A}$ receptor ($A_{2A}R$) antagonist, is effective as an adjunct to levodopa and can alleviate "off" time and motor symptoms in patients with PD (Takahashi et al., 2018). Our previous exploratory study demonstrated that istradefylline could be efficacious for postural abnormalities evaluated by Unified Dystonia Rating Scale (UDRS) (Takahashi et al., 2022) (Fig. 1) and was generally well tolerated in patients with PD experiencing the wearing-off phenomenon with levodopa-containing therapies (Takahashi et al., 2018; Takahashi et al., 2022). A subset of PD patients in the advanced stage shows more severe postural

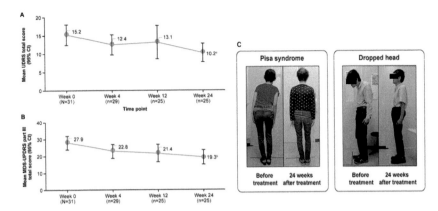

Fig. 1 Time course of (A) UDRS score and (B) MDS-UPDRS (part III) score by the treatment with Istradefylline in PD patients with camptocormia. Antagonising $A_{2A}R$ by istradefylline improved both dystonia and Parkinsonism, which is evaluated by UDRS and MDS-UPDRS(Part III). (A) Regression coefficient: −4.85 ($P=0.001$); P-value was calculated using fixed effect in the mixed-effects model. (B) Regression coefficient: −6.129 ($P<0.001$); P-value was calculated using fixed effect in the mixed-effects model. (C) Representative images of patients with postural abnormalities during the course of the study. (Courtesy of Takahashi, M. (2022). *Journal* of the *Neurological Sciences, 432*:120078.) Abbreviations; CI, confidence interval; MDS-UPDRS, Movement Disorder Society-Unified Parkinson's Disease Rating Scale; UDRS, Unified Dystonia Rating Scale.

abnormalities leading to significant disabilities, such as camptocormia, anterocollis, Pisa syndrome, and scoliosis (Bloch et al., 2006). Although the pathophysiology of these postural abnormalities is largely unknown, several factors have been suggested, which include disease progression of PD, dystonia, drug-induced factors, proprioceptive impairment, and paraspinal myopathy (Srivanitchapoom, & Hallett, 2016). The PET study measuring adenosine $A_{2A}$ receptors(R) revealed that $A_{2A}R$ expression in the striatum increases in accordance with the disease progression (Mishina et al., 2011). The treatment of postural abnormalities/truncal dystonia remains challenging; however, it is important to prevent secondary skeletal deformities, and the treatment of dystonia is one such approach.

According to the several studies of PD mouse model, adenosine $A_{2A}$ receptors colocalize with dopamine D2 receptors in D2-specific spiny projections neurons in the striatum. By way of indirect pathway in basal ganglia circuit, D2R signal is thought to be inhibited by Adenosine $A_{2A}R$ signal. The reduced D2R signal in indirect pathway causes enhanced GABAergic inhibitory input to the GPe, eventually enhances inhibitory output to the thalamus, leading to the decrease of excitatory output to the cortex and provoke hypokinesis. Therefore, antagonizing adenosine $A_{2A}$ receptor can reverse the reduced dopamine D2 signal, leading to the reverse of hypokinesis (Mori, 2020) (Fig. 2).

In addition, striatal cholinergic dysfunction is a common phenotype associated with various forms of dystonia in which anti-cholinergic drugs have some therapeutic benefits. However, the underlying substrate of striatal cholinergic defects in dystonia remain poorly understood. Striatal cholinergic interneurons are thought to project to virtually all subtypes of striatal spiny projection neurons and involved in dopamine D2/Adenosine $A_{2A}$ signals (Tozzi et al., 2011). The co-localization of $A_{2A}$ was studied in striatal

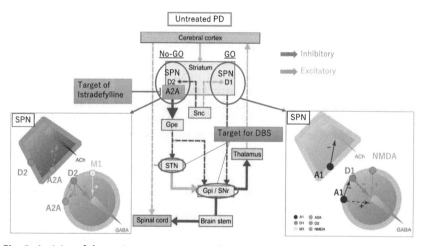

**Fig. 2** Activity of the major connections in the basal ganglia in untreated PD and the striatopallidal neuronal functions regulated by adenosine $A_{2A}$ receptors. In untreated PD, decreased levels of dopamine induces decreased activity of excitatory D1-specific spiny projection neurons (medium spiny neuron; MSN) and decreased activity of inhibitory D2-specific MSN, which eventually causes dominancy of inhibitory indirect pathway (No-Go) over excitatory direct pathway (Go), leading to hypokinesis. Adenosine $A_{2A}$ receptor (R) is colocalized with D2R expressing SPN, and $A_{2A}R$ disturbed D2R signals. D2, D1 (D1 and D2-like dopaminergic receptors); Gpe (external segment of globus pallidus); Gpi (internal segment of globus pallidus); Snc (compact segment of substantia nigra); SNr (reticular segment of substantia nigra); STN (subthalamic nucleus). The striatopallidal neuronal function seems to be regulated by $A_{2A}$ receptors both at the postsynaptic level and at the $A_{2A}$ presynaptic level (by an $A_{2A}$ receptor-mediated regulation of acetylcholine release). The striatonigral and striatoentopeduncular neuronal function seems to be mainly regulated by $A_1$ receptors both at the postsynaptic level and at the presynaptic level (by an $A_1$ receptor-mediated inhibition of glutamate release). Abbreviation: $M_1$, muscarinic $M_1$ receptors.

cholinergic interneurons identified by anti-choline-acetyltransferase (ChAT) antibody (D'Angelo et al., 2021). The combined activation of both $A_{2A}$ and D2 receptors on cholinergic interneurons may decrease the release of acetylcholine at the synaptic sites of D1- and D2-R- expressing MSNs and trigger a $Ca^{2+}$ inflow by relieving the cholinergic inhibition of L-type $Ca^{2+}$ channels mediated by M1 muscarinic receptors. (Tozzi et al., 2011) (Fig. 2).

Since the functional abnormality in basal ganglia is not fully understood, many animal models of dystonia were created and studied. In DYT1 (Tor1a+/−) mouse model, enhanced adenosine $A_{2A}$ receptor expression was found in the striatopallidal complex. In this model, opposite changes of $A_{2A}$ receptors' expression in the striatal-pallidal complex and the entopeduncular nucleus were also observed as well. Adenosine $A_{2A}$ receptor antagonism recovers bidirectional plasticity in rodent models of DYT1 (Napolitano et al., 2010), DYT25 (Yu-Taeger et al., 2020) and DYT11 (Maltese et al., 2017), but may trigger dystonia in some models (Richter & Hamann, 2001).

Taken together with these evidences, the pathophysiology of dystonia is critically dependent on a composite functional imbalance of the indirect over the direct pathway in basal ganglia. Antagonizing adenosine $A_{2A}$ receptor, by using istradefylline eg., could be a potential therapeutic option for generalized dystonia/camptocormia, in addition to levodopa and/or anti-cholinergic agents.

## References

Albanese, A., Di Giovanni, M., & Lalli, S. (2019). Dystonia: Diagnosis and management. *European Journal of Neurology: The Official Journal of the European Federation of Neurological Societies, 26*(1), 5–17.

Bhatia, K. P., & Marsden, C. D. (1994). The behavioural and motor consequences of focal lesions of the basal ganglia in man. *Brain, 117*(Pt 4), 859–876.

Bloch, F., Houeto, J. L., Tezenas du Montcel, S., Bonneville, F., Etchepare, F., Welter, M. L., et al. (2006). Parkinson's disease with camptocormia. *Journal of Neurology, Neurosurgery, and Psychiatry, 77*(11), 1223–1228.

D'Angelo, V., Giorgi, M., Paldino, E., Cardarelli, S., Fusco, F. R., Saverioni, I., et al. (2021). $A_{2A}$ receptor dysregulation in dystonia DYT1 knock-out mice. *International Journal of Molecular Sciences, 22*(5), 2691.

Ferré, S., Fredholm, B. B., Morelli, M., Popoli, P., & Fuxe, K. (1997). Adenosine-dopamine receptor-receptor interactions as an integrative mechanism in the basal ganglia. *Trends in Neurosciences, 20*(10), 482–487.

Frucht, S. J. (2013). The definition of dystonia: Current concepts and controversies. *Movement Disorders: Official Journal of the Movement Disorder Society, 28*(7), 884–888.

Kanda, T., Tashiro, T., Kuwana, Y., & Jenner, P. (1998). Adenosine A2A receptors modify motor function in MPTP-treated common marmosets. *Neuroreport, 9*(12), 2857–2860.

Maltese, M., et al. (2017). Abnormal striatal plasticity in a DYT11/SGCE myoclonus dystonia mouse model is reversed by adenosine A2A receptor inhibition. *Neurobiology of Disease, 108*, 128–139.

Mishina, M., Ishiwata, K., Naganawa, M., Kimura, Y., Kitamura, S., Suzuki, M., et al. (2011). Adenosine A(2A) receptors measured with [C]TMSX PET in the striata of Parkinson's disease patients. *PLoS One, 6*(2), e17338.

Mori, A. (2020). How do adenosine $A_{2A}$ receptors regulate motor function? *Parkinsonism & Related Disorders, 80*(Suppl 1), S13–S20.

Napolitano, F., Pasqualetti, M., Usiello, A., Santini, E., Pacini, G., Sciamanna, G., et al. (2010). Dopamine D2 receptor dysfunction is rescued by adenosine A2A receptor antagonism in a model of DYT1 dystonia. *Neurobiology of Disease, 38*(3), 434–445.

Richter, A., & Hamann, M. (2001). Effects of adenosine receptor agonists and antagonists in a genetic animal model of primary paroxysmal dystonia. *British Journal of Pharmacology, 134*(2), 343–352.

Scarduzio, M., Hess, E. J., Standaert, D. G., & Eskow Jaunarajs, K. L. (2022). Striatal synaptic dysfunction in dystonia and levodopa-induced dyskinesia. *Neurobiology of Disease, 166*, 105650.

Schulz-Schaeffer, W. J., Margraf, N. G., Munser, S., Wrede, A., Buhmann, C., Deuschl, G., et al. (2015). Effect of neurostimulation on camptocormia in Parkinson's disease depends on symptom duration. *Movement Disorders: Official Journal of the Movement Disorder Society, 30*(3), 368–372.

Srivanitchapoom, P., & Hallett, M. (2016). Camptocormia in Parkinson's disease: Definition, epidemiology, pathogenesis and treatment modalities. *Journal of Neurology, Neurosurgery, and Psychiatry, 87*(1), 75–85.

Takahashi, M., Ito, S., Tsuji, Y., & Horiguchi, S. (2022). Safety and effectiveness of istradefylline as add-on therapy to levodopa in patients with Parkinson's disease: Final report of a post-marketing surveillance study in Japan. *Journal of the Neurological Sciences, 15*(443), 120479.

Takahashi, M., Fujita, M., Asai, N., Saki, M., & Mori, A. (2018). Safety and effectiveness of istradefylline in patients with Parkinson's disease: Interim analysis of a post-marketing surveillance study in Japan. *Expert Opinion on Pharmacotherapy, 19*(15), 1635–1642.

Takahashi, M., Shimokawa, T., Koh, J., Takeshima, T., Yamashita, H., Kajimoto, Y., et al. (2022). Efficacy and safety of istradefylline in patients with Parkinson's disease presenting with postural abnormalities: Results from a multicenter, prospective, and open-label exploratory study in Japan. *Journal of the Neurological Sciences, 432*, 120078.

Tozzi, A., de Iure, A., Di Filippo, M., Tantucci, M., Costa, C., Borsini, F., et al. (2011). The distinct role of medium spiny neurons and cholinergic interneurons in the $D_2/A_2A$ receptor interaction in the striatum: implications for Parkinson's disease. *The Journal of Neuroscience, 31*(5), 1850–1862.

Yu-Taeger, L., Ott, T., Bonsi, P., Tomczak, C., Wassouf, Z., Martella, G., et al. (2020). Impaired dopamine- and adenosine-mediated signaling and plasticity in a novel rodent model for DYT25 dystonia. *Neurobiology of Disease, 134*, 104634.

CHAPTER NINE

# $A_{2A}R$ antagonist treatment for multiple sclerosis: Current progress and future prospects

Chenxing Qi[a,b,1], Yijia Feng[c,1], Yiwei Jiang[d], Wangchao Chen[a], Serhii Vakal[e], Jiang-Fan Chen[a,b], and Wu Zheng[a,b,*]

[a]State Key Laboratory of Ophthalmology, Optometry and Visual Science, Eye Hospital, Wenzhou Medical University, Wenzhou, P.R. China
[b]Oujiang Laboratory (Zhejiang Laboratory for Regenerative Medicine, Vision and Brain Health), School of Ophthalmology & Optometry and Eye Hospital, Wenzhou Medical University, Wenzhou, P.R. China
[c]Key Laboratory of Alzheimer's Disease of Zhejiang Province, Institute of Aging, Wenzhou Medical University, Wenzhou, Zhejiang, P.R. China
[d]Alberta Institute, Wenzhou Medical University, Wenzhou, P.R. China
[e]Structural Bioinformatics Laboratory, Biochemistry, Faculty of Science and Engineering, Åbo Akademi University, Turku, Finland
*Corresponding author. e-mail address: zhengwu@wmu.edu.cn

## Contents

1. Introduction — 187
2. Adenosine metabolism — 188
3. $A_{2A}R$ distribution in MS/EAE pathology — 190
4. The confusing effect of $A_{2A}R$ signaling in MS/EAE — 191
5. $A_{2A}R$ signaling in macrophages and dendritic cells — 193
6. The complex effects of $A_{2A}R$ signaling in T cells — 195
7. $A_{2A}R$ signaling is essential for maintaining the suppressive capacity of Tregs — 198
8. $A_{2A}R$ antagonist attenuates the pro-inflammation phenotype of microglia — 199
9. $A_{2A}R$ antagonist inhibits the activation of astrocytes — 201
10. The $A_{2A}R$ antagonist protects oligodendrocytes from damage — 203
11. $A_{2A}R$ antagonist protects against lymphocyte infiltration into the CNS across the BBB — 205
12. $A_{2A}R$ antagonist suppresses T-cell infiltration by decreasing the ChP gateway activity — 207
13. Safety of the $A_{2A}R$ antagonist in clinical trials — 209
14. The limitations of current studies on the potential efficacy of $A_{2A}R$ antagonists in MS pathology — 213
15. Conclusion — 214
Acknowledgments — 214
Declarations — 214
Ethics approval and consent to participate — 214
Consent for publication — 214

[1] These authors contributed equally to this work.

| | |
|---|---|
| Funding | 215 |
| Competing interests | 215 |
| References | 215 |

## Abstract

Emerging evidence suggests that both selective and non-selective Adenosine $A_{2A}$ receptor ($A_{2A}$R) antagonists could effectively protect mice from experimental autoimmune encephalomyelitis (EAE), which is the most commonly used animal model for multiple sclerosis (MS) research. Meanwhile, the recent FDA approval of Nourianz® (istradefylline) in 2019 as an add-on treatment to levodopa in Parkinson's disease (PD) with "OFF" episodes, along with its proven clinical safety, has prompted us to explore the potential of $A_{2A}$R antagonists in treating multiple sclerosis (MS) through clinical trials. However, despite promising findings in experimental autoimmune encephalomyelitis (EAE), the complex and contradictory role of $A_{2A}$R signaling in EAE pathology has raised concerns about the feasibility of using $A_{2A}$R antagonists as a therapeutic approach for MS. This review addresses the potential effect of $A_{2A}$R antagonists on EAE/MS in both the peripheral immune system (PIS) and the central nervous system (CNS). In brief, $A_{2A}$R antagonists had a moderate effect on the proliferation and inflammatory response, while exhibiting a potent anti-inflammatory effect in the CNS through their impact on microglia, astrocytes, and the endothelial cells/epithelium of the blood–brain barrier. Consequently, $A_{2A}$R signaling remains an essential immunomodulator in EAE/MS, suggesting that $A_{2A}$R antagonists hold promise as a drug class for treating MS.

## Abbreviation

| | |
|---|---|
| $A_{2A}$R | Adenosine $A_{2A}$ receptor |
| BBB | blood-brain barrier |
| CNS | central nervous system |
| MS | multiple sclerosis |
| EAE | experimental autoimmune encephalomyelitis |
| PD | Parkinson's disease |
| CSF | cerebrospinal fluid |
| ChP | choroid plexus |
| PIS | peripheral immune system |
| DMT | disease modification therapy |
| IFN-β | interferonβ |
| PML | progressive multifocal leukoencephalopathy |
| CAM | schemotaxis and adhesion molecules |
| ADK | adenosine kinase |
| MDMs | monocyte-derived macrophages |
| PKA | protein kinase A |
| ADP | adenosine diphosphate |
| ROP | retinopathy of prematurity |
| ENTs | equilibrium nucleoside transporters |
| ADA | adenosine deaminase |

| | |
|---|---|
| MAPK | mitogen-activated protein kinase |
| ECM | extracellular matrix |
| GPCR | G-protein-coupled receptors |
| SAH | S-adenosyl-L-homocysteine |

## 1. Introduction

Multiple sclerosis (MS) is an autoimmune disease characterized by an overactive immune system, lymphocyte infiltration, gliosis, and demyelination, which results in paralysis, cognitive decline, vision loss, and sensory sensitivity, among other symptoms (Inojosa, Schriefer, & Ziemssen, 2020; Koch-Henriksen & Magyari, 2021). MS affects approximately 2.5 million people worldwide, with the majority of cases occurring in the Western world. Along with the leakage of the blood–brain barrier (BBB), the infiltration of immune cells, such as T cells, B cells, and other monocytes, into the central nervous system (CNS) can cause inflammation, demyelination, and impaired signaling in damaged nerves (Correale, Gaitán, Ysrraelit, & Fiol, 2017). Although the exact cause of MS remains unclear, high-risk factors are believed to include a combination of genetic and environmental factors, such as genetic polymorphisms (including HLA, IL2, and IL7R), smoking, and viral infections.

MS is clinically divided into four types based on disease course: relapsing-remitting MS (RRMS), secondary-progressive MS (SPMS), primary-progressive MS (PPMS), and progressive-relapsing MS (PRMS). Approximately 70–80% of MS patients have RRMS, and about half of RRMS patients experience a slow progressive exacerbation without relapsing remission within 15 years (Gholamzad et al., 2019; Lassmann, 2019). PPMS accounts for 10–15% of MS patients and presents a slow exacerbation progression without relapsing-remitting. PRMS affects 5% of patients and is characterized by gradual deterioration with superimposed relapses (Nicholas & Rashid, 2013). Experimental autoimmune encephalomyelitis (EAE) is the dominant animal model used to explore the immunopathogenic injury mechanisms of MS. EAE is mainly mediated by specifically priming $CD4^+$ T cells and simulates the processing and presentation of autoantigen, activation, and infiltration of $CD4^+$ T helper cells, neuroinflammation, demyelination, and hindlimb paralysis in MS pathology (Constantinescu, Farooqi, O'Brien, & Gran, 2011).

Therapeutic schedules for MS patients are generally decided according to the pathologic phase of the disease. During the acute phase, high-dose corticosteroids are the first-line treatment, regardless of MS type, and are widely used clinically. In the remission phase, disease modification therapy

(DMT) is recommended (Rae-Grant et al., 2018a; Spelman et al., 2021; Wingerchuk & Carter, 2014). DMTs can be categorized into four types based on their mechanisms of action: (1) Inflammatory mediators, such as interferon β (IFN-β), which was approved by the FDA for MS treatment in 1993; (2) Inhibitors of immune cell migration, such as Fingolimod and Natalizumab; (3) Immune cell depletion, such as Ocrelizumab and Ublituximab (Steinman et al., 2022); and (4) immunotolerance therapy, represented by Glatiramer acetate. However, approved DMTs have distinct risk profiles, and most result in varying degrees of immunosuppression, inhibiting the beneficial effect of immune cells. So, MS patients receiving DMTs need to be monitored for immunodeficiency-associated risks, such as infections or progressive multifocal leukoencephalopathy (PML) (Maillart et al., 2017). As a result, designing new therapeutic drugs with fewer unwanted side effects is a trend in MS treatment.

Adenosine is considered a significant immunomodulator under both pathological and physiological conditions (Boison & Yegutkin, 2019; Kammer, 1986). In inflammatory conditions, extracellular adenosine is rapidly released and performs anti-inflammatory or proinflammatory functions by activating four adenosine receptors ($A_1$, $A_{2A}$, $A_{2B}$, and $A_3$) that are widely expressed in the central nervous system (CNS) and peripheral tissues and organs. Among them, the Adenosine $A_{2A}$ receptor ($A_{2A}R$) has been identified as an attractive target for the treatment of several neurodegenerative disorders, such as Parkinson's disease (PD), Huntington's disease, and MS (Chen, Eltzschig, & Fredholm, 2013; Ren & Chen, 2020). Istradefylline, a selective and orally active $A_{2A}R$ antagonist, was recently approved by the US FDA as an adjunctive therapy for PD (Dungo & Deeks, 2013; Hauser et al., 2021). This exciting achievement has encouraged further exploration of the potential effects of $A_{2A}R$ antagonists on MS, as previous studies have demonstrated their ability to protect animals from experimental autoimmune encephalomyelitis (EAE), a dominant animal model of MS. In this review, we summarize the effect or potential effect of $A_{2A}R$ antagonists on the peripheral and central immune system in MS pathology and propose that $A_{2A}R$ antagonists hold promise as a therapeutic drug for MS.

## 2. Adenosine metabolism

The transduction process of an adenine energy signal is of great significance, and extracellular adenosine accumulates via the breakdown of

ATP through both intracellular and extracellular pathways. Intracellular ATP and adenosine diphosphate (ADP) are transferred from the intracellular space to the extracellular matrix (ECM) via pannexin-1 channels. The main source of extracellular adenosine (eADO) is ATP, which is hydrolyzed to adenosine (ADO) by a series of membrane-localized enzymes of several cell types, including extracellular nucleoside triphosphate diphosphohydrolase 1 (CD39), ecto-5'-nucleotidase (CD73), ectonucleotide pyrophosphatase/phosphodiesterase (ENPP), and prostatic acid phosphatase (PAP). The eADO production may also occur through intrinsic metabolic pathways, mainly involving the $NAD^+$ recovery pathway of adenosine kinase (ADK), S-adenosylhomocysteine hydrolase (SAHH), cytoplasmic 5'-nucleotidase-I (cN-I), and cell surface cyclic ADP ribose hydrolase (CD38) (Nakanishi, 2007). CD39, along with CD73, is responsible for the cascade of ATP conversion to ADP and cyclic adenosine phosphate (cAMP), ultimately resulting in eADO. The nature of ADO release plays a critical role in determining the beneficial or harmful effects of P1R signaling pathways. Specifically, an acute increase in ADO tone or extracellular ADO levels is protective, inhibiting the immune response and harmful effects of an overactive immune response, promoting tissue homeostasis and healing. However, chronically elevated extracellular ADO levels are associated with harmful effects and pathological conditions such as inflammatory tissue trauma, inflammation, hypoxia, and cancer. CD39 is a cell membrane protein that belongs to the extracellular nucleotide triphosphate diphosphohydrolase family and has ATPase and ADPase enzyme activity. CD39 hydrolyzes extracellular ATP and ADP to adenosine monophosphate (AMP), thereby limiting the immune response (Moesta, Li, & Smyth, 2020). CD73, also known as Ecto-5'-Nucleotidase, is a cell surface protein encoded by the *NT5E* gene and is the main enzyme that catalyzes AMP breakdown to form extracellular adenosine (Minor, Alcedo, Battaglia, & Snider, 2019). After being released into the extracellular space, ATP and ADP are converted to AMP by the exoenzyme CD39. AMP is then converted to ADO by the enzyme CD73. The CD39 adenosine pathway occurs through purinergic receptors type 1 ($A_1$, $A_{2A}$, $A_{2B}$, $A_3$), which are G-protein-coupled receptors (GPCR). After binding to the receptor, ADO triggers a complex signaling cascade. The $A_1$ and $A_3$ receptors inhibit adenylate cyclase and cAMP production, often described as immune-promoting adenosine receptors. In contrast, the $A_{2A}$ and $A_{2B}$ receptors are generally associated with high levels of immunosuppression, triggering intracellular cAMP accumulation. cAMP production activates downstream protein kinase A (PKA), regulating gene expression. On the other hand, $G_{i/o}$ ($A_1R$ and $A_3R$) inhibits AC and cAMP production. Notably, mitogen-activated protein kinase (MAPK) activation is a downstream step common to all adenosine receptors. Both $A_{2A}$ and $A_{2B}$ receptors are highly expressed in various immune cells, including myeloid cells and lymphocytes. After binding to the receptor and triggering intracellular effects, ADO is captured by equilibrium nucleoside transporters (ENTs) 1 and 2 and transported to cells, where it is either converted to inosine by adenosine deaminase (ADA) or to AMP by ADK.

Intracellular ATP is converted to ADP by ATPase and subsequently to AMP, which can be further broken down to adenosine by 5'-nucleotidase enzymes. Another pathway of adenosine synthesis is the decomposition of S-adenosyl-L-homocysteine (SAH) by SAH hydrolase (SAHH) into adenosine and homocysteine following S-adenosyl-L-homocysteine-dependent methylation (Burgos, Gulab, Cassera, & Schramm, 2012). Similarly, adenosine can be converted back to S-adenosyl-L-homocysteine through reversible production from ADO and homocysteine, catalyzed by SAHH. Intracellular decomposition of ADO can also occur through the phosphorylation of AMP by ADK.

## 3. $A_{2A}R$ distribution in MS/EAE pathology

Under normal physiological conditions, $A_{2A}R$ is primarily localized in certain regions of the brain, such as the striatum, olfactory tubercle, entorhinal cortex, and nucleus accumbens (Liu et al., 2019). In the peripheral system, $A_{2A}R$ is abundant in immune cells such as T lymphocytes, macrophages, and platelets (Rajasundaram, 2018). However, in acute injury or ischemia and hypoxia, ATP energy metabolism is accelerated to counteract local inflammation which results in a rapid amplification of eADO signaling. Of note, only expression of $A_{2A}R$, but not the other three ARs, was markedly increased in lymphocytes from 46 MS patients, compared to 50 healthy subjects (Vincenzi et al., 2013). The combination of positron emission tomography (PET) imaging using [$^{11}$C]TMSX (a selective radioligand that binds to $A_{2A}R$) and magnetic resonance imaging (MRI) has also demonstrated that $A_{2A}R$ expression is positively correlated with higher expanded disability status scale (EDSS) scores in SPMS patients. Furthermore, $A_{2A}R$ protein levels are increased in the cerebrovascular endothelial cells of 28 SPMS patients (Liu, Alahiri, Ulloa, Xie, & Sadiq, 2018), and CD73 is enriched in the microvasculature of MS patients (Airas, Niemelä, Yegutkin, & Jalkanen, 2007). In the MS animal model, an elevated $A_{2A}R$ level was also observed in the peripheral lymphocytes and brain parenchyma during EAE pathology (Ingwersen et al., 2016). These findings suggest that abnormal $A_{2A}R$ signaling may play a critical role in the infiltration of immune cells and autoimmune injury in MS, making $A_{2A}R$ antagonists a promising therapeutic approach for MS.

## 4. The confusing effect of $A_{2A}R$ signaling in MS/EAE

Direct evidence for $A_{2A}R$ involvement in MS pathogenesis was obtained through the use of $A_{2A}R$-KO ($A_{2A}R^{-/-}$) mice in the EAE animal model. The $A_{2A}R$-KO mice demonstrated more severe EAE pathology than WT mice, including greater infiltration of $CD4^+$ T lymphocytes into the brain parenchyma, stronger neuroinflammation, more serious demyelination, and increased motor paralysis (Ingwersen et al., 2016; Yao et al., 2012). This suggests that $A_{2A}R$ signaling plays a protective role in EAE. However, several pharmacological studies have repeatedly confirmed that $A_{2A}R$ antagonists, such as SCH58261 and KW6002, can significantly improve EAE pathological symptoms (Chen et al., 2019; Zheng et al., 2022). Similarly, caffeine, a non-selective $A_{2A}R$ antagonist, has also been shown to attenuate EAE pathology (Wang et al., 2014). These opposing outcomes suggest that $A_{2A}R$ antagonist-mediated neuroprotection in EAE occurs through a mechanism distinct from that observed in $A_{2A}R$-KO mice. Following EAE induction, the $CD4^+$ T lymphocytes previously activated in $A_{2A}R^{-/-}$ mice showed an increased ability to proliferate and produce a higher level of IFN-γ when stimulated with $MOG_{35-55}$ in vitro (Mills, Kim, Krenz, Chen, & Bynoe, 2012). To elucidate the role of $A_{2A}R$ signaling in different cell types in MS/EAE pathology, this contradiction was further explored in a series of adoptive transfer experiments. A study using a radiation bone marrow chimera model found that $A_{2A}R^{-/-}$ mice, which were irradiated and then transplanted with bone marrow from WT donor mice, were protected against EAE and developed mild disease compared to irradiated WT mice that were transferred with bone marrow from WT donor mice (Mills, Kim, et al., 2012). As a result, the irradiated WT mice reconstituted with $A_{2A}R^{-/-}$ bone marrow developed more severe EAE pathology than the control group, indicating that $A_{2A}R$

signaling on hematopoietic cells is indispensable for limiting MS progression, while inactivation of $A_{2A}R$ in non-hematopoietic cells contributes to the protective scheme. Interestingly, the $A_{2A}R$ antagonist SCH58261 alleviated the pathological symptoms of $TCR\alpha^{-/-}$ mice reconstituted with $CD4^+$ T lymphocytes from $A_{2A}R^{-/-}$ donors compared to vehicle controls (Mills, Kim, et al., 2012). This suggests that $A_{2A}R$ antagonist-mediated neuroprotection in EAE is independent of T lymphocytes.

Another puzzle is the conflicting effects of $A_{2A}R$ agonists and antagonists. Surprisingly, treatment with the $A_{2A}R$ agonist CGS21680 in the pre-onset and early stages significantly delayed disease onset and mitigated pathological lesions in mice. However, administering CGS21680 in the late stages of EAR resulted in more severe neuroinflammation (Friberg, 1989). In contrast, it has been found that an $A_{2A}R$ antagonist is not effective when administered before disease onset, but it does show a protective effect when given after EAE onset (Friberg, 1989). During EAE pathology, it is commonly believed that T lymphocytes rapidly proliferate during the pre-onset stage and subsequently infiltrate the CNS during the peak stage. Based on the evidence that $A_{2A}R$ agonists effectively inhibit the proliferation and migration of T lymphocytes in vitro, it is plausible to speculate that $A_{2A}R$ agonists protect mice from the advancement of EAE in the early stages by restraining the proliferation and infiltration of pathogenic T lymphocytes. $A_{2A}R$ antagonists, on the other hand, can effectively block T lymphocyte infiltration into the brain (Chen et al., 2019). Recent studies have shown that the $A_{2A}R$ antagonist KW6002 also prevents T lymphocytes from infiltrating the cerebrospinal fluid (CSF) by controlling the expression of chemotaxis and adhesion molecules (CAMs) in the choroid plexus (ChP) (Zheng et al., 2022). Thus, while both $A_{2A}R$ agonists and antagonists can exert protective effects on EAE, their pharmacological mechanisms are quite different. During EAE pathology, the $A_{2A}R$ agonist appears to act by activating $A_{2A}R$ signaling in immune cells, whereas the $A_{2A}R$ antagonist prefers to downregulate $A_{2A}R$ signals in non-immune cells, such as the BBB and ChP.

In light of the conflicting results regarding the efficacy of $A_{2A}R$ agonists in the late stages of EAE and the difficulty of distinguishing between different stages of MS in patients, caution should be exercised when considering the use of $A_{2A}R$ agonists for MS treatment (Ingwersen et al., 2016). In contrast, the consistent protective effect of $A_{2A}R$ antagonists throughout EAE progress suggests that they may be a more feasible and safe option for treating MS patients.

## 5. A$_{2A}$R signaling in macrophages and dendritic cells

As early as the last century, the potential role of macrophages in human MS disease was observed. Autopsy studies of MS patients have shown a positive correlation between the degree of demyelinating lesions and macrophage infiltration (Ferguson, Matyszak, Esiri, & Perry, 1997). In MS intracranial inflammation, the infiltrated macrophages can be divided into two categories: peripherally infiltrated monocyte-derived macrophages (MDMs) and centrally resident microglia-derived macrophages (MiDMs). Immunohistochemical analysis of the central nervous system of EAE animal models has revealed that A$_{2A}$R is expressed in both MDMs and MiDMs clustered near inflammatory lesions (Ingwersen et al., 2016), so the role of A$_{2A}$Rs in disease has also been studied in macrophages. Although both play similar roles in phagocytosis during immune responses, recent studies have found that they have different morphological structures and physiological characteristics, which support the conclusion that they have specific functions at different stages of the disease (Yamasaki et al., 2014).

Bone MDMs are immune cells that originate from hematopoietic stem cells in the bone marrow and play a crucial role in the pathogenesis of MS, an autoimmune disease. These cells are known for their ability to present antigens and initiate an immune response. In the context of MS, macrophages respond directly to central inflammation and play a significant role in phagocytosing necrotic tissues and antigens, leading to the production of proinflammatory cytokines and amplification of the immune response. First, studies have revealed that myelin debris can rapidly up-regulate A$_{2A}$R expression in macrophages in vitro (Ingwersen et al., 2016). The A$_{2A}$R signal is known to inhibit immune cells' function in promoting inflammation, so the use of A$_{2A}$R agonist CGS21680 can restrain macrophages' phagocytosis and migration, which is unfavorable for their proinflammatory activity (Ingwersen et al., 2016). In several common neurodegenerative diseases, myelin debris removal and demyelination of neurons may not occur promptly, which may hinder the repair and regeneration of myelin sheaths (Neumann, Kotter, & Franklin, 2009). Given that the A$_{2A}$R signals can regulate the phagocytosis ability of macrophages, its involvement may impact the clinical prognosis of MS. This highlights the importance of understanding A$_{2A}$R signaling in macrophages in MS pathology.

Extensive research has proved that A$_{2A}$R expression on immune cells is involved in suppressing inflammation, and the activation of A$_{2A}$R signaling on lymphocytes can up-regulate the expression of anti-inflammatory cytokines while downregulating the release of proinflammatory cytokines. Similarly, the same anti-inflammatory effect has been observed in macrophages with $A_{2A}R$ signaling, which is of particular interest given the strong association of TNF-α with MS disease severity. Studies on transgenic mice have shown that overexpression of TNF-α can lead to demyelinating lesions similar to MS, while TNF-α-specific antibodies protected against EAE (Probert, Akassoglou, Pasparakis, Kontogeorgos, & Kollias, 1995; Selmaj, Papierz, Glabiński, & Kohno, 1995). Further research has shown that knocking out $A_{2A}R$ on mouse monocytes significantly increases TNF-α expression, whereas $A_{2A}R$ activation reduces its expression (Zhang, Hepburn, Cruz, Borman, & Clark, 2005). However, the results of phase II clinical trials of the TNF-α receptor blocker Lenercept and studies targeting different TNF receptor subtypes on oligodendrocytes have found that TNF-α plays a more complex role in brain inflammation. Therefore, altering TNF-α as a therapeutic strategy for MS based on existing studies is not feasible (Rajasundaram, 2018). However, understanding the potential adverse effects of changes in TNF-α levels is a prerequisite for the use of $A_{2A}R$-targeted therapies.

Macrophages also act as antigen-presenting cells in MS pathogenesis, indirectly inducing autoimmune T lymphocytes. Studies have shown that $A_{2A}Rs$ regulate macrophage antigen-presenting ability, and in the presence of adenosine, the ability of T lymphocytes to respond to LPS-induced antigen presentation by macrophages is diminished, leading to a weakening of the Th1 polarization of activated T cells (Linden & Cekic, 2012).

Dendritic cells (DCs) are highly specialized antigen-presenting cells that play a significant role in the immune response to MS. $A_{2A}R$ is highly expressed in DCs and is known to mediate lymphocyte response. Several studies have suggested that $A_{2A}R$ is the main ADO receptor that regulates the immune response in EAE models of MS (Alam et al., 2009; Blackburn, Vance, Morschl, & Wilson, 2009). Additionally, adenosine has been found to inhibit the production of cytokines, such as IFN-a, IL-6, and IL-12, by activating plasmacytoid dendritic cells through $A_{2A}Rs$ (Schnurr et al., 2004).

The $A_{2A}R$ agonist, ATL313, has been shown to inhibit lymphocyte proliferation and IFN-γ release in a dose-dependent manner through its effects on T lymphocytes and antigen-presenting cells (APCs), according to previous studies (Sevigny et al., 2007). Plasmacytoid DCs are known to regulate both the innate immune response by producing high levels of IFN-1, as well as the adaptive immune response by acting as a powerful regulator of T-cell responses (Cella, Facchetti, Lanzavecchia, & Colonna, 2000; Fonteneau et al., 2003). Adenosine signaling promotes a tolerogenic phenotype of DCs, characterized by an increase in IL-10 and a decrease in IFN-γ expression, resulting in functional immunosuppressive effects (Panther et al., 2001; Wilson et al., 2011). The expression of $A_{2A}Rs$ is increased during DC maturation, and activation of $A_{2A}Rs$ can stimulate DCs to produce inhibitory cytokines (Schnurr et al., 2004; Cekic & Linden, 2016). In line with these results, studies have shown that $A_{2A}R$ antagonists can reduce DC maturation and IL-10 expression (Cekic & Linden, 2016). Additionally, DCs are targeted by Treg cells for immunosuppression, as Treg cells attract DCs by activating the cAMP signaling pathway (Ring et al., 2015). Therefore, the protective mechanism of the targeted $A_{2A}R$ pathway against EAE models of MS may be partly achieved through the regulation of DCs themselves and the interaction with T cells.

## 6. The complex effects of $A_{2A}R$ signaling in T cells

The T-cell hypothesis for MS was first proposed when Kabat immunized monkeys with genetically homologous monkey brain myelin antigen, leading to inflammation of the brain and spinal cord with similar pathological features to MS (Kabat, Wolf, & Bezer, 1947). For many years, the inflammatory response produced by T lymphocytes has been regarded as the primary pathological mechanism of MS, playing a central role in mediating the immunopathological process of MS. Helper T cells, which differentiate from naive $CD4^+$ T cells, can mainly be divided into Th1, Th2, Th17, and Treg lineages based on secreted cytokines, specific transcription factors, and differences in function. They play distinct roles in recruiting and chemoattracting other immune cells and participate in adaptive immune responses (Zhu & Paul, 2010).

The onset of MS is marked by the activation of effector T cells in the peripheral immune system, followed by their infiltration into the central nervous system through various chemokines, leading to characteristic pathological changes in MS. The T cells specific to the myelin that are present in the CNS of individuals with MS exhibit a Th1 phenotype and secrete proinflammatory cytokines such as IFN-γ and TNF-α, resulting in a state of inflammation in the brain that contributes to MS development and is considered harmful (Allegretta, Nicklas, Sriram, & Albertini, 1990; Fletcher, Lalor, Sweeney, Tubridy, & Mills, 2010).

Th17-secreted IL-17 is a crucial proinflammatory cytokine in MS pathology due to its broad receptor distribution. IL-17 specifically promotes the activation state of CNS astrocytes (Elain, Jeanneau, Rutkowska, Mir, & Dev, 2014), leading to the release of other proinflammatory factors, chemokines, and effector proteins to promote a pathological immune response (Onishi & Gaffen, 2010). Additionally, it can activate microglia and recruit other immune cells to gather at inflammation sites (Steinman, 2007). While $CD4^+$ T cells have long been the focus of MS research, recent studies suggest that $CD8^+$ T cells also play a significant role in MS neuroinflammation pathological changes. $CD8^+$ T cells recognize upregulated MHC Class I molecules in the brains of MS patients and mediate demyelination and neuronal damage through cytolytic action. The inflammatory microenvironment of the CNS also promotes the function of $CD8^+$ T cells, which may be involved in MS pathogenesis through other pathways such as cytokine secretion (Kaskow & Baecher-Allan, 2018).

Recent studies have focused on the effect of $A_{2A}R$ signaling on immune cells in MS, with adenosine serving as an endogenous antiinflammatory agent that modulates the onset and progression of inflammation in the body via adenosine receptors. However, the effect of $A_{2A}R$ signals on immune cells in animal models of EAE can be complex and even contradictory (Rajasundaram, 2018).

According to Vincenzi F's study, $A_{2A}R$ expression was found to be upregulated in the lymphocytes of MS patients compared to healthy individuals. Activation of $A_{2A}R$ on Th1 cells significantly inhibits the secretion and proliferation of cytokines such as TNF-$\alpha$, IFN-$\gamma$, IL-6, IL-1$\beta$, and IL-1 (Vincenzi et al., 2013). Moreover, the study also revealed that $A_{2A}R$ activation reduces VLA-4 expression and NF-κB activation, the latter being a crucial intracellular signaling pathway associated with inflammatory and immune responses. Overactivation of NF-κB is often linked to various autoimmune diseases in humans. Cytokines can roughly be divided into Th1 and Th2 types according to their different effects on inflammation, and the correct Th1/Th2 ratio is essential for proper immune function. In an animal study, $A_{2A}R$ knockout mice induced by $MOG_{35-55}$ developed more severe EAE disease and had more inflammatory cell infiltration in the brain and spinal cord compared to wild-type mice. The researchers analyzed the expression of key cytokines and found that IFN-$\gamma$, IL-17, and IL-6 mRNAs were significantly elevated in the brain and spinal cord of $A_{2A}R$-KO mice, while TGF-$\beta$1 and IL-10 mRNA were significantly decreased, compared to WT mice. These findings suggest that the absence of $A_{2A}R$ signaling increases the expression of proinflammatory cytokines and reduces the expression of anti-inflammatory cytokines at the transcriptional level. As a result, the Th1/Th2 balance is shifted towards Th1, leading to a poorer disease prognosis. Furthermore, the inactivation of $A_{2A}Rs$ selectively increased $CD4^+$ T cells in EAE disease, indicating that $A_{2A}Rs$ may partially protect against the disease by inhibiting Th1 production (Yao et al., 2012).

Th17 cells have gained significant research attention in MS pathogenesis due to the widespread distribution of IL-17 receptors throughout the body. A recent study by Meiko Tokano and his team investigated the role of adenosine-induced T cells during EAE. They demonstrated, for the first time, that inhibitors of both $A_{2A}R$ and the $G_s$ downstream pathway AC-PKA-CREB inhibit adenosine-mediated IL-17A release from Th17 cells after $CD4^+$ T cell differentiation. Follow-up experiments showed that $A_{2A}R$ antagonists also inhibited the overproduction of Th17 cell-specific IL-17A caused by PLP peptide. Abnormal IL-17A concentration plays a significant role in the pathogenesis of autoimmune diseases by inducing neutrophil aggregation in local inflammatory lesions through stimulating stromal and epithelial cells to produce neutrophil chemokines (Tokano, Matsushita, Takagi, Yamamoto, & Kawano, 2022). Thus, the use of $A_{2A}R$ antagonists can potentially reduce the migration of autoimmune cells from draining lymph nodes to inflammatory lesions, thus reducing IL-17A-mediated EAE symptoms. Additionally, $A_{2A}R$ agonists have been found to promote $A_{2A}R$ signaling by reducing IL-6 and increasing IL-10 expression, leading to the development of Th1, Th2, and Th17 effector cells in both in vivo and in vitro studies involving T cell development mixed with APC (Linden & Cekic, 2012).

EAE, which mimics the neuroinflammation and demyelination pathology of MS, is mainly mediated by $CD4^+$ T cells, although this was later found to be different from the actual pathogenesis of MS. In MS patients, postmortem analysis revealed a higher number of $CD8^+$ T cells than $CD4^+$ T cells in the lymphocytic perivascular cuffs (Hauser et al., 1986), suggesting that $CD8^+$ T cells may play a dominant role in the pathogenesis of human MS. A study by Akio Ohta certified that $A_{2A}R$ agonists can inhibit the cytotoxicity of T cells, and the inhibition of cytotoxicity remained significant even after the signal was removed. The effect of $A_{2A}R$ agonists on T-cell proliferation was found to be insignificant in several experiments. The increased cAMP level in T cells upon agonist exposure leads to downstream immunosuppressant effects of $A_{2A}R$

activation, which are observed in both $CD4^+$ and $CD8^+$ T cells. Moreover, $A_{2A}R$ signaling activates T lymphocytes to develop into regulatory T cells, which can further produce eADO through CD39 and CD73. This, in turn, locally inhibits the activation and inflammatory effects of other immune cells (Linden & Cekic, 2012).

## 7. $A_{2A}R$ signaling is essential for maintaining the suppressive capacity of Tregs

Treg cells (Tregs) play a critical role in the immune system by suppressing immune responses to self-antigens and exogenous antigens. These cells are characterized by the expression of CD4 and CD25 markers and constitutively express the transcription factor forkhead box P3 (FoxP3) (Sakaguchi, Yamaguchi, Nomura, & Ono, 2008). However, under inflammatory conditions, the expression of FoxP3 can be reduced, and Tregs may acquire effector T cell characteristics (Zhou et al., 2009). In this state, Tregs lose their suppressive function and secrete proinflammatory cytokines such as IL-17 and/or IFN-γ. Although meta-analysis showed no significant difference in the number of $CD4^+CD25^+$ Tregs between MS patients and healthy controls, a decrease in the proportion of $CD4^+CD25^+FoxP3^+$ Tregs was observed in MS patients (Li et al., 2019). Similarly, in children with MS, while the overall frequencies of Tregs ($CD4^+CD25^{hi}CD127^{low}$) did not differ, FoxP3 expression and suppressive capacity of Tregs were reduced (Mexhitaj et al., 2019). Of note, Treg subsets expressing CD39 were also decreased in MS children (Mexhitaj et al., 2019).

CD39 and CD73 are surface markers expressed on Tregs, and their interaction with $A_{2A}R$ on activated T effector cells generates an immunosuppressive loop. Studies have shown that $CD39^{-/-}$ Tregs exhibit impaired suppressive properties and cannot prevent allograft rejection in vivo (Deaglio et al., 2007). With sorting, only $CD4^+CD25^{high}CD39^+$ suppressed IL-17 production, whereas $CD4^+CD25^{high}CD39^-$ T cells produced IL-17. This suggests that CD39 plays a crucial role in Treg-mediated immunosuppression, and the interaction between CD39 and adenosine may be a crucial mechanism for Tregs to regulate immune responses (Fletcher et al., 2009). Studies have reported that knockout of $A_{2A}R$ does not affect the proportion of Tregs or the ability of T-cell proliferation to suppress the immune response in EAE (35). However, $A_{2A}R$ stimulation is required for the emergence of Tregs after autoimmune disease resolution by upregulating the expression of a tyrosine-based inhibitory motif (TIGIT) (Muhammad et al., 2020). To date, no studies have been conducted to investigate the effect of $A_{2A}R$ antagonists on Treg in MS/EAE. However, in a sepsis model induced by LPS, the administration of the $A_{2A}R$ antagonist ZM241385 resulted in the inhibition of Treg immunosuppressive activity in T lymphocyte proliferation (Zhang, Zhao, Fu, Chen, & Ma, 2022). Further investigation is necessary to evaluate the potential effect of $A_{2A}R$ antagonists on Tregs in MS/EAE.

## 8. $A_{2A}R$ antagonist attenuates the pro-inflammation phenotype of microglia

Microglia are innate immune cells of the CNS that play a role similar to macrophages and are recognized as essential factors in tissue repair and regeneration, as well as anti-infection immunity. In the course of MS, in addition to the characteristic peripheral immune cell infiltration, and microglial cell proliferation also occurs. In MS tissues, macrophage accumulation is associated with demyelination (Ferguson et al., 1997), and nearly half of these macrophages are derived from microglia, while the other half originates from monocytes (Brück et al., 1995). Although microglia and mononuclear macrophages have similar functions, they have different origins during embryogenesis. Microglia develop from yolk capsule progenitor cells and self-proliferate independently of bone marrow hematopoietic stem cells (Gomez Perdiguero, Schulz, & Geissmann, 2013), whereas mononuclear macrophages are derived from bone marrow hematopoietic stem cells (Ginhoux et al., 2010). Therefore, while microglia exhibit similar behavior to mononuclear macrophages in the nervous system, it has been suggested that they play different roles in human MS and EAE models. Mononuclear macrophages cause tissue damage with their powerful phagocytosis and inflammatory effects, whereas microglia surprisingly exhibit immunosuppressive properties (Yamasaki et al., 2014). At the same time, microglia have a bidirectional effect on promoting tissue repair and exacerbating the injury, and the $A_{2A}R$ regulates this contrasting effect (Rajasundaram, 2018). However, the role of $A_{2A}R$ signaling in this process is complex. Studies have shown that the $A_{2A}R$ signal in non-immune cells that make up the brain barrier mediates immune cell entry into the CNS. Conversely, the $A_{2A}R$ signal in immune cells leads to a phenotypic transformation of immune cells that inhibits inflammation.

During the acute phase of the EAE model, microglia rapidly differentiate into the M1 phenotype and secrete Th1 cytokines, leading to an inflammatory response and causing tissue damage, such as demyelination and neuronal death. However, during the late stage of EAE, the microglia in the nervous system mainly exhibit the M2 phenotype, which secretes Th2-type inflammatory cytokines and inhibits the progression of EAE (Chu et al., 2018). $A_{2A}R$ signaling has been demonstrated to regulate this phenotypic transition, and its activation can promote the transformation of classically activated microglia into an M2-like phenotype through an IL-4Rα-independent pathway (Ferrante et al., 2013).

In vitro experiments have shown that $A_{2A}R$ expression is rapidly upregulated in microglia treated with a myelin sheath. Subsequently, $A_{2A}R$ activator CGS21680 was found to reduce microglia phagocytosis activity and migration ability (Ingwersen et al., 2016). Furthermore, this study found that $A_{2A}R$ activator treatment had protective effects when administered in the early stage of EAE, while administration during the later stage of the disease worsened the symptoms. One possible explanation for the protective effect of early $A_{2A}R$ signaling in EAE is that it inhibits the phagocytosis and migration of microglia cells, which limits their invasion and destruction of the CNS. However, in advanced stages, microglia have weak phagocytosis and reduced action, which hinders myelin regeneration due to impaired uptake of myelin fragments (Neumann et al., 2009). Therefore, the timing of microglia activation associated with the $A_{2A}R$ signal can have opposite effects on the disease. Moreover, the bidirectional effect of $A_{2A}R$ signaling may also be influenced by the local microenvironment and dosage (Loram et al., 2015).

Experiments have shown that the $A_{2A}R$ antagonist SCH58261 can effectively reduce microglia infiltration in the CNS, reduce demyelinating areas in lysolecithin (LPC)-induced inflammation, improve spatial memory (Akbari, Khalili-Fomeshi, Ashrafpour, Moghadamnia, & Ghasemi-Kasman, 2018), and prevent neurobehavioral defects in the $MOG_{35-55}$-induced EAE model, providing protective effects for EAE (Chen et al., 2019). However, the absence of $A_{2A}R$ signaling may lead to more severe EAE disease (Wang et al., 2014) because the microglia of $A_{2A}R^{-/-}$ mice change to a more proinflammatory phenotype, which negates any barrier protective effect of the CNS conferred by the absence of $A_{2A}R$ signaling.

In conclusion, $A_{2A}R$ signaling is implicated in both the protective and detrimental effects of microglia in MS and EAE models. $A_{2A}R$ ligands influence microglia behavior by several factors, such as treatment timing, drug dosage, microenvironment, and other factors. Therefore, identifying the actual effect of the $A_{2A}R$ signal on microglia under different conditions is crucial for the clinical application of $A_{2A}R$ antagonists in MS treatment.

## 9. $A_{2A}R$ antagonist inhibits the activation of astrocytes

Astrocytes are glial cells found throughout the brain. They play a crucial role in the formation and maintenance of the BBB, as well as the regulation of local ion levels and various physiological processes in neurons (Kwon & Koh, 2020). MS is a disease that can be categorized into two types: relapsing MS and progressive MS. The former is characterized by peripheral immune cells transiently infiltrating, and peripheral immunosuppressive therapy is an effective treatment strategy. In contrast, the immune damage effect of the latter is mainly mediated by CNS resident cells such as astrocytes and microglia, making peripheral immunosuppressive therapy ineffective (Healy, Stratton, Kuhlmann, & Antel, 2022; Katz Sand, 2015). Therefore, a deeper understanding of the role of CNS resident cells in neuroinflammation is needed to develop effective therapeutic strategies for progressive MS.

Astrocytes can be activated and proliferated by $A_{2A}R$ signaling through glutamic acid uptake regulation. In addition, $A_{2A}R$ signaling can promote the release of neurotrophic factors such as GDNF and BDNF by astrocytes, which can contribute to myelin repair and regeneration (Merighi et al., 2022). However, during chronic neuroinflammation in EAE mice, astrocytes produce LacCer, which can lead to neurodegeneration (Mayo et al., 2014). A recent study showed that $A_{2A}R$ knockout increased disease deterioration in EAE mice (Ingwersen et al., 2016), but interestingly, the team also found higher levels of the astrocyte activation marker GFAP in KO mice than in WT mice. The authors explained that astrocyte proliferation in KO mice may have been a compensatory response to inflammation. However, due to the dual nature of astrocytes, which can have both cytotoxic and neuroprotective effects (Kwon & Koh, 2020), it cannot be definitively determined whether the increased astrocyte proliferation is a compensatory protective response against disease without further studies. Additionally, inhibiting $A_{2A}R$ signaling can decrease the activation of astrocytes and microglia, allowing for comparison in the number of activations between the two cell types, ultimately leading to a neuroprotective effect (Lambertucci et al., 2022). In conclusion, $A_{2A}R$ is involved in various pathways related to astrocytes and contributes to neuroinflammation, neurodegeneration, and other pathological processes in MS. However, experimental results have shown that the protective effect of $A_{2A}R$-mediated astrocyte activation is not always absolute and depends on multiple factors, such as the disease stage and interactions with microglia (Liddelow et al., 2017). Further research is needed to fully understand the role of $A_{2A}R$ signaling in astrocytes and develop effective therapeutic strategies for MS.

Many studies have demonstrated the protective effect of blocking or inactivating $A_{2A}R$ on neurodegenerative diseases. A potential therapeutic direction for MS treatment is the use of the selective $A_{2A}R$ inhibitor SCH58261, which has been shown to reduce LPS-induced demyelination (Akbari et al., 2018). In vitro experiments have also proved that SCH58261 can block the $A_{2A}R$ signal and reverse transcriptional dysregulation caused by $A_{2A}R$ overexpression in astrocytes. Dysregulated genes are mainly associated with immune response, angiogenesis, and cell activation (Paiva et al., 2019). Numerous clinical data have demonstrated that caffeine, a non-selective antagonist of $A_{2A}R$, is strongly associated with a reduced risk of MS (Ruggiero et al., 2022).

Although numerous studies have demonstrated the correlation between astrocytes and $A_{2A}R$ signaling in MS pathology, there is still a lack of research on the specific effect of $A_{2A}R$ knockout in EAE. Investigating the pathogenic mechanism of $A_{2A}$ receptors on astrocytes in MS is essential to gain a better understanding of the role of $A_{2A}R$ signaling on astrocytes in this disease. It will also help determine the feasibility and effectiveness of treating astrocytes with $A_{2A}R$ antagonists.

Astrocyte-specific deletion of $A_{2A}R$ has been found to affect the pathological process of the CNS in other diseases. For instance, Sandhoff disease is a CNS lysosomal storage disorder that triggers intense CNS inflammation and is characterized by the activation of astrocytes and microglia due to a *HEXB* gene mutation. The interactions between astrocytes and microglia have been demonstrated to play a key role in disease development (Jha, Jo, Kim, & Suk, 2019). Based on the previous conclusion, inhibition of $A_{2A}R$ on astrocytes can reduce the activation of microglia during the late inflammatory period of Sandhoff disease (Ogawa et al., 2018), suggesting that $A_{2A}R$ antagonists may play a neuroprotective role indirectly by reducing the communication between these two types of glial cells. Furthermore, a previous study on schizophrenia found that knocking out $A_{2A}R$ in astrocytes disrupts glutamate homeostasis via the

GLU-1 pathway, leading to the psychomotor enhancement and cognitive impairment (Matos et al., 2015). These studies suggest that the $A_{2A}R$ signal on astrocytes may also be involved in the pathogenesis of chronic nerve inflammation and neuronal degeneration related to cognitive function in MS, but further studies are needed to confirm this.

Overall, there is increasing evidence to support the application of $A_{2A}R$ antagonists to inhibit astrocyte activation in MS. This provides direct or indirect protection for MS and offers promising treatment options for this debilitating disease. However, the role of astrocytes in MS disease is complex, and there are still many unanswered questions. For example, numerous studies have shown that EAE model mice with whole-body $A_{2A}R$ knockout have a poorer prognosis than WT mice despite having more astrocytes in the brain (Kwon & Koh, 2020). Additionally, the heterogeneity of astrocytes, which is difficult to distinguish due to the lack of specific cell markers, may play a significant role in the progression of MS disease. There is growing evidence that different astrocyte subtypes have distinct roles in MS, and more targeted studies are needed to develop effective astrocyte-targeted MS therapies.

## 10. The $A_{2A}R$ antagonist protects oligodendrocytes from damage

Oligodendrocytes (OL) are a type of glial cells found throughout the CNS. They are named for their relatively small number of dendrites visible in silver staining. OLs are responsible for wrapping neuronal axons with multiple layers of the myelin sheath, which insulates the axons and facilitates rapid jump conduction of bioelectrical signals (Moore et al., 2020). At the same time, OLs also play a crucial role in nourishing and supporting neurons, helping to maintain their normal physiological functions and protecting them from damage. The dysfunction of oligodendrocytes is linked to various neurodegenerative and demyelinating disorders such as MS and Alzheimer's disease.

OLs are derived from oligodendrocyte progenitor cells (OPCs), which are resident stem cells in the CNS that retain the ability to differentiate and proliferate during embryonic development and adulthood. OPCs protect the normal operation of the CNS by differentiating into OLs that form the myelin sheath (Nave & Werner, 2014). When this process is interrupted, it can cause CNS damage and the aforementioned conditions.

In a typical demyelinating disease, the myelin sheath surrounding neurons' axons is degraded due to various pathological factors, requiring OPCs differentiation to form new OLs to facilitate myelin sheath repair. However, the unfavorable microenvironment near the lesion becomes a major obstacle to this process, leading to myelin regeneration disorders, nerve conduction obstructions, and clinical symptoms of the disease. While the molecular signals that regulate OPC proliferation and differentiation during embryonic development are well understood, the regulatory mechanisms behind adult OPC-mediated myelin regeneration remain unclear. However, there is a growing body of evidence indicating that therapies targeting OPC-mediated myelin regeneration could be feasible and effective in the clinical treatment of demyelinating diseases (Skaper, 2019). Among them, $A_{2A}R$ has emerged as a promising target for such therapies, given its ability to regulate OPC differentiation, proliferation, and migration.

Current research indicates that $A_{2A}R$ signaling may prevent OPC maturation in vitro. In a study conducted by Coppi and colleagues, OPC cultures exposed to the $A_{2A}R$ agonist CGS21680 demonstrated a higher percentage of naive OPCs in 12-day cell cultures. Additional experiments utilizing tetraethylammonium to block the I(K) current provide further evidence that $A_{2A}R$ signaling delays the differentiation of OPCs in vitro by inhibiting the I(K) current on OPC cell membranes. Conversely, the $A_{2A}R$ antagonist SCH58261 can negate the above effects of $A_{2A}R$ activation (Coppi, Cellai, Maraula, Pugliese, & Pedata, 2013). The same research team has shown a direct relationship between cAMP and I(K) current in recent years. They observed a reduction in I(K) current with the exogenous AC agonist forskolin (Coppi et al., 2020). Moreover, the protective effect of the $A_{2A}R$ antagonist on oligodendrocytes was confirmed by the finding that SCH58261 could prevent oligodendrocyte damage by reducing the activation of JNK/MAPK signaling in oligodendrocytes in the MCAO mode (Melani et al., 2009).

However, there are some paradoxical phenomena observed in the development of OPCs mediated by the $A_{2A}R$ signal. For example, in a study on the association between oligodendrocyte differentiation and $A_{2A}R$ signaling in Niemann Pick type C (NPC) disease, De Nuccio et al. observed a dual effect of $A_{2A}R$ signaling on OL differentiation (De Nuccio et al., 2019). NPC is a rare hereditary central nervous system disorder characterized by abnormal myelin development and OPC maturation disorders. In the study, the team induced an NPC phenotype in OPCs using the cholesterol transport inhibitor u18666a and treated it with the $A_{2A}R$ agonist CGS21680. Experimental results showed that the $A_{2A}R$ signal could eliminate NPC characteristics of OPCs, promote the differentiation of OPCs, and provide protection against the disease by reducing cholesterol accumulation and changes in mitochondrial membrane potential in u18666a-treated OPCs. However, the study also revealed that CGS21680 inhibited normal OPC differentiation in the control group, which is consistent with previous in vitro experiments involving OPCs. Hence, it is believed that the $A_{2A}R$ signal exhibits a dual effect on regulating OPC differentiation, as observed in studies on NPC and Huntington's disease (Martire et al., 2007). Although the specific mechanism behind this dual regulation remains unclear, it is thought that the $A_{2A}R$ signal satisfies the physiological demands for maintaining a stable state of OPC differentiation and proliferation in both normal and pathological states, leading to positive outcomes (De Nuccio et al., 2019).

The $A_{2A}R$ antagonist SCH58261 has been found to have a distinct effect on zebrafish central OPCs, unlike other adenosine receptor antagonists. SCH58261 was found to enable these OPCs to migrate to the periphery, which is necessary for proper myelination. However, in contrast to the previous conclusion that the $A_{2A}R$ signal inhibited OPC maturation, this experiment demonstrated that SCH58261 not only increased OPC migration by reducing neuronal activity but also decreased their ability to differentiate and form myelin (Fontenas et al., 2019).

The focus of OLs research has primarily been on inducing remyelination in response to MS as a demyelinating disease. However, with an improved understanding of MS pathogenesis, researchers have realized that phenotypic changes in OPCs may also play a role. In vitro experiments have shown that IFN-γ can induce OPCs to assume an immune-like state and activate T cells via the expression of major histocompatibility complex I and II. This heterogeneity of OPCs was further confirmed through single-cell RNA sequencing in EAE mice (Falcão et al., 2018).

## 11. $A_{2A}R$ antagonist protects against lymphocyte infiltration into the CNS across the BBB

The CNS blood vessels are distinct due to the BBB, a structure composed of vascular endothelium, pericytes, perivascular astrocytes, and basal lamina (Ortiz et al., 2014). The BBB maintains brain homeostasis by restricting biomacromolecules and immune cells from entering the CNS through tight junctions (TJs) between the monolayer endothelial cells (ECs) of the blood vessels. This strict control protects neural tissue from toxins or pathogens. However, the BBB properties are affected by neuroinflammation, which can be a crucial factor in the progression of various neurological diseases.

The breakdown of the BBB and the subsequent infiltration of immune cells into the CNS is a well-established hallmark of MS pathogenesis. Numerous MRI imaging studies have observed BBB leakage and indicated that BBB dysfunction may precede myelin lesions in MS (Bastianello et al., 1990; Kermode et al., 1990; Mattioli et al., 1993). Histopathological studies on postmortem MS brain tissues also provide direct evidence of BBB dysfunction at different MS phases. During MS/EAE pathology, the expression of adhesion molecules, such as ICAM-1 (intercellular cell adhesion molecule-1) and VCAM-1 (vascular cell adhesion molecule-1), significantly increases in ECs. The upregulated adhesion molecules on endothelial cells facilitate the binding and infiltration of activated immune cells through the BBB via leukocyte integrins. Currently, effective MS therapy involves inhibiting activated lymphocyte infiltration across the BBB. Natalizumab, a recombinant humanized monoclonal antibody targeting VLA-4 (very late antigen-4), blocks the entry of leukocytes into the CNS via the BBB, reducing the MS relapse rate by approximately 70% (Polman et al., 2006).

The evidence supporting the role of $A_{2A}R$ signaling in mediating BBB permeability is strong. Injection of an $A_{2A}R$ agonist (Lexiscan) in mice resulted in rapid detection of dextran (10 kDa) or β-amyloid antibody in the brain within 5 min (Carman, Mills, Krenz, Kim, & Bynoe, 2011). Maximum BBB permeability was observed at around 30–60 min both in vitro and in vivo (Carman, Mills, Krenz, Kim, & Bynoe, 2011; Gao et al., 2014). Furthermore, it was discovered that the $A_{2A}R$ agonist Lexiscan can modulate the P-glycoprotein (P-gp) expressed on brain endothelial cells, thereby regulating BBB permeability. Based on its biological properties, $A_{2A}R$ nano agonists have been developed to induce cell contraction and paracellular TJ opening for drug delivery (Gao et al., 2017). While studies have shown that treatment with $A_{2A}R$ antagonists can reduce the invasion of immune cells into the CNS in EAE, it is difficult to determine the specific contribution of $A_{2A}R$ signaling in ECs to EAE pathology, as current observations are limited to pharmacological or knockout studies. However, a study on mice specifically deficient in endothelial $A_{2A}R$

($A_{2A}R^{\Delta VEC}$) found that they were resistant to stroke, with reduced leukocyte infiltration and alleviative BBB leakage (Zhou et al., 2019). This suggests that inhibiting the infiltration of immune cells across the BBB is one of the protective mechanisms of $A_{2A}R$ antagonists.

## 12. $A_{2A}R$ antagonist suppresses T-cell infiltration by decreasing the ChP gateway activity

The choroid plexus (ChP) is composed mainly of porous vascular endothelial cells, epithelial cells, a few fibroblasts, and resident immune cells in the stroma. Its primary biological function is to secrete CSF to remove harmful substances from the brain and maintain intracranial pressure (Benarroch, 2016; Lun, Monuki, & Lehtinen, 2015). In addition to its cleansing function, the ChP also secretes several neurotrophins, including brain-derived neurotrophic factor (BDNF), nerve growth factor (NGF), transforming growth factor (TGF), and glial cell line-derived neurotrophic factor (GDNF) (Cruz et al., 2018), to nourish the brain. Furthermore, the ChP is responsible for maintaining the blood–CSF barrier, where the monolayer epithelium acts as a gatekeeper to prevent infiltration of peripheric immune cells into the CSF during infections or tissue damage.

The ChP is a pivotal structure located at the nexus between the CSF, brain parenchyma, and systemic circulation. Traditionally, the blood–brain barrier has been viewed as the primary route for leukocyte infiltration into the brain in MS/EAE. However, recent studies have found that $CD4^+$ T cells, including $Th1^+$ and $Th17^+$ T cells, are significantly enriched in the CSF of MS patients (Schafflick et al., 2020; Schneider-Hohendorf et al., 2021; van Langelaar et al., 2018). In addition, under EAE/MS pathology, chemotaxis, and adhesion molecules (CAMs), such as ICAM-1, VCAM-1, and CCL5 are highly expressed in the choroid plexus tissue (Steffen, Breier, Butcher, Schulz, & Engelhardt, 1996). The evidence indicates that $CD4^+$ T cells infiltrate the CSF through the choroid plexus. This was directly demonstrated in a T-cell tracing experiment, where fluorescently labeled T lymphocytes were transferred into mice with EAE induction, showing a "two-wave" pattern of infiltration (Reboldi et al., 2009). The "first wave" of T lymphocytes migrates into the CSF through the choroid plexus at the early stage of EAE, followed by the "second wave" invasion of T cells into the brain through the BBB at subsequent stages. It is believed that the "first wave" T lymphocytes infiltrate the CSF and disperse around the perivascular space, which results in the release of proinflammatory factors, such as TNF-α and IFN-γ, from activated lymphocytes. This, in turn, leads to BBB integrity disruption and a strong invasion of immune cells across breached BBB endothelial cells. To validate the role of the choroid plexus in the pathogenesis of MS/EAE, researchers employed a pharmacological approach and gene knockout technology (Breuer et al., 2018; Jovanova-Nesic et al., 2009).

However, the systemic effects of drugs and knockout technology were insufficient to determine the exact role of the choroid plexus in MS/EAE. In the EAE model, it was found that $CD4^+$ T cells isolated from $CD73^{-/-}$ mice produced more proinflammatory cytokines than those from WT mice and induced more severe EAE pathology when transferred to T cell-deficient recipients. Unexpectedly, $CD73^{-/-}$ mice were resistant to EAE. It was initially hypothesized that the BBB ECs might be potential candidates for the role of CD73 signaling since CD73 is often enriched in peripheral vessels (Koszalka et al., 2004). Immunohistochemistry results demonstrated that CD73 was expressed at a high level in the ChP epithelium, while it was almost undetectable in the endothelial cells of the BBB, providing the first indication of the role of adenosine signaling in regulating ChP permeability (Mills et al., 2008). Similarly, CD73 antagonists effectively inhibit immune infiltration across the ChP in spinal cord injury (SCI) models (Shechter et al., 2013). In parallel, it was observed that the $A_{2A}R$ agonist CGS21680 induced the expression of chemokines in vivo and suggested that $A_{2A}R$ could regulate T lymphocyte trafficking by modulating CX3CL1 (Mills, Alabanza, Mahamed, & Bynoe, 2012). Our recent study aimed to eliminate interferences from other brain regions by performing a ChP-specific knockdown of $A_{2A}R$ by ICV injection of TAT-Cre recombinant protein. Upon $MOG_{35-55}$ immunization, we found that specific knockdown of $A_{2A}R$ in the ChP inhibited the infiltration of $Th17^+$ T cells, reduced the degree of pathological damage, and delayed disease onset by downregulating the CCR6-CCL20 axis (Zheng et al., 2022). This study provides direct evidence that increased $A_{2A}R$ signaling in the ChP contributes to MS pathology. In normal conditions, bulk RNA-seq data showed that administering the $A_{2A}R$ antagonist KW6002 via intraperitoneal injection could decrease the expression of TJ and CAM genes in the ChP (Ye et al., 2023). Additionally, in vivo experiments demonstrated that KW6002 reduced the expression of CAMs as well as prevented the infiltration of $CD3^+$ T lymphocytes into the CNS through the ChP. Of note, the treatment of $MOG_{35-55}$ immunized mice with KW6002 during the immunoinfiltrating phase was found to impede the infiltration of T cells across the ChP and relieve pathological symptoms, as demonstrated in a recent study (Zheng et al., 2022). Therefore, one way in which $A_{2A}R$ antagonists are protective in MS/EAE is by blocking $A_{2A}R$ signaling in the ChP and preventing the infiltration of "first wave" T lymphocytes across the ChP route. This conclusion is consistent with the hypothesis that $A_{2A}R$ antagonists preferentially target $A_{2A}R$ in non-hematopoietic cells.

## 13. Safety of the $A_{2A}R$ antagonist in clinical trials

Multiple sclerosis (MS) is an autoimmune disease that affects the central nervous system, characterized by demyelination and synaptic lesions in the brain and spinal cord. Various drugs are currently used to treat MS, such as glucocorticoids, interferon-β (IFN-β), glatiramer acetate, teriflunomide, natalizumab, fingolimod, and others (Rae-Grant et al., 2018). These drugs work primarily by regulating the immune response of lymphocytes to slow the disease progression and decrease the recurrence rate. However, they cannot completely cure the disease and often have adverse reactions. Progressive multifocal leukoencephalopathy is the most specific adverse effect of these drugs because lymphopenia, caused by the drugs, blocks the migration of T cells to the CNS, thereby interfering with the normal monitoring function of T cells in the CNS (Berger, 2017). Therefore, conducting clinical trials to identify low-toxicity, effective, and safe treatment strategies for MS is of paramount importance.

Over the past two decades, studies have shown that targeting the $A_{2A}R$ is a promising approach to treating brain damage and neuroinflammation in various animal models of neurodegeneration, including MS (Chen et al., 2001; Ning et al., 2013; Rajasundaram, 2018a; Zheng et al., 2022a). This is supported by several studies that have demonstrated the neuroprotective and anti-inflammatory effects of $A_{2A}R$ antagonists. The molecular mechanism underlying these effects is thought to involve the regulation of proinflammatory cytokine production and proliferation. Moreover, the successful use of $A_{2A}R$ antagonists in treating Parkinson's disease has provided valuable insights into the safety of these drugs in the context of drug development for MS patients (Table 1) (Takahashi, Fujita, Asai, Saki, & Mori, 2018).

In a placebo-controlled, double-blind trial, Istradefylline showed a clinically meaningful improvement in motor symptoms and was well-tolerated

Table 1 Safety studies of $A_{2A}R$ antagonists conducted with Parkinson's disease patients.

| $A_{2A}R$ antagonists | Time | Phase of clinical development | Number of subjects | Safety evaluation | References |
|---|---|---|---|---|---|
| Istradefylline | March 2003 | Phase II | 15 | No medically important drug toxicity occurred. | Bara-Jimenez et al. (2003b) |
| Istradefylline | May 2003 | Phase II | 83 | Nausea was the most common adverse event. | Hauser, Hubble, Truong, and Istradefylline US-001 Study Group (2003) |
| Istradefylline | Feb 2008 | Phase III | 196 | Treatment-emergent adverse effects were generally mild. | LeWitt et al. (2008) |
| Istradefylline | June 2008 | Phase III | 395 | Among the common adverse events were dyskinesia, nausea, dizziness, and hallucinations. | Hauser et al. (2008) |
| Istradefylline | October 2008 | Phase III | 231 | Dyskinesia, lightheadedness, tremor, constipation, and weight decrease were reported more often with istradefylline than with placebo. | Liu et al. (2019) |

| | | | | | |
|---|---|---|---|---|---|
| Preladenant | June 2013 | Phase II | 140 | Adverse events, reported by ≥15% of subjects, were dyskinesia and constipation. | Factor et al. (2013) |
| Tozadenant | Dec 2010 | Phase II | 21 | Dizziness, nausea, mood anxiety | Black, Koller, Campbell, Gusnard, and Bandak (2010) |
| Preladenant | March 2011 | Phase II | 253 | Preladenant showed no difference in adverse effects compared with the placebo. | Hauser et al. (2011) |
| Preladenant | Nov 2016 | Phase II | 450 | Preladenant was well tolerated, and the frequency of adverse events appeared to be dose-related. | Hattori et al. (2016) |

as a levodopa adjunct in Parkinson's disease. Nausea, dizziness, and hallucinations were the most frequent adverse drug reactions observed (Stacy et al., 2008). Hauser et al. conducted a Phase III randomized study of Istradefylline, which included 231 subjects with PD, and showed a significant reduction in OFF time with good tolerability and mild treatment-related adverse effect (Hauser et al., 2008). Bara-Jimenez et al. reported that the selective $A_{2A}R$ antagonist KW6002 could alleviate symptoms in patients with PD who were treated with levodopa, with no occurrence of medically significant drug toxicity (Bara-Jimenez et al., 2003a). Phase III clinical trials of the selective $A_{2A}R$ antagonist Istradefylline (KW6002) and the competitive $A_{2A}R$ antagonist Preladenant (SCH420814) have shown that they are highly effective and safe in PD patients without any apparent adverse drug reactions (Hauser et al., 2011; Takahashi, Fujita, Asai, Saki, & Mori, 2018a). The US Food and Drug Administration (FDA) has approved KW6002 for the treatment of PD patients as an adjunct to levodopa drugs after clinical studies in more than 4000 PD patients (Ren & Chen, 2020). KW6002 was launched in Japan in 2013. A systematic review and meta-analysis were conducted by searching electronic databases up to November 2021, which integrated qualified studies on efficacy, tolerance, and incidence of adverse drug events. In total, 2727 subjects in nine clinical studies were treated with Istradefylline for Parkinson's disease, and the results showed a statistically significant difference in efficacy compared to a placebo. Moreover, there was no significant difference in tolerability or adverse effects between Istradefylline and placebo (Wang et al., 2022). Animal and clinical trials have demonstrated that $A_{2A}R$ antagonists are well-tolerated, safe, and have shown sustained improvement in patients with neurological injury. Nausea is the most commonly reported adverse effect associated with Istradefylline use. Due to its selective adenosine $A_{2A}$ receptor antagonist activity, Istradefylline can block these receptors without stimulating other receptors, which may offer a safe and beneficial treatment option for MS patients with dyskinesia. $A_{2A}R$ antagonists have also demonstrated significant therapeutic effects in retinopathy of prematurity (ROP) by reducing pathological neovascularization. Zhou et al. conducted a study showing that long-term use of caffeine or KW6002 did not affect physiological retinal vascular development during postnatal development in newborns (Zhou et al., 2015). Some researchers have expressed concerns about the potential neurodevelopmental effects of using $A_{2A}R$ antagonists to prevent ROP. However, these concerns do not apply to the use of $A_{2A}R$ antagonists in MS patients. Several studies have shown that the inactivation of $A_{2A}R$ reduces cognitive dysfunction

(Chen & Pedata, 2008; Mills et al., 2008), which is particularly important for the 65% of MS patients who experience cognitive impairment (Guilloton et al., 2020).

The studies conducted on $A_{2A}R$ antagonists and their effectiveness in treating various neurological diseases provide a sound biological basis and safety guarantee for the forthcoming clinical trials on MS. It is crucial to investigate the molecular targets of $A_{2A}R$ antagonists for shielding EAE and the optimal timing of treatment to achieve maximum therapeutic efficacy and minimize adverse reactions for safety reasons.

## 14. The limitations of current studies on the potential efficacy of $A_{2A}R$ antagonists in MS pathology

It is worth noting that while the EAE animal model is widely used to study MS, it only partially replicates the clinical symptoms observed in MS patients (Table 2). Therefore, when evaluating the potential therapeutic efficacy

Table 2 Limitations of the experimental autoimmune encephalomyelitis model compared to multiple sclerosis.

|  | MS | EAE |
|---|---|---|
| Predisposing factors | Autoimmune antigens, hereditary susceptibility (HLA polymorphism), bacterial and viral infections, environment (smoking, chill, et al.) | Active immunization induced with $MOG_{35-55}$ or proteolipid protein $(PLP_{139-151})$ |
| Immune cells involved in the pathological process | $CD4^+/CD8^+$ T lymphocytes and B lymphocytes involved in MS pathology | EAE is mainly driven by $CD4^+$ T cells. |
| Immune response | Various extents of lymphocyte infiltration or no immune response | Strong infiltration of lymphocytes |
| Classification of the disease course | Relapsing-remitting MS, secondary-progressive MS, primary-progressive MS, progressive-relapsing MS | EAE animal model recapitulates the relapsing-remitting MS. |

of candidate drugs, it is crucial to carefully consider the differences in symptoms and mechanisms between EAE and MS. For instance, the EAE model does not fully capture secondary-progressive disability, axonal degeneration, and remyelination that are hallmarks of MS. The EAE model is primarily focused on the activation and infiltration of T lymphocytes, but it should be noted that over 40% of MS patients do not exhibit immune cell infiltration (Behan & Chaudhuri, 2014). Moreover, the potential effects of $A_{2A}R$ antagonists on other immune cell types require further investigation. Recent studies have highlighted the crucial role of B cells in MS and EAE (Comi et al., 2021; Kim et al., 2021; Ramesh et al., 2020). Therefore, it is important to carefully consider the limitations and possible biases of animal models in the development of therapeutic strategies for MS. As far as we know, there are no studies investigating the potential role of $A_{2A}R$ signaling in B cells involved in MS/EAE pathology. Thus, more systematic studies are needed to elucidate the function of $A_{2A}R$ signaling in the immune system. While the EAE model cannot replicate all aspects of MS pathology, it has been crucial in promoting the development of several DMTs approved for MS treatment, such as glatiramer acetate and natalizumab. Therefore, despite its limitations, the EAE model remains an indispensable tool for studying MS pathology.

## 15. Conclusion

While the effects of $A_{2A}R$ signaling are complex and ambiguous, $A_{2A}R$ antagonists have consistently demonstrated protective effects in EAE pathology. Furthermore, considering the preferential actions on non-hematopoietic cells and their safety in clinical trials for other diseases, $A_{2A}R$ antagonists remain excellent candidates for the treatment of multiple sclerosis.

## Acknowledgments
Not applicable.

## Declarations
Not applicable.

## Ethics approval and consent to participate
Not applicable.

## Consent for publication
All authors have approved the contents of this manuscript and provided consent for publication.

## Funding

This work was supported by the Research Fund for International Senior Scientists (Grant # 82150710558), and the National Natural Science Foundation of China (Grants 81630040, 31771178 to Jiang-fan Chen, grants 31800903 to Wu Zheng).

## Competing interests

The authors declare that they have no competing interests.

## References

Airas, L., Niemelä, J., Yegutkin, G., & Jalkanen, S. (2007). Mechanism of action of IFN-beta in the treatment of multiple sclerosis: A special reference to CD73 and adenosine. *Annals of the New York Academy of Sciences, 1110*, 641–648.

Akbari, A., Khalili-Fomeshi, M., Ashrafpour, M., Moghadamnia, A. A., & Ghasemi-Kasman, M. (2018). Adenosine A2A receptor blockade attenuates spatial memory deficit and extent of demyelination areas in lyolecithin-induced demyelination model. *Life Sciences, 205*, 63–72.

Alam, M. S., Kurtz, C. C., Wilson, J. M., Burnette, B. R., Wiznerowicz, E. B., Ross, W. G., ... Ernst, P. B. (2009). A2A adenosine receptor (AR) activation inhibits pro-inflammatory cytokine production by human CD4+ helper T cells and regulates Helicobacter-induced gastritis and bacterial persistence. *Mucosal Immunology, 2*(3), 232–242.

Allegretta, M., Nicklas, J. A., Sriram, S., & Albertini, R. J. (1990). T cells responsive to myelin basic protein in patients with multiple sclerosis. *Science (New York, N. Y.), 247*(4943), 718–721.

Bara-Jimenez, W., Sherzai, A., Dimitrova, T., Favit, A., Bibbiani, F., Gillespie, M., ... Chase, T. N. (2003a). Adenosine A(2A) receptor antagonist treatment of Parkinson's disease. *Neurology, 61*(3), 293–296.

Bara-Jimenez, W., Sherzai, A., Dimitrova, T., Favit, A., Bibbiani, F., Gillespie, M., ... Chase, T. N. (2003b). Adenosine A(2A) receptor antagonist treatment of Parkinson's disease. *Neurology, 61*(3), 293–296.

Bastianello, S., Pozzilli, C., Bernardi, S., Bozzao, L., Fantozzi, L. M., Buttinelli, C., & Fieschi, C. (1990). Serial study of gadolinium-DTPA MRI enhancement in multiple sclerosis. *Neurology, 10*(4), 591–595.

Behan, P. O., & Chaudhuri, A. (2014). EAE is not a useful model for demyelinating disease. *Multiple Sclerosis and Related Disorders, 3*(5), 565–574.

Benarroch, E. E. (2016). Choroid plexus–CSF system: Recent developments and clinical correlations. *Neurology, 86*(3), 286–296.

Berger, J. R. (2017). Classifying PML risk with disease modifying therapies. *Multiple Sclerosis and Related Disorders, 12*, 59–63.

Black, K. J., Koller, J. M., Campbell, M. C., Gusnard, D. A., & Bandak, S. I. (2010). Quantification of indirect pathway inhibition by the adenosine A2a antagonist SYN115 in Parkinson disease. *The Journal of Neuroscience: The Official Journal of the Society for Neuroscience, 30*(48), 16284–16292.

Blackburn, M. R., Vance, C. O., Morschl, E., & Wilson, C. N. (2009). Adenosine receptors and inflammation. *Handbook of Experimental Pharmacology, 193*, 215–269.

Boison, D., & Yegutkin, G. G. (2019). Adenosine metabolism: Emerging concepts for cancer therapy. *Cancer cell, 36*(6), 582–596.

Breuer, J., Korpos, E., Hannocks, M. J., Schneider-Hohendorf, T., Song, J., Zondler, L., ... Schwab, N. (2018). Blockade of MCAM/CD146 impedes CNS infiltration of T cells over the choroid plexus. *Journal of neuroinflammation, 15*(1), 236.

Brück, W., Porada, P., Poser, S., Rieckmann, P., Hanefeld, F., Kretzschmar, H. A., & Lassmann, H. (1995). Monocyte/macrophage differentiation in early multiple sclerosis lesions. *Annals of Neurology, 38*(5), 788–796.

Burgos, E. S., Gulab, S. A., Cassera, M. B., & Schramm, V. L. (2012). Luciferase-based assay for adenosine: Application to S-adenosyl-L-homocysteine hydrolase. *Analytical Chemistry, 84*(8), 3593–3598.

Carman, A. J., Mills, J. H., Krenz, A., Kim, D. G., & Bynoe, M. S. (2011). Adenosine receptor signaling modulates permeability of the blood-brain barrier. *The Journal of Neuroscience: The Official Journal of the Society for Neuroscience, 31*(37), 13272–13280.

Cekic, C., & Linden, J. (2016). Purinergic regulation of the immune system. *Nature Reviews. Immunology, 16*(3), 177–192.

Cella, M., Facchetti, F., Lanzavecchia, A., & Colonna, M. (2000). Plasmacytoid dendritic cells activated by influenza virus and CD40L drive a potent TH1 polarization. *Nature Immunology, 1*(4), 305–310.

Chen, J. F., & Pedata, F. (2008). Modulation of ischemic brain injury and neuroinflammation by adenosine A2A receptors. *Current Pharmaceutical Design, 14*(15), 1490–1499.

Chen, J. F., Eltzschig, H. K., & Fredholm, B. B. (2013). Adenosine receptors as drug targets—What are the challenges? *Nature Reviews. Drug Discovery, 12*(4), 265–286.

Chen, J. F., Xu, K., Petzer, J. P., Staal, R., Xu, Y. H., Beilstein, M., ... Schwarzschild, M. A. (2001). Neuroprotection by caffeine and A(2A) adenosine receptor inactivation in a model of Parkinson's disease. *The Journal of neuroscience : the official journal of the Society for Neuroscience, 21*(10), RC143.

Chen, Y., Zhang, Z. X., Zheng, L. P., Wang, L., Liu, Y. F., Yin, W. Y., ... Zheng, R. Y. (2019). The adenosine A2A receptor antagonist SCH58261 reduces macrophage/microglia activation and protects against experimental autoimmune encephalomyelitis in mice. *Neurochemistry International, 129*, 104490.

Chu, F., Shi, M., Zheng, C., Shen, D., Zhu, J., Zheng, X., & Cui, L. (2018). The roles of macrophages and microglia in multiple sclerosis and experimental autoimmune encephalomyelitis. *Journal of Neuroimmunology, 318*, 1–7.

Comi, G., Bar-Or, A., Lassmann, H., Uccelli, A., Hartung, H. P., Montalban, X., ... Expert Panel of the 27th Annual Meeting of the European Charcot Foundation (2021). Role of B cells in multiple sclerosis and related disorders. *Annals of Neurology, 89*(1), 13–23.

Constantinescu, C. S., Farooqi, N., O'Brien, K., & Gran, B. (2011). Experimental autoimmune encephalomyelitis (EAE) as a model for multiple sclerosis (MS). *British Journal of Pharmacology, 164*(4), 1079–1106.

Coppi, E., Cellai, L., Maraula, G., Pugliese, A. M., & Pedata, F. (2013). Adenosine $A_2A$ receptors inhibit delayed rectifier potassium currents and cell differentiation in primary purified oligodendrocyte cultures. *Neuropharmacology, 73*, 301–310.

Coppi, E., Cherchi, F., Fusco, I., Dettori, I., Gaviano, L., Magni, G., ... Pugliese, A. M. (2020). Adenosine A2B receptors inhibit K+ currents and cell differentiation in cultured oligodendrocyte precursor cells and modulate sphingosine-1-phosphate signaling pathway. *Biochemical Pharmacology, 177*, 113956.

Correale, J., Gaitán, M. I., Ysrraelit, M. C., & Fiol, M. P. (2017). Progressive multiple sclerosis: From pathogenic mechanisms to treatment. *Brain: A Journal of Neurology, 140*(3), 527–546. https://doi.org/10.1093/brain/aww258

Cruz, Y., García, E. E., Gálvez, J. V., Arias-Santiago, S. V., Carvajal, H. G., Silva-García, R., ... Ibarra, A. (2018). Release of interleukin-10 and neurotrophic factors in the choroid plexus: Possible inductors of neurogenesis following copolymer-1 immunization after cerebral ischemia. *Neural Regeneration Research, 13*(10), 1743–1752.

De Nuccio, C., Bernardo, A., Ferrante, A., Pepponi, R., Martire, A., Falchi, M., ... Minghetti, L. (2019). Adenosine A2A receptor stimulation restores cell functions and differentiation in Niemann-Pick type C-like oligodendrocytes. *Scientific Reports, 9*(1), 9782.

Deaglio, S., Dwyer, K. M., Gao, W., Friedman, D., Usheva, A., Erat, A., ... Robson, S. C. (2007). Adenosine generation catalyzed by CD39 and CD73 expressed on regulatory T cells mediates immune suppression. *The Journal of Experimental Medicine, 204*(6), 1257–1265.

Dungo, R., & Deeks, E. D. (2013). Istradefylline: First global approval. *Drugs, 73*(8), 875–882.

Elain, G., Jeanneau, K., Rutkowska, A., Mir, A. K., & Dev, K. K. (2014). The selective anti-IL17A monoclonal antibody secukinumab (AIN457) attenuates IL17A-induced levels of IL6 in human astrocytes. *Glia, 62*(5), 725–735.

Factor, S. A., Wolski, K., Togasaki, D. M., Huyck, S., Cantillon, M., Ho, T. W., ... Pourcher, E. (2013). Long-term safety and efficacy of preladenant in subjects with fluctuating Parkinson's disease. *Movement Disorders: Official Journal of the Movement Disorder Society, 28*(6), 817–820.

Falcão, A. M., van Bruggen, D., Marques, S., Meijer, M., Jäkel, S., Agirre, E., ... Castelo-Branco, G. (2018). Disease-specific oligodendrocyte lineage cells arise in multiple sclerosis. *Nature Medicine, 24*(12), 1837–1844.

Ferguson, B., Matyszak, M. K., Esiri, M. M., & Perry, V. H. (1997). Axonal damage in acute multiple sclerosis lesions. *Brain: A Journal of Neurology, 120*(Pt 3), 393–399.

Ferrante, C. J., Pinhal-Enfield, G., Elson, G., Cronstein, B. N., Hasko, G., Outram, S., & Leibovich, S. J. (2013). The adenosine-dependent angiogenic switch of macrophages to an M2-like phenotype is independent of interleukin-4 receptor alpha (IL-4Rα) signaling. *Inflammation, 36*(4), 921–931.

Fletcher, J. M., Lalor, S. J., Sweeney, C. M., Tubridy, N., & Mills, K. H. (2010). T cells in multiple sclerosis and experimental autoimmune encephalomyelitis. *Clinical and Experimental Immunology, 162*(1), 1–11. https://doi.org/10.1111/j.1365-2249

Fletcher, J. M., Lonergan, R., Costelloe, L., Kinsella, K., Moran, B., O'Farrelly, C., ... Mills, K. H. (2009). CD39+Foxp3+ regulatory T cells suppress pathogenic Th17 cells and are impaired in multiple sclerosis. *Journal of Immunology (Baltimore, Md.: 1950), 183*(11), 7602–7610.

Fontenas, L., Welsh, T. G., Piller, M., Coughenour, P., Gandhi, A. V., Prober, D. A., & Kucenas, S. (2019). The neuromodulator adenosine regulates oligodendrocyte migration at motor exit point transition zones. *Cell Reports, 27*(1), 115–128.e5.

Fonteneau, J. F., Gilliet, M., Larsson, M., Dasilva, I., Münz, C., Liu, Y. J., & Bhardwaj, N. (2003). Activation of influenza virus-specific CD4+ and CD8+ T cells: A new role for plasmacytoid dendritic cells in adaptive immunity. *Blood, 101*(9), 3520–3526.

Friberg, L. (1989). Regional cerebral blood flow in psychiatry as measured by single photon emission computed tomography. *Psychiatry Research, 29*(3), 319–322.

Gao, X., Qian, J., Zheng, S., Changyi, Y., Zhang, J., Ju, S., ... Li, C. (2014). Overcoming the blood-brain barrier for delivering drugs into the brain by using adenosine receptor nanoagonist. *ACS Nano, 8*(4), 3678–3689.

Gao, X., Wang, Y. C., Liu, Y., Yue, Q., Liu, Z., Ke, M., ... Li, C. (2017). Nanoagonist-mediated endothelial tight junction opening: A strategy for safely increasing brain drug delivery in mice. *Journal of Cerebral Blood Flow and Metabolism: Official Journal of the International Society of Cerebral Blood Flow and Metabolism, 37*(4), 1410–1424.

Gholamzad, M., Ebtekar, M., Ardestani, M. S., Azimi, M., Mahmodi, Z., Mousavi, M. J., & Aslani, S. (2019). A comprehensive review on the treatment approaches of multiple sclerosis: Currently and in the future. *Inflammation Research: Official Journal of the European Histamine Research Society, 68*(1), 25–38. https://doi.org/10.1007/s00011-018-1185-0

Ginhoux, F., Greter, M., Leboeuf, M., Nandi, S., See, P., Gokhan, S., ... Merad, M. (2010). Fate mapping analysis reveals that adult microglia derive from primitive macrophages. *Science (New York, N. Y.), 330*(6005), 841–845.

Gomez Perdiguero, E., Schulz, C., & Geissmann, F. (2013). Development and homeostasis of "resident" myeloid cells: The case of the microglia. *Glia, 61*(1), 112–120.

Guilloton, L., Camdessanche, J. P., Latombe, D., Neuschwander, P., Cantalloube, S., Thomas-Anterion, C., ... Cognition-MS group of the Rhône-Alpes MS network (2020). A clinical screening tool for objective and subjective cognitive disorders in multiple sclerosis. *Annals of Physical and Rehabilitation Medicine, 63*(2), 116–122.

Hattori, N., Kikuchi, M., Adachi, N., Hewitt, D., Huyck, S., & Saito, T. (2016). Adjunctive preladenant: A placebo-controlled, dose-finding study in Japanese patients with Parkinson's disease. *Parkinsonism & Related Disorders, 32*, 73–79.

Hauser, R. A., Cantillon, M., Pourcher, E., Micheli, F., Mok, V., Onofrj, M., ... Wolski, K. (2011). Preladenant in patients with Parkinson's disease and motor fluctuations: A phase 2, double-blind, randomised trial. The. *Lancet. Neurology, 10*(3), 221–229.

Hauser, R. A., Hattori, N., Fernandez, H., Isaacson, S. H., Mochizuki, H., Rascol, O., ... LeWitt, P. (2021). Efficacy of istradefylline, an adenosine A2A receptor antagonist, as adjunctive therapy to levodopa in Parkinson's disease: A pooled analysis of 8 phase 2b/3 trials. *Journal of Parkinson's Disease, 11*(4), 1663–1675.

Hauser, R. A., Hubble, J. P., Truong, D. D., & Istradefylline US-001 Study Group (2003). Randomized trial of the adenosine A(2A) receptor antagonist istradefylline in advanced PD. *Neurology, 61*(3), 297–303.

Hauser, R. A., Shulman, L. M., Trugman, J. M., Roberts, J. W., Mori, A., Ballerini, R., ... Istradefylline 6002-US-013 Study Group (2008a). Study of istradefylline in patients with Parkinson's disease on levodopa with motor fluctuations. *Movement Disorders: Official Journal of the Movement Disorder Society, 23*(15), 2177–2185.

Hauser, S. L., Bhan, A. K., Gilles, F., Kemp, M., Kerr, C., & Weiner, H. L. (1986). Immunohistochemical analysis of the cellular infiltrate in multiple sclerosis lesions. *Annals of Neurology, 19*(6), 578–587.

Healy, L. M., Stratton, J. A., Kuhlmann, T., & Antel, J. (2022). The role of glial cells in multiple sclerosis disease progression. *Nature Reviews. Neurology, 18*(4), 237–248.

Ingwersen, J., Wingerath, B., Graf, J., Lepka, K., Hofrichter, M., Schröter, F., ... Aktas, O. (2016). Dual roles of the adenosine A2a receptor in autoimmune neuroinflammation. *Journal of Neuroinflammation, 13*, 48.

Inojosa, H., Schriefer, D., & Ziemssen, T. (2020). Clinical outcome measures in multiple sclerosis: A review. *Autoimmunity Reviews, 19*(5), 102512. https://doi.org/10.1016/j.autrev.2020.102512

Jha, M. K., Jo, M., Kim, J. H., & Suk, K. (2019). Microglia-astrocyte crosstalk: An intimate molecular conversation. *The Neuroscientist: A Review Journal Bringing Neurobiology, Neurology and Psychiatry, 25*(3), 227–240.

Jovanova-Nesic, K., Koruga, D., Kojic, D., Kostic, V., Rakic, L., & Shoenfeld, Y. (2009). Choroid plexus connexin 43 expression and gap junction flexibility are associated with clinical features of acute EAE. *Annals of the New York Academy of Sciences, 1173*, 75–82. https://doi.org/10.1111/j.1749-6632

Kabat, E. A., Wolf, A., & Bezer, A. E. (1947). The rapid production of acute disseminated encephalomyelitis in rhesus monkeys by injection of heterologous and homologous brain tissue with adjuvants. *The Journal of Experimental Medicine, 85*(1), 117–130.

Kammer, G. M. (1986). Adenosine. *Lancet (London, England), 1*(8471), 50.

Kaskow, B. J., & Baecher-Allan, C. (2018). Effector T cells in multiple sclerosis. *Cold Spring Harbor perspectives in Medicine, 8*(4), a029025.

Katz Sand, I. (2015). Classification, diagnosis, and differential diagnosis of multiple sclerosis. *Current Opinion in Neurology, 28*(3), 193–205.

Kermode, A. G., Thompson, A. J., Tofts, P., MacManus, D. G., Kendall, B. E., Kingsley, D. P., ... McDonald, W. I. (1990). Breakdown of the blood-brain barrier precedes symptoms and other MRI signs of new lesions in multiple sclerosis. Pathogenetic and clinical implications. *Brain: A Journal of Neurology, 113*(Pt 5), 1477–1489.

Kim, J., Islam, S. M. T., Qiao, F., Singh, A. K., Khan, M., Won, J., & Singh, I. (2021). Regulation of B cell functions by S-nitrosoglutathione in the EAE model. *Redox biology, 45*, 102053.

Koch-Henriksen, N., & Magyari, M. (2021). Apparent changes in the epidemiology and severity of multiple sclerosis. *Nature Reviews. Neurology, 17*(11), 676–688. https://doi.org/10.1038/s41582-021-00556-y

Koszalka, P., Ozüyaman, B., Huo, Y., Zernecke, A., Flögel, U., Braun, N., ... Schrader, J. (2004). Targeted disruption of cd73/ecto-5'-nucleotidase alters thromboregulation and augments vascular inflammatory response. *Circulation Research, 95*(8), 814–821.

Kwon, H. S., & Koh, S. H. (2020). Neuroinflammation in neurodegenerative disorders: The roles of microglia and astrocytes. *Translational Neurodegeneration, 9*(1), 42.

Lambertucci, C., Marucci, G., Catarzi, D., Colotta, V., Francucci, B., Spinaci, A., ... Volpini, R. (2022). A2A adenosine receptor antagonists and their potential in neurological disorders. *Current Medicinal Chemistry, 29*(28), 4780–4795.

Lassmann, H. (2019). Pathogenic mechanisms associated with different clinical courses of multiple sclerosis. *Frontiers in Immunology, 9*, 3116. https://doi.org/10.3389/fimmu.2018.03116

LeWitt, P. A., Guttman, M., Tetrud, J. W., Tuite, P. J., Mori, A., Chaikin, P., ... 6002-US-005 Study Group (2008). Adenosine A2A receptor antagonist istradefylline (KW-6002) reduces "off" time in Parkinson's disease: A double-blind, randomized, multicenter clinical trial (6002-US-005). *Annals of Neurology, 63*(3), 295–302.

Li, Y. F., Zhang, S. X., Ma, X. W., Xue, Y. L., Gao, C., Li, X. Y., & Xu, A. D. (2019). The proportion of peripheral regulatory T cells in patients with multiple sclerosis: A meta-analysis. *Multiple Sclerosis and Related Disorders, 28*, 75–80.

Liddelow, S. A., Guttenplan, K. A., Clarke, L. E., Bennett, F. C., Bohlen, C. J., Schirmer, L., ... Barres, B. A. (2017). Neurotoxic reactive astrocytes are induced by activated microglia. *Nature, 541*(7638), 481–487.

Linden, J., & Cekic, C. (2012). Regulation of lymphocyte function by adenosine. *Arteriosclerosis, Thrombosis, and Vascular Biology, 32*(9), 2097–2103.

Liu, Y. J., Chen, J., Li, X., Zhou, X., Hu, Y. M., Chu, S. F., ... Chen, N. H. (2019). Research progress on adenosine in central nervous system diseases. *CNS Neuroscience & Therapeutics, 25*(9), 899–910.

Liu, Y., Alahiri, M., Ulloa, B., Xie, B., & Sadiq, S. A. (2018). Adenosine A2A receptor agonist ameliorates EAE and correlates with Th1 cytokine-induced blood brain barrier dysfunction via suppression of MLCK signaling pathway. *Immunity, Inflammation and Disease, 6*(1), 72–80.

Loram, L. C., Strand, K. A., Taylor, F. R., Sloane, E., Van Dam, A. M., Rieger, J., & Watkins, L. R. (2015). Adenosine 2A receptor agonism: A single intrathecal administration attenuates motor paralysis in experimental autoimmune encephalopathy in rats. *Brain, Behavior, and Immunity, 46*, 50–54.

Lun, M. P., Monuki, E. S., & Lehtinen, M. K. (2015). Development and functions of the choroid plexus-cerebrospinal fluid system. *Nature Reviews. Neuroscience, 16*(8), 445–457.

Maillart, E., Vidal, J. S., Brassat, D., Stankoff, B., Fromont, A., de Sèze, J., ... Papeix, C. (2017). Natalizumab-PML survivors with subsequent MS treatment: Clinico-radiologic outcome. *Neurology(R) Neuroimmunology & Neuroinflammation, 4*(3), e346.

Martire, A., Calamandrei, G., Felici, F., Scattoni, M. L., Lastoria, G., Domenici, M. R., ... Popoli, P. (2007). Opposite effects of the A2A receptor agonist CGS21680 in the striatum of Huntington's disease versus wild-type mice. *Neuroscience Letters, 417*(1), 78–83.

Matos, M., Shen, H. Y., Augusto, E., Wang, Y., Wei, C. J., Wang, Y. T., ... Chen, J. F. (2015). Deletion of adenosine A2A receptors from astrocytes disrupts glutamate homeostasis leading to psychomotor and cognitive impairment: relevance to schizophrenia. *Biological Psychiatry, 78*(11), 763–774.

Mattioli, F., Cappa, S. F., Cominelli, C., Capra, R., Marcianoc, N., & Gasparotti, R. (1993). Serial study of neuropsychological performance and gadolinium-enhanced MRI in MS. *Acta neurologica Scandinavica, 87*(6), 465–468.

Mayo, L., Trauger, S. A., Blain, M., Nadeau, M., Patel, B., Alvarez, J. I., ... Quintana, F. J. (2014). Regulation of astrocyte activation by glycolipids drives chronic CNS inflammation. *Nature Medicine, 20*(10), 1147–1156.

Melani, A., Cipriani, S., Vannucchi, M. G., Nosi, D., Donati, C., Bruni, P., ... Pedata, F. (2009). Selective adenosine A2a receptor antagonism reduces JNK activation in oligodendrocytes after cerebral ischaemia. *Brain: A Journal of Neurology, 132*(Pt 6), 1480–1495.

Merighi, S., Borea, P. A., Varani, K., Vincenzi, F., Travagli, A., Nigro, M., ... Gessi, S. (2022). Pathophysiological role and medicinal chemistry of A2A adenosine receptor antagonists in Alzheimer's disease. *Molecules (Basel, Switzerland), 27*(9), 2680.

Mexhitaj, I., Nyirenda, M. H., Li, R., O'Mahony, J., Rezk, A., Rozenberg, A., ... Bar-Or, A. (2019). Abnormal effector and regulatory T cell subsets in paediatric-onset multiple sclerosis. *Brain: A Journal of Neurology, 142*(3), 617–632.

Mills, J. H., Alabanza, L. M., Mahamed, D. A., & Bynoe, M. S. (2012). Extracellular adenosine signaling induces CX3CL1 expression in the brain to promote experimental autoimmune encephalomyelitis. *Journal of Neuroinflammation, 9*, 193.

Mills, J. H., Kim, D. G., Krenz, A., Chen, J. F., & Bynoe, M. S. (2012a). A2A adenosine receptor signaling in lymphocytes and the central nervous system regulates inflammation during experimental autoimmune encephalomyelitis. *Journal of Immunology (Baltimore, Md.: 1950), 188*(11), 5713–5722.

Mills, J. H., Thompson, L. F., Mueller, C., Waickman, A. T., Jalkanen, S., Niemela, J., ... Bynoe, M. S. (2008). CD73 is required for efficient entry of lymphocytes into the central nervous system during experimental autoimmune encephalomyelitis. *Proceedings of the National Academy of Sciences of the United States of America, 105*(27), 9325–9330.

Minor, M., Alcedo, K. P., Battaglia, R. A., & Snider, N. T. (2019). Cell type- and tissue-specific functions of ecto-5'-nucleotidase (CD73). *American Journal of Physiology. Cell Physiology, 317*(6), C1079–C1092.

Moesta, A. K., Li, X. Y., & Smyth, M. J. (2020). Targeting CD39 in cancer. *Nature Reviews. Immunology, 20*(12), 739–755.

Moore, S., Meschkat, M., Ruhwedel, T., Trevisiol, A., Tzvetanova, I. D., Battefeld, A., ... Nave, K. A. (2020). A role of oligodendrocytes in information processing. *Nature Communications, 11*(1), 5497.

Muhammad, F., Wang, D., McDonald, T., Walsh, M., Drenen, K., Montieth, A., ... Lee, D. J. (2020). TIGIT+ A2Ar-dependent anti-uveitic Treg cells are a novel subset of Tregs associated with resolution of autoimmune uveitis. *Journal of Autoimmunity, 111*, 102441.

Nakanishi, M. (2007). S-Adenosyl-L-homocysteine hydrolase as an attractive target for antimicrobial drugs. *Yakugaku Zasshi: Journal of the Pharmaceutical Society of Japan, 127*(6), 977–982.

Nave, K. A., & Werner, H. B. (2014). Myelination of the nervous system: Mechanisms and functions. *Annual Review of Cell and Developmental Biology, 30*, 503–533.

Neumann, H., Kotter, M. R., & Franklin, R. J. (2009). Debris clearance by microglia: An essential link between degeneration and regeneration. *Brain: A Journal of Neurology, 132*(Pt 2), 288–295.

Nicholas, R., & Rashid, W. (2013). Multiple sclerosis. *American Family Physician, 87*(10), 712–714.

Ning, Y. L., Yang, N., Chen, X., Xiong, R. P., Zhang, X. Z., Li, P., ... Zhou, Y. G. (2013). Adenosine A2A receptor deficiency alleviates blast-induced cognitive dysfunction. *Journal of cerebral blood flow and metabolism : official journal of the International Society of Cerebral Blood Flow and Metabolism, 33*(11), 1789–1798.

Ogawa, Y., Furusawa, E., Saitoh, T., Sugimoto, H., Omori, T., Shimizu, S., ... Oishi, K. (2018). Inhibition of astrocytic adenosine receptor A2A attenuates microglial activation in a mouse model of Sandhoff disease. *Neurobiology of Disease, 118*, 142–154.

Onishi, R. M., & Gaffen, S. L. (2010). Interleukin-17 and its target genes: Mechanisms of interleukin-17 function in disease. *Immunology, 129*(3), 311–321.

Ortiz, G. G., Pacheco-Moisés, F. P., Macías-Islas, M.Á., Flores-Alvarado, L. J., Mireles-Ramírez, M. A., González-Renovato, E. D., ... Alatorre-Jiménez, M. A. (2014). Role of the blood-brain barrier in multiple sclerosis. *Archives of Medical Research, 45*(8), 687–697.

Paiva, I., Carvalho, K., Santos, P., Cellai, L., Pavlou, M. A. S., Jain, G., ... Blum, D. (2019). A2A R-induced transcriptional deregulation in astrocytes: An in vitro study. *Glia, 67*(12), 2329–2342.

Panther, E., Idzko, M., Herouy, Y., Rheinen, H., Gebicke-Haerter, P. J., Mrowietz, U., ... Norgauer, J. (2001). Expression and function of adenosine receptors in human dendritic cells. *FASEB Journal: Official Publication of the Federation of American Societies for Experimental Biology, 15*(11), 1963–1970.

Polman, C. H., O'Connor, P. W., Havrdova, E., Hutchinson, M., Kappos, L., Miller, D. H., ... AFFIRM Investigators (2006). A randomized, placebo-controlled trial of natalizumab for relapsing multiple sclerosis. *The New England Journal of Medicine, 354*(9), 899–910.

Probert, L., Akassoglou, K., Pasparakis, M., Kontogeorgos, G., & Kollias, G. (1995). Spontaneous inflammatory demyelinating disease in transgenic mice showing central nervous system-specific expression of tumor necrosis factor alpha. *Proceedings of the National Academy of Sciences of the United States of America, 92*(24), 11294–11298.

Rae-Grant, A., Day, G. S., Marrie, R. A., Rabinstein, A., Cree, B. A. C., Gronseth, G. S., ... Pringsheim, T. (2018a). Practice guideline recommendations summary: Disease-modifying therapies for adults with multiple sclerosis: Report of the Guideline Development, Dissemination, and Implementation Subcommittee of the American Academy of Neurology. *Neurology, 90*(17), 777–788.

Rajasundaram, S. (2018a). Adenosine A2A receptor signaling in the immunopathogenesis of experimental autoimmune encephalomyelitis. *Frontiers in Immunology, 9*, 402.

Ramesh, A., Schubert, R. D., Greenfield, A. L., Dandekar, R., Loudermilk, R., Sabatino, J. J., ... Wilson, M. R. (2020). A pathogenic and clonally expanded B cell transcriptome in active multiple sclerosis. *Proceedings of the National Academy of Sciences of the United States of America, 117*(37), 22932–22943.

Reboldi, A., Coisne, C., Baumjohann, D., Benvenuto, F., Bottinelli, D., Lira, S., ... Sallusto, F. (2009). C-C chemokine receptor 6-regulated entry of TH-17 cells into the CNS through the choroid plexus is required for the initiation of EAE. *Nature Immunology, 10*(5), 514–523.

Ren, X., & Chen, J. F. (2020a). Caffeine and Parkinson's disease: Multiple benefits and emerging mechanisms. *Frontiers in Neuroscience, 14*, 602697.

Ring, S., Pushkarevskaya, A., Schild, H., Probst, H. C., Jendrossek, V., Wirsdörfer, F., ... Mahnke, K. (2015). Regulatory T cell-derived adenosine induces dendritic cell migration through the Epac-Rap1 pathway. *Journal of Immunology (Baltimore, Md.: 1950), 194*(8), 3735–3744.

Ruggiero, M., Calvello, R., Porro, C., Messina, G., Cianciulli, A., & Panaro, M. A. (2022). Neurodegenerative diseases: Can caffeine be a powerful ally to weaken neuroinflammation? *International Journal of Molecular Sciences, 23*(21), 12958.

Sakaguchi, S., Yamaguchi, T., Nomura, T., & Ono, M. (2008). Regulatory T cells and immune tolerance. *Cell, 133*(5), 775–787.

Schafflick, D., Xu, C. A., Hartlehnert, M., Cole, M., Schulte-Mecklenbeck, A., Lautwein, T., ... Meyer Zu Horste, G. (2020). Integrated single cell analysis of blood and cerebrospinal fluid leukocytes in multiple sclerosis. *Nature communications, 11*(1), 247.

Schneider-Hohendorf, T., Schulte-Mecklenbeck, A., Ostkamp, P., Janoschka, C., Pawlitzki, M., Luessi, F., ... Schwab, N. (2021). High anti-JCPyV serum titers coincide with high CSF cell counts in RRMS patients. Multiple sclerosis (Houndmills, Basingstoke. England), 27(10), 1491–1496.

Schnurr, M., Toy, T., Shin, A., Hartmann, G., Rothenfusser, S., Soellner, J., ... Maraskovsky, E. (2004). Role of adenosine receptors in regulating chemotaxis and cytokine production of plasmacytoid dendritic cells. Blood, 103(4), 1391–1397.

Selmaj, K., Papierz, W., Glabiński, A., & Kohno, T. (1995). Prevention of chronic relapsing experimental autoimmune encephalomyelitis by soluble tumor necrosis factor receptor I. Journal of Neuroimmunology, 56(2), 135–141.

Sevigny, C. P., Li, L., Awad, A. S., Huang, L., McDuffie, M., Linden, J., ... Okusa, M. D. (2007). Activation of adenosine 2A receptors attenuates allograft rejection and alloantigen recognition. Journal of Immunology (Baltimore, Md.: 1950), 178(7), 4240–4249.

Shechter, R., Miller, O., Yovel, G., Rosenzweig, N., London, A., Ruckh, J., ... Schwartz, M. (2013). Recruitment of beneficial M2 macrophages to injured spinal cord is orchestrated by remote brain choroid plexus. Immunity, 38(3), 555–569.

Skaper, S. D. (2019). Oligodendrocyte precursor cells as a therapeutic target for demyelinating diseases. Progress in Brain Research, 245, 119–144.

Spelman, T., Magyari, M., Piehl, F., Svenningsson, A., Rasmussen, P. V., Kant, M., ... Lycke, J. (2021). Treatment escalation vs immediate initiation of highly effective treatment for patients with relapsing-remitting multiple sclerosis: Data from 2 different national strategies. JAMA Neurology, 78(10), 1197–1204.

Stacy, M., Silver, D., Mendis, T., Sutton, J., Mori, A., Chaikin, P., & Sussman, N. M. (2008). A 12-week, placebo-controlled study (6002-US-006) of istradefylline in Parkinson disease. Neurology, 70(23), 2233–2240.

Steffen, B. J., Breier, G., Butcher, E. C., Schulz, M., & Engelhardt, B. (1996). ICAM-1, VCAM-1, and MAdCAM-1 are expressed on choroid plexus epithelium but not endothelium and mediate binding of lymphocytes in vitro. The American Journal of Pathology, 148(6), 1819–1838.

Steinman, L. (2007). A brief history of T(H)17, the first major revision in the T(H)1/T(H)2 hypothesis of T cell-mediated tissue damage. Nature Medicine, 13(2), 139–145.

Steinman, L., Fox, E., Hartung, H. P., Alvarez, E., Qian, P., Wray, S., ... ULTIMATE I and ULTIMATE II Investigators (2022). Ublituximab versus Teriflunomide in Relapsing Multiple Sclerosis. The New England Journal of Medicine, 387(8), 704–714.

Takahashi, M., Fujita, M., Asai, N., Saki, M., & Mori, A. (2018a). Safety and effectiveness of istradefylline in patients with Parkinson's disease: Interim analysis of a post-marketing surveillance study in Japan. Expert Opinion on Pharmacotherapy, 19(15), 1635–1642.

Tokano, M., Matsushita, S., Takagi, R., Yamamoto, T., & Kawano, M. (2022). Extracellular adenosine induces hypersecretion of IL-17A by T-helper 17 cells through the adenosine A2a receptor. Brain, Behavior, & Immunity - Health, 26, 100544.

van Langelaar, J., van der Vuurst de Vries, R. M., Janssen, M., Wierenga-Wolf, A. F., Spilt, I. M., Siepman, T. A., ... van Luijn, M. M. (2018). T helper 17.1 cells associate with multiple sclerosis disease activity: Perspectives for early intervention. Brain: A Journal of Neurology, 141(5), 1334–1349.

Vincenzi, F., Corciulo, C., Targa, M., Merighi, S., Gessi, S., Casetta, I., ... Varani, K. (2013). Multiple sclerosis lymphocytes upregulate A2A adenosine receptors that are antiinflammatory when stimulated. European Journal of Immunology, 43(8), 2206–2216.

Wang, T., Xi, N. N., Chen, Y., Shang, X. F., Hu, Q., Chen, J. F., & Zheng, R. Y. (2014). Chronic caffeine treatment protects against experimental autoimmune encephalomyelitis in mice: Therapeutic window and receptor subtype mechanism. Neuropharmacology, 86, 203–211.

Wang, X. L., Feng, S. T., Chen, B., Hu, D., Wang, Z. Z., & Zhang, Y. (2022). Efficacy and safety of istradefylline for Parkinson's disease: A systematic review and meta-analysis. *Neuroscience Letters, 774*, 136515.

Wilson, J. M., Kurtz, C. C., Black, S. G., Ross, W. G., Alam, M. S., Linden, J., & Ernst, P. B. (2011). The A2B adenosine receptor promotes Th17 differentiation via stimulation of dendritic cell IL-6. *Journal of Immunology (Baltimore, Md.: 1950), 186*(12), 6746–6752.

Wingerchuk, D. M., & Carter, J. L. (2014). Multiple sclerosis: Current and emerging disease-modifying therapies and treatment strategies. *Mayo Clinic Proceedings, 89*(2), 225–240.

Yamasaki, R., Lu, H., Butovsky, O., Ohno, N., Rietsch, A. M., Cialic, R., ... Ransohoff, R. M. (2014). Differential roles of microglia and monocytes in the inflamed central nervous system. *The Journal of Experimental Medicine, 211*(8), 1533–1549.

Yao, S. Q., Li, Z. Z., Huang, Q. Y., Li, F., Wang, Z. W., Augusto, E., ... Zheng, R. Y. (2012). Genetic inactivation of the adenosine A(2A) receptor exacerbates brain damage in mice with experimental autoimmune encephalomyelitis. *Journal of Neurochemistry, 123*(1), 100–112.

Ye, M., Wang, M., Feng, Y., Shang, H., Yang, Y., Hu, L., ... Zheng, W. (2023). Adenosine A2A receptor controls the gateway of the choroid plexus. *Purinergic Signalling, 19*(1), 135–144.

Zhang, J. G., Hepburn, L., Cruz, G., Borman, R. A., & Clark, K. L. (2005). The role of adenosine A2A and A2B receptors in the regulation of TNF-alpha production by human monocytes. *Biochemical Pharmacology, 69*(6), 883–889.

Zhang, T., Zhao, J., Fu, J., Chen, G., & Ma, T. (2022). Improvement of the sepsis survival rate by adenosine 2a receptor antagonists depends on immune regulatory functions of regulatory T-cells. *Frontiers in Immunology, 13*, 996446.

Zheng, W., Feng, Y., Zeng, Z., Ye, M., Wang, M., Liu, X., ... Chen, J. F. (2022a). Choroid plexus-selective inactivation of adenosine A2A receptors protects against T cell infiltration and experimental autoimmune encephalomyelitis. *Journal of Neuroinflammation, 19*(1), 52.

Zhou, R., Wang, C., Lin, B., Wang, Y., Chen, J., & Liu, X. (2015). Association for Research in Vision and Ophthalmology. (p. 214).

Zhou, X., Bailey-Bucktrout, S. L., Jeker, L. T., Penaranda, C., Martínez-Llordella, M., Ashby, M., ... Bluestone, J. A. (2009). Instability of the transcription factor Foxp3 leads to the generation of pathogenic memory T cells in vivo. *Nature Immunology, 10*(9), 1000–1007.

Zhou, Y., Zeng, X., Li, G., Yang, Q., Xu, J., Zhang, M., ... Huo, Y. (2019). Inactivation of endothelial adenosine A2A receptors protects mice from cerebral ischaemia-induced brain injury. *British Journal of Pharmacology, 176*(13), 2250–2263.

Zhu, J., & Paul, W. E. (2010). Heterogeneity and plasticity of T helper cells. *Cell Research, 20*(1), 4–12.

CHAPTER TEN

# $A_{2A}R$ and traumatic brain injury

Yan Zhao[a,b,1], Ya-Lei Ning[a,b,1], and Yuan-Guo Zhou[a,b,*]

[a]Department of Army Occupational Disease, State Key Laboratory of Trauma and Chemical Poisoning, Research Institute of Surgery and Daping Hospital, Army Medical University, P.R. China
[b]Institute of Brain and Intelligence, Army Medical University, Chongqing, P.R. China
*Corresponding author. e-mail address: ygzhou@tmmu.edu.cn

## Contents

1. Preface — 226
2. $A_{2A}R$ modulation of neuroinflammation at the early stage of TBI — 228
   2.1 Regulation of glial cells by $A_{2A}R$ antagonists — 228
   2.2 Regulation of neuroinflammation by BMDC $A_{2A}R$ — 231
   2.3 Regulation of NLRP3 inflammasome activation by $A_{2A}R$ — 231
   2.4 Bidirectional modulation of neuroinflammation by $A_{2A}R$ — 232
3. $A_{2A}R$ modulation of glutamate excitotoxicity at the early stage of TBI — 235
   3.1 Potentiation of presynaptic glutamate release by $A_{2A}R$ activation — 236
   3.2 Exacerbation of postsynaptic glutamate excitotoxicity by $A_{2A}R$ activation — 236
   3.3 Control of cerebral glutamate homeostasis by astroglial $A_{2A}R$ — 237
4. $A_{2A}R$ modulation of aberrant proteins — 240
   4.1 Modulation of tau hyperphosphorylation after TBI — 240
   4.2 Modulation of autophagy after TBI — 241
5. $A_{2A}R$ modulation of cognitive dysfunction after TBI — 241
   5.1 Alteration of adenosine and its receptors during a chronic period of TBI — 242
   5.2 $A_{2A}R$ regulation in pathological proteins — 244
   5.3 $A_{2A}R$ and delayed cognitive dysfunction of TBI — 245
6. $A_{2A}R$ modulation of chronic neuroinflammation in TBI — 247
   6.1 Long-lasting neuroinflammation — 247
   6.2 Impairment of synaptic structure and function — 249
7. $A_{2A}R$ modulation of other long-lasting consequences of TBI — 250
   7.1 Sleep disorders — 250
   7.2 PTSD — 251
   7.3 Posttraumatic seizures and epilepsy — 252
References — 253

## Abstract

Accumulating evidence has revealed the adenosine 2A receptor is a key tuner for neuropathological and neurobehavioral changes following traumatic brain injury by experimental animal models and a few clinical trials. Here, we highlight recent data involving acute/sub-acute and chronic alterations of adenosine and adenosine 2A

---

[1] These authors contributed equally to this work.

receptor-associated signaling in pathological conditions after trauma, with an emphasis of traumatic brain injury, including neuroinflammation, cognitive and psychiatric disorders, and other severe consequences. We expect this would lead to the development of therapeutic strategies for trauma-related disorders with novel mechanisms of action.

## 1. Preface

Adenosine, an important intermediate in energy metabolism, plays an important regulatory role by binding its four G-protein-coupled receptors ($A_1R$, $A_{2A}R$, $A_{2B}R$, and $A_3R$) in various physiological activities such as sleep, learning, immune function and neural function (Dunwiddie & Masino, 2001). Moreover, it is also involved in traumatic pathological changes such as traumatic brain injury (TBI), spinal cord injury, and their sequelae (Vlajkovic, Housley, & Thorne, 2009).

Among them, TBI is the leading cause for high disability and mortality after trauma and leads to long-lasting cognitive and emotional disorders. With the ongoing improvement in medical treatment in ICU, the mortality caused by trauma has significantly reduced. However, patients with severe trauma who survive early treatment still face serious threats of various complications during hospital treatment due to the complex pathophysiology of TBI. TBI triggers an acute surge in adenosine, most likely as a result of ATP release (Latini & Pedata, 2001; Ribeiro, 2005). Since extracellular adenosine concentrations increase sharply during brain damage, adenosine and its receptors, especially adenosine 2A receptor ($A_{2A}R$), have been demonstrated to play multifaceted roles in the progression and outcome of TBI and long-term sequelae.

In addition to the primary injury, brain trauma also leads to secondary brain injury, which occurs within minutes and evolves over a period of hours, days and even months (Aires et al., 2019) after the original insult. The level of secondary injury closely correlates with the clinical prognosis. Multiple disturbed pathways have been reported to contribute to secondary injury, leading to further brain tissue damage and the development of the neurological impairments after TBI. The mechanisms of secondary brain injury include neuroinflammation (Jassam, Izzy, Whalen, Mcgavern, & El Khoury, 2017; O'brien et al., 2020; Paudel et al., 2018), glutamate excitotoxicity (Dorsett et al., 2017), oxidative stress (Anthonymuthu, Kenny, & Bayir, 2016), blood–brain barrier (BBB) disruption (Van Vliet et al., 2020) and cell death. The primary (mechanical) injury, which occurs at the time of trauma to the cerebral tissue, is irreversible. The delayed nature of secondary injury suggests its potential reversibility, providing a window for therapeutic intervention.

While the roles of $A_{2A}R$ in central nervous system (CNS) appear to be controversial, most studies agree that $A_{2A}R$ blockade confers neuroprotective effects against a broad spectrum of neuropathological processes associated with TBI (Chen & Pedata, 2008; Chen et al., 2007; Cunha, 2005; Dai et al., 2008; Li et al., 2009; Ning et al., 2013; Zhao Li et al., 2017), such as neuroinflammation, excitotoxicity, autophagy blocking, tauopathy, and neurobehavioral disorders.

Studies by Li et al. suggested that inactivation of $A_{2A}R$, either by genetic knockout (Li et al., 2009) or by chronically consumption of $A_{2A}R$ antagonist, caffeine (Li et al., 2008), attenuated the consequences of TBI in the acute stage in a mouse cortical impact model. According to clinical research, higher levels of caffeine in cerebrospinal fluid (CSF) are associated with improved clinical outcomes in patients with severe TBI (Sachse et al., 2008). In a blast-induced TBI model, Ning et al. found that genetic inactivation of $A_{2A}R$ alleviated both early brain injury and prolonged cognitive dysfunction induced by blast overpressure (Ning et al., 2013). A selective $A_{2A}R$ antagonist, SCH58261, attenuated hyperlocomotion in a lateral fluid percussion (LFP)-induced brain injury model in Mongolian Gerbils (Mullah, Inaji, Nariai, Ishibashi, & Ohno, 2013).

There are also several literatures showing that $A_{2A}R$ activation plays a neuroprotective role. Accordingly, available evidence has suggested that the neuromodulatory effects of $A_{2A}R$ are complex in the progression of TBI, and that the neuroprotective effects of $A_{2A}R$ vary depending on the local environment, stages of pathological processes, and the nature and severity of brain damage.

More importantly, TBI patients also show decreased executive function (Krawczyk et al., 2019), impaired attention (Cusimano et al., 2021), social dysfunction (Mcdonald, Dalton, Rushby, & Landin-Romero, 2019), affective disorders (Howlett, Nelson, & Stein, 2022; Medeiros et al., 2022), and posttraumatic stress disorder (PTSD). The persistence of various sequelae of trauma seriously affects the quality of life and causes a heavy burden on society.

This chapter mainly focuses on the acute, sub-acute, and chronic changes of $A_{2A}R$ in TBI, its adverse consequences, and the underlying mechanisms, as well as various intervention strategies targeting $A_{2A}R$ in the context of TBI.

## 2. A$_{2A}$R modulation of neuroinflammation at the early stage of TBI

Neuroinflammation is defined as the activation of the brain's innate immune system in response to an inflammatory challenge and is characterized by a host of cellular and molecular changes within the brain (Dejan, Snjezana, Breyer, Aschner, & Montine, 2017). The neuroinflammatory response following TBI is a highly complex cascade of events and is regarded as a key secondary injury mechanism, involving the activation of glial cells, recruitment of peripheral immune cells such as lymphocytes, neutrophils, and monocytes, release of danger signals such as damage-associated molecular patterns (DAMPs) (Roth et al., 2014; Shi, Zhang, Dong, & Shi, 2019), and upregulation of inflammatory mediators such as pro- and anti-inflammatory cytokines and chemokines, growth factors, and proteinases (Kumar & Loane, 2012; Morganti-Kossmann, Satgunaseelan, Bye, & Kossmann, 2007).

Both endogenous brain cells and circulating inflammatory cells can activate immune response after brain injury. In the brain context, glial cells, neurons, and endothelial cells all produce signals to coordinate these reactions. The brain's indigenous immune cells, microglia and astrocytes, are the predominant cells responsible for neuroinflammatory cascade. Complement mediates neuroinflammation post TBI and causes aberrant microglial activation (Mallah et al., 2021). Activated microglia produce interleukin-1β (IL-1β), tumor necrosis factor α (TNF-α), and complement 1q, which can lead to neurotoxic A$_1$ astrocyte activation (Shinozaki et al., 2017). On the other hand, reactive astrocytes facilitate the activation of microglia and immune cells, resulting in persistent neuroinflammation. Astrocytes can also produce IL-33, which promotes the recruitment of microglia/macrophages in TBI mice (Michinaga & Koyama, 2021; Wicher et al., 2017). Then these cells produce more aggressive substances, such as proinflammatory cytokines, chemokines, or reactive oxygen species (ROS), which are not typically produced under normal circumstances (Patraca et al., 2017).

### 2.1 Regulation of glial cells by A$_{2A}$R antagonists

Extracellular adenosine concentrations in physiological conditions are around the nanomolar levels. In inflammatory conditions, the levels of adenosine are dramatically increased. The main actions of adenosine on the CNS are mediated through its A$_1$ and A$_{2A}$ receptors. As a result, it has a variety of effects, some of which protect neuronal integrity while others worsen neuronal injury. But accumulating evidence suggests that in a number of inflammatory conditions, $A_{2A}R$ is the key regulator of adenosine-mediated inflammatory effects (Lambertucci et al., 2022; Marucci et al., 2021). $A_{2A}R$ has been shown to modulate various aspects of neuroinflammation in a variety of animal models of both acute and chronic CNS disorders, including TBI (Aires et al., 2019; Dai & Zhou, 2011). $A_{2A}R$ therefore has long been considered an attractive therapeutic target for modulating neuroinflammation and for drug development in the above disorders (Chen, Eltzschig, & Fredholm, 2013).

The expression of $A_{2A}R$ in glial cells is usually low under physiological conditions. Brain injuries cause an elevation in the level of $A_{2A}R$ in microglia, resulting in signal transductions that do not take place in cells that have the level of the receptors at a normal level, such as promoting the release of cytokines (Marti Navia et al., 2020). $A_{2A}R$ antagonists, on the other hand, prevent microglia from becoming activated in both in vitro and in vivo tests (Aires et al., 2019; Chen et al., 2018; Gomes et al., 2013; Rebola et al., 2011). Studies have shown that activation of adenosine $A_{2A}$ receptors exacerbates neuroinflammation and neurodegeneration after TBI. For instance, primary cultures of mixed glial cells derived from mouse cortex exposed to lipopolysaccharide (LPS) induced production of $A_{2A}R$ mRNA and protein. The $A_{2A}R$ agonist CGS21680 promoted LPS-induced release of nitrous oxide (NO) and NO synthase-II expression in a time- and concentration-dependent manner. The effect was inhibited by the $A_{2A}R$ antagonist ZM241385 and was absent on mixed glial cultures from $A_{2A}R$ knockout mice (Saura et al., 2005). $A_{2A}R$ agonist CGS21680 and non-specific agonist NECA can promote the secretion of cyclooxygenase 2 and prostaglandin E2 in rat microglia cells, and aggravate Alzheimer's disease (Marcoli et al., 2004). The expression and activation of potassium (K) ion channels are closely related with the activation of microglia cells. Kust et al. found that activation of $A_{2A}R$ on microglia cells by CGS21680 can up-regulate the expression of potassium channel Kv1.3 mRNA and protein via the cAMP-protein kinase A (PKA) pathway. Additionally, it can also enhance the expression of ROMK1 mRNA through the protein kinase C (PKC) pathway. In some animal models of neuroinflammation mediated by LPS, there is a rise in $A_{2A}R$ mRNA expression in a nuclear factor kappa-light-chain-enhancer of activated B cells (NF-κB)-dependent way (Latini & Pedata, 2001). According to a study by Chen et al. (Chen, Lee, & Chern, 2014), mice lacking the $A_{2A}R$

displayed lower inflammation, increased neuronal survival, and enhanced neurological function following TBI, compared to wild-type mice. The study also showed that administration of an $A_{2A}R$ antagonist reduced neuroinflammation and improved neurological function in wild-type mice. Another study by Muzzi et al. demonstrated that treatment with an $A_{2A}R$ antagonist reduced neuroinflammation and improved cognitive function in rats after TBI. The study also showed that $A_{2A}R$ antagonist reduced the expression of proinflammatory cytokines and chemokines.

In addition, $A_{2A}R$ antagonist has also been reported to ameliorate neuroinflammation by suppressing the activation of glial cells in peripheral nerves. $A_{2A}R$ antagonist ZM241385 is reported to reduce the activation of microglia and downregulate the proinflammatory cytokines under the conditions of chronic ocular hypertension in experimental glaucoma (Liu, Huang et al., 2016). Intravitreal injection of adenosine $A_{2A}R$ antagonist SCH58261 was able to control microglial reactivity and halt neuroinflammation, thus affords protection against cell loss and reduces vascular leakage in the retina of diabetic mice (Aires et al., 2019).

The induction of $A_{2A}Rs$ in glial cells and the inflammatory signals triggered by cerebral insults, coupled with the local increase in adenosine and proinflammatory cytokine levels, may serve as an important feed-forward mechanism to locally regulate neuroinflammatory responses in the brain (Boison, Chen, & Fredholm, 2010; Nishizaki et al., 2002). In addition, $A_{2A}R$ establish a network of interactions with many other receptors, such as glutamate, dopamine, cannabinoid, and adenosine $A_1$ receptors. Therefore, activation of $A_{2A}R$ plays complex roles in the function of glial cells. Activation of $A_{2A}Rs$ in microglial cells has mixed effects on microglial proliferation, but has clear facilitating effects on cytokine release, including upregulation of cyclooxygenase 2 and release of prostaglandin E2 (PGE2), as well as increases in nitric oxide synthase (NOS) activity and NO release, and nerve growth factor expression (Boison et al., 2010; Hasko, Pacher, Vizi, & Illes, 2005). While in astrocytes, activation of $A_{2A}Rs$ by stimulates astrocyte proliferation and activation (Brambilla, Cottini, Fumagalli, Ceruti, & Abbracchio, 2003; Hindley, Herman, & Rathbone, 1994), but suppresses inducible NOS (iNOS) expression and the NO production (Brodie, Blumberg, & Jacobson, 1998), and controls astrocytes glutamate efflux (Boison et al., 2010).

Accordingly, to limit excessive neuroinflammation and prevent neurological impairment, $A_{2A}R$ antagonists can preferentially constrain the progression of microglial activation towards a fully activated state (Yu et al., 2008).

These antagonists have the potential to decrease the threshold for microglial reactivity by inhibiting cAMP accumulation, while also constraining microglial activation by blocking p38 MAPK signaling (Dai & Zhou, 2011).

## 2.2 Regulation of neuroinflammation by BMDC $A_{2A}R$

Immune activation can be initiated by a combination of endogenous brain cells and circulating inflammatory cells. In addition to the involvement of endogenous brain cells, $A_{2A}Rs$ on bone marrow-derived cells (BMDC) also play a role in regulating neuroinflammation and influencing outcomes in several CNS injuries. Yu et al. (2004) demonstrated that BMDC $A_{2A}R$ inactivation significantly reduced ischemic brain injury by preventing neuroinflammation. This study provides concrete proof indicating that $A_{2A}Rs$ on BMDC, in addition to neurons and glia cells, contribute to the modulation of neuroinflammatory responses. They demonstrated that the selective inactivation of BMDC $A_{2A}Rs$ reduced infarct volumes and the expression of multiple proinflammatory cytokines induced by ischemia in ischemic brain injury. In a mouse model of compression spinal cord injury, Li et al. (2006) showed that activating BMDC $A_{2A}Rs$ or deleting non-BMDC $A_{2A}Rs$ exerted protective effects. Additionally, in a mouse cortical impact model, Dai, Li, An et al. (2010) demonstrated that selectively inactivating BMDC $A_{2A}Rs$ significantly attenuated brain damage and inhibited the expressions of inflammatory cytokine TNF-a and IL-1b. Furthermore, they found that selective inactivation of BMDC $A_{2A}Rs$ showed better protection against neurological deficits and stronger inhibition of neuroinflammation compared to the selective inactivation of non-BMDC $A_{2A}Rs$.

These findings suggest that BMDC $A_{2A}Rs$ play an indispensably role in the development of brain injury. The underlying mechanism of $A_{2A}R$-mediated protection involves inhibition of proinflammatory responses in BMDCs, such as platelets, monocytes, mast cells, neutrophils, and T cells.

## 2.3 Regulation of NLRP3 inflammasome activation by $A_{2A}R$

The NOD-, LRR and pyrin domain-containing protein 3 (NLRP3) inflammasome is a multiprotein complex consisting of NLRP3, ASC and pro-caspase 1, which is mostly located in microglia and is essential for neuroinflammation (Swanson, Deng, & Ting, 2019). When the NLRP3 inflammasome is activated, it cleaves pro-caspase 1 and releases the subunit caspase 1 p20, which converts IL-1β and IL-18 precursors into mature forms and cleaves gasdermin D (GSDMD) to release the N-terminal part (N-GSDMD) (Shi et al., 2015). It has been demonstrated that the microglial NLRP3 inflammasome plays a critical role in the secondary injury after TBI (Du et al., 2022; Swanson et al., 2019) and the consequent cognitive impairment (Chakraborty, Tabassum, & Parvez, 2023; Tan Zhao et al., 2021), and neurodegenerative diseases (Anderson, Biggs, Rankin, & Havrda, 2023; Du et al., 2022; Duan, Kelley, & He, 2020). Inhibiting the activation of the NLRP3 inflammasome has shown neuroprotective effects (Anderson et al., 2023; Duan et al., 2020; Wu et al., 2021).

The $A_{2A}R$ signaling has been linked to the regulation of NLRP3 inflammasome activation. In a cerebral ischemia brain injury model, both endothelial $A_{2A}R$ inactivation and $A_{2A}R$ antagonist KW 6002 reduced post-injury outcomes through antiinflammatory effects via suppression of NLRP3 inflammasome activation, down-regulating cleaved caspase 1 and IL-1beta expression (Hindley et al., 1994). Caffeine inhibited NLRP3 inflammasome activation by suppressing $A_{2A}R$-associated ROS production in LPS-induced THP-1 macrophages (Zhao, Ma, Cai, & Gong, 2019). $A_{2A}R$ activation was also demonstrated to facilitate the sustained activation of the NLRP3 inflammasome in murine peritoneal macrophages (Ouyang et al., 2013).

Recent findings by Du et al. demonstrated that, in primary microglia, pharmacological activation of $A_{2A}R$ by CGS21680 exacerbated NLRP3 inflammasome activation and downstream inflammatory cytokine (IL-1β, IL-18) release under high glutamate concentrations by promoting NLRP3 assembly. This detrimental effect of CGS21680 on NLRP3 inflammasome activation was counteracted by the treatment of $A_{2A}R$ antagonist ZM241385 (Du et al., 2022).

## 2.4 Bidirectional modulation of neuroinflammation by $A_{2A}R$

Despite the antiinflammatory effects of $A_{2A}R$ inactivation mentioned above, an antiinflammatory and beneficial effect of $A_{2A}R$ activation has also been observed. Increased extracellular adenosine can profoundly reduce LPS/interferon-γ-mediated NO production and induce iNOS and TNF-α gene expression in glial cells (Chakraborty et al., 2023). The $A_{2A}R$ agonist CGS21680 has been shown to reduce TNF-α production and neutrophil infiltration linked to intracerebral hemorrhage (Mayne et al., 2001). Similarly, the $A_{2A}R$ agonist ATL146e, given during rabbit spinal cord reperfusion, significantly inhibited the levels of proinflammatory cytokines like TNF-α and IL-1β induced by traumatic/ischemic spinal cord injury through the NF-κB pathway and improved the spinal function, and reduced paralysis and apoptosis (Cassada et al., 2002). These results showed antiinflammatory properties of $A_{2A}R$ activation. In a mouse spinal cord injury (SCI) model, $A_{2A}R$ agonists has been shown to give strong neuroprotection against compression SCI by reducing inflammation and this protection was eliminated by specifically inactivating $A_{2A}R$ in bone marrow cells (Li et al., 2006).

The activation of $A_{2A}R$ plays a protective role in peripheral tissue injuries, such as ischemia, trauma, and hemorrhage of the liver, lung, or kidney. However, it exerts a bidirectional effect in certain CNS injuries, both protective and deleterious. The research conducted by Dai, Zhou, Li et al. (2010) in a mouse model of TBI provides new evidence supporting the dual role of $A_{2A}R$ in neuroinflammation and acute neurological injury prognosis. This study hypothesized that the levels of local glutamate in the brain dictates the effects of $A_{2A}R$ on neuroinflammation and brain damage outcome. Firstly, the researchers demonstrated that in the presence of low concentration of glutamate, CGS21680, which activates $A_{2A}Rs$ on microglia, inhibited the NOS activity of microglia induced by LPS through the cAMP-PKA pathway in pure microglia cultures (microglia > 95%). On the other hand, in high concentrations of glutamate, CGS21680 promoted the LPS-induced NOS activity of microglia via the PKC signaling pathway. These in vitro results were then confirmed in vivo using a mouse TBI model. When the glutamate level in the local environment is low, activation of $A_{2A}R$ attenuates brain damage by inhibiting neuroinflammation and reducing glutamate outflow. However, when the glutamate level in the local environment is high, activation of $A_{2A}R$ aggravates brain damage by promoting neuroinflammation and increasing glutamate outflow. These findings suggest that the level of glutamate in the local environment influences the effects of $A_{2A}R$ activation on neuroinflammation and outcome of acute neurological injury. A high level of local glutamate redirects $A_{2A}R$ signaling from the PKA to the PKC pathway, resulting in a shift from antiinflammatory to proinflammatory effects.

These discoveries reveal a new role for glutamate in regulating neuroinflammation and brain injury via the adenosine-$A_{2A}R$ system and presents a plausible explanation for the contradictory effects of $A_{2A}R$ agonists and antagonists observed in various CNS injury models. For instance, in a study by Blum et al. (Dai, Zhou, Li et al., 2010), $A_{2A}R$ activation had a detrimental effect when a high dose of 3-nitro-propionic acid was used to induce high glutamate outflow in the striatum. However, at a moderate dose and thus reduced outflow of glutamate, $A_{2A}R$ activation had a protective effect. Jones, Smith, and Stone (1998a, 1998b) discovered that central injection of the $A_{2A}R$ antagonist ZM241385 or the peripheral delivery of the $A_{2A}R$ agonist CGS21680 protected the hippocampus against kainate-induced excitotoxicity. It is tempting to hypothesize that this injection site-dependent variation in the effects of $A_{2A}R$ ligand may be caused by varying local glutamate levels in the hippocampus (high glutamate) and peripheral circulation (relatively low glutamate levels). In a rat experimental TBI model, the $A_{2A}R$ agonist exhibited a beneficial neuroprotective effect at the early stages (15 min and 1 h) of TBI, while the $A_{2A}R$ antagonist showed a benefit at the later stages (12 and 24 h) after TBI, suggesting that the extracellular glutamate concentrations may regulate the time-based switching of the $A_{2A}R$'s antiinflammatory and proinflammatory actions (Velayutham, Murthy, & Babu, 2019). Similarly, in a SCI model, the administration of $A_{2A}R$ agonists provided neuroprotection at the early stages post-injury, whereas $A_{2A}R$ inhibition was neuroprotective at later stages (Li et al., 2006). This time-dependent impact of $A_{2A}R$ agonists and antagonists could be attributed to changes in the local glutamate concentration in the spinal cord at different stages of the injury response. According to the study by Du et al. (2022), $A_{2A}R$ activation ameliorated the activation of NLRP3 inflammasome by inhibiting inflammasome assembly under normal or low glutamate concentrations, while high glutamate concentration reversed this inhibition. The results highlight the significance of local glutamate concentrations in brain injury and contribute the understanding of the role of $A_{2A}R$ in neuroinflammation.

Having a better understanding of the cellular basis and factors that contribute to $A_{2A}R$'s bidirectional action may help us to improve the design of clinical trials for $A_{2A}R$ agonists or antagonists in acute neurological injuries. Firstly, the conclusion that glutamate levels in local environments influence the bidirectional effect of $A_{2A}R$ on neuroinflammation and outcome addresses the site- and time-dependent action of $A_{2A}R$, providing new evidence for clinical treatment strategies for acute CNS injuries using $A_{2A}R$ agonists or antagonists. Thus, in order to achieve the best neuroprotection, $A_{2A}R$ agonists or antagonists can be used at various disease stages, and different dose regimens of $A_{2A}R$ antagonists can be used depending on the glutamate concentration. Additionally, various adjunctive agents, such as glutamate release inhibitors or glutamate receptor modulators, can be combined with $A_{2A}R$ agonists or antagonists. Secondly, $A_{2A}R$ has distinct effects on different cells, highlighting the need to develop cell type-specific adenosine receptor agents that specifically target the receptors where and when they are important in the course of acute brain injury. Finally, although the evidence presented here clearly highlights the complexity of using $A_{2A}R$ agents therapeutically in acute CNS injuries and demonstrates many areas for further investigation. These studies also confirm many promising characteristics of $A_{2A}R$ ligands that encourage the pursuit of their therapeutic potential in clinical treatment of acute neurological injuries.

## 3. $A_{2A}R$ modulation of glutamate excitotoxicity at the early stage of TBI

Glutamate is widely recognized as the most important excitatory neurotransmitter in the CNS and is released from all excitatory synapses. Excitotoxicity, predominantly mediated by glutamate, is a significant pathogenesis of TBI. It is characterized as a process in which neurons are injured by excessive stimulation of excitatory glutamate neurotransmitter receptors, such as the N-methyl-D-aspartic acid (NMDA) and AMPA receptors (Kaur & Sharma, 2018; Lewen et al., 2001). Experimental and clinical studies demonstrate that extracellular glutamate levels rise sharply shortly after TBI. The substantial depolarization of neuron leads to elevated extracellular glutamate levels due to energy failure and damage. The excessive intracellular $Ca^{2+}$ contributes to increased influx of $Na^+$ and $Ca^{2+}$ into the cells, ultimately triggering pathways that cause cell damage. These mechanisms of cellular damage result in apoptosis, which occurs as a consequence of caspase activation (Bullock et al., 1998). When astrocytes exhibit central resistance through glutamate absorption, glutamate excitotoxicity first manifests in the neurons (Kaur & Sharma, 2018; Robertson et al., 2001). Astrocytes are known to transport glutamate due to the presence of glutamate-aspartate transporter (GLAST) and glutamate transporter-1 (GLT-1). The cells then convert glutamate to glutamine and provide significant defense alongside other interconnected astrocytes through gap junction channels. Excessive extracellular glutamate also initiates astrocyte pathology through overuptake of $Na^+$ through sodium-glutamate co-transporters. Even these events cause catabolic reactions including blood brain barrier (BBB) disintegration, the body's attempt to compensate for ionic gradients, an increase in $Na^+/K^+$-ATPase activity, and ultimately a rise in metabolic demand, lead to a severe loop in which flow-metabolism uncouples from the cell (Floyd, Gorin, & Lyeth, 2005; Kaur & Sharma, 2018).

Several lines of evidence have suggested that the deleterious effects of $A_{2A}R$ activation in acute neurological injuries could be largely due to aggravated glutamate excitotoxicity via both neuronal and glial mechanisms, and both presynaptic and postsynaptic mechanisms. Additionally, it could disrupt the balance of the intraparenchymal-blood concentration gradient of glutamate. In a TBI animal model, genetic inactivation of $A_{2A}R$ was reported to reduce levels of glutamate significantly in the CSF at 24 h post TBI (Dai et al., 2008; Li et al., 2009; Ning et al., 2013). It was also reported that chronic caffeine consumption decreased the glutamate level in CSF in severe TBI patients (Li et al., 2008).

## 3.1 Potentiation of presynaptic glutamate release by $A_{2A}R$ activation

Increased release of excitatory amino acid after acute CNS injuries, including TBI, is thought to prime an excitotoxic cascade that eventually results in cell death (Bruce-Keller, 1999; Faden, Demediuk, Panter, & Vink, 1989). Numerous studies have shown that $A_{2A}R$ is abundant in the striatum, primarily found on GABAergic neurons (Schiffmann, Jacobs, & Vanderhaeghen, 1991). Besides, $A_{2A}R$ is also found presynaptically on glutamatergic terminals, where it can regulate glutamate release by interacting with other neurotransmitter receptors, including cannabinoid CB1 receptor, dopamine D2 receptor (Ferre, Sarasola, Quiroz, & Ciruela, 2023; Kofalvi et al., 2020; Marcoli et al., 2003; Pinna, Serra, Marongiu, & Morelli, 2020). $A_{2A}R$ activation has been shown to increase glutamate release by a PKA-dependent signaling pathway, resulting in enhanced $Ca^{2+}$ influx (Chen & Chern, 2011; Higley & Sabatini, 2010). As a result, blockade of $A_{2A}R$ either by antagonist or by genetic inactivation may play a neuroprotective role by alleviating excitatory toxicity.

## 3.2 Exacerbation of postsynaptic glutamate excitotoxicity by $A_{2A}R$ activation

Released glutamate induces excitotoxicity through activating its postsynaptic ionotropic receptors, including NMDA and AMPA. Hippocampal $A_{2A}R$ activation increases postsynaptic AMPA-evoked currents, as well as the phosphorylation and surface expression of GluR1, the subunit of AMPA receptor (Dias, Ribeiro, & Sebastiao, 2012). Also, synergistic interaction between $A_{2A}R$ and metabotropic glutamate receptor 5 (mGluR5) is observed in vitro and in vivo (Ferre et al., 2002; Hamor, Gobin, & Schwendt, 2020). The expression and activation of mGluR5, a member of the group I mGluR, can encourage postsynaptic influx of $Ca^{2+}$ to increase the excitotoxic activity of glutamate (Lea, Custer, Vicini, & Faden, 2002; Skeberdis et al., 2001). Accordingly, activation of $A_{2A}R$ can promote glutamate excitotoxicity via these postsynaptic mechanisms.

It has been suggested that $A_{2A}R$ activation could facilitate excitatory postsynaptic current (Mouro, Rombo, Dias, Ribeiro, & Sebastiao, 2018). The frequency of glutamate released from the presynaptic membranes of hippocampal neurons may be regulated by $A_{2A}R$, which then modify the frequency of the calcium channels associated with NMDA receptor (NMDAR) opening at the postsynaptic membrane (Rebola, Lujan, Cunha, & Mulle, 2008). However, there are studies demonstrated that postsynaptic $A_{2A}R$ activation inhibited NMDAR-mediated synaptic currents. Consequently, $A_{2A}R$ antagonists decrease neurotransmitter release while amplifying the excitotoxic effect of direct NMDAR stimulation (Chen et al., 2014). Moreover, some findings concluded that $A_{2A}R$ activation impaired synaptic plasticity in animal models (Costenla et al., 2011; Laurent et al., 2016). Recently, Tan et al. confirmed that $A_{2A}R$ and NMDAR interact to form heteromeric complexes in hippocampal synaptic structures. Their findings support the notion that the $A_{2A}R$–NMDAR heteromers exhibit distinct subcellular localizations in different brain regions, potentially contributing to the bidirectional regulation of NMDAR function by $A_{2A}R$. Specifically, a presynaptic distribution of $A_{2A}R$–NMDAR facilitated $A_{2A}R$-mediated promotion of long-term depression (LTD), whereas a postsynaptic distribution facilitated $A_{2A}R$-mediated long-term potentiation (LTP) (Tan, Li et al., 2021). These results may enhance our understanding of the effects of $A_{2A}R$-mediated postsynaptic glutamate excitotoxicity via NMDAR.

### 3.3 Control of cerebral glutamate homeostasis by astroglial $A_{2A}R$

A substantial body of evidence has demonstrated the crucial role of astrocytes in the regulation of glutamate homeostasis and the brain's vulnerability to excitotoxic damage (Lopes, Cunha, & Agostinho, 2021). It has also been documented that $A_{2A}R$ can act as an astrocytic modulator. And astrocytic $A_{2A}R$ dysfunction has been found to impair glutamate homeostasis.

$A_{2A}R$ can regulate many aspects of astrocytic functions as shown below, especially in pathological conditions, which all contributes to the astrocyte-related glutamate excitotoxicity.

**1.** $A_{2A}R$ can induce abnormal transcription and trigger astrocyte reactivation (Brambilla et al., 2003; Paiva et al., 2019). The antagonists of

$A_{2A}R$ have exhibited potential beneficial effects on cerebral glutamate homeostasis. For instance, the $A_{2A}R$ antagonist SCH58261 restored the expression levels of genes related to astrocytic activation- and consequently improved astrocytic dysfunctions (Paiva et al., 2019). In addition, treatment with $A_{2A}R$ antagonist SCH58261 and KW6002 counteracted the formation of reactive astrocytes induced by basic fibroblast growth factor in rat striatal primary astrocytes in a concentration-dependent manner (Brambilla et al., 2003).

2. $A_{2A}R$ regulates glutamate release and, consequently, synaptic transmission (Amato et al., 2022; Cervetto et al., 2017; Liu et al., 2021; Nishizaki et al., 2002). A study by Nishizaki T et al. showed that astrocytic $A_{2A}R$ promoted glutamate release in a GLT-1-independent but $Ca^{2+}$-dependent manner, via PKA pathway, thereby increasing the glutamate concentration in the synaptic cleft (Nishizaki et al., 2002). $A_{2A}R$ is also suggested to regulate astrocytic glutamate release through receptor–receptor interaction (RRI) with other CNS receptors. Amato S et al. reported that $A_{2A}R$ and oxytocin receptor (OTR) co-localized in the same striatal astrocyte and that $A_{2A}R$ activation abolished OT-mediated inhibition of glutamate release from astrocyte processes, resulting in modulation of glutamatergic synapse functioning (Amato et al., 2022). $A_{2A}R$ was also shown to potentiate astrocytic glutamate release through $A_{2A}R$–D2R interaction (Cervetto et al., 2017).

3. $A_{2A}R$ regulates glutamate uptake by controlling the levels or function of glutamate transporters (GLT) (Constantino et al., 2021; Hou, Li, Huang, Zhao, & Gui, 2020; Matos, Augusto, Agostinho, Cunha, & Chen, 2013; Nishizaki et al., 2002; Pintor et al., 2004; Yu et al., 2012). Rapid clearance of extracellular glutamate contributes greatly to the maintaining of glutamate homeostasis. Astrocytes play a major role in the uptake of extracellular glutamate, which relies on excitatory amino acid transporters (EAAT), also known as GLT. Several lines of evidence have shown that $A_{2A}R$ activation increases extracellular glutamate by suppressing the levels or the functions of the GLT and the resultant glutamate uptake. For example, in rat retinal Muller cells under elevated hydrostatic pressure in vitro, the $A_{2A}R$ antagonist SCH442416 increased the levels of glutamine synthetase (GS) and GLAST, leading to accelerated clearance of extracellular glutamate (Yu et al., 2012). Adenosine, act at $A_{2A}R$, was reported to prevent glutamate uptake by inhibiting the function of the glial GLT-1 (Nishizaki et al., 2002). Recently, Hou X et al. reported that (Hou et al., 2020) adenosine $A_1R$–$A_{2A}R$ heteromers

regulated GLT-1 expression and glutamate uptake under oxygen-glucose deprivation (OGD) conditions. Inactivation of $A_{2A}R$ by SCH58251 reversed OGD-mediated glutamate uptake dysfunction and elevated GLT-1 level through YY1-induced repression of PPAR gamma transcription. $A_{2A}R$ was also reported to inhibit glutamate uptake by decreasing the activity of $Na^+/K^+$-ATPase (NKA) selectively in astrocytes in the cerebral cortex and striatum through modulation of NKA-α2 activity (Matos et al., 2013).

4. $A_{2A}R$ plays a significant role in modulating endothelial EAAT function. In addition to uptake by astrocytes wrapped around the synapse, the EAAT on abluminal endothelial membranes, primarily GLAST and GLT-1, continuously transport extracellular glutamate into endothelial cells and then into the blood through facilitated transport, playing a crucial role in maintaining glutamate homeostasis in the brain (Ballabh, Braun, & Nedergaard, 2004; Zhao, Nelson, Betsholtz, & Zlokovic, 2015). After acute brain injury, such as stroke and TBI (Nicholls & Attwell, 1990; Rossi, Oshima, & Attwell, 2000), endothelial EAAT dysfunction or even blood–brain reverse transportation may be another important factor contributing to the robust increase in glutamate levels in the brain (Bai & Zhou, 2017). A study by Bai W et al. showed that under OGD conditions, activation of $A_{2A}R$ reduces PKA- and glutamate level-dependent strengthening of the interaction between NKA-α1 and the FXYD1 subunit, leading to a decline in the activity of NKA. Activation of $A_{2A}R$ also increases its interaction with EAAT, which worsens the reverse transport function of endothelial EAATs. By rectifying the decreased connection between NKA-α1 and FXYD1, two NKA subunits, and regaining the NKA activity, the $A_{2A}R$ antagonist ZM241385 restores the normal transport function of EAAT, increases endothelial cell absorption of glutamate, and lowers extracellular glutamate concentrations (Bai et al., 2018). $A_{2A}R$ antagonists provide protection against severe TBI in animal models as well, and this protection is linked to chronically high glutamate levels after severe TBI.

Altogether, the aforementioned findings suggest that the deleterious effects of $A_{2A}R$ activation in acute neurological injuries could be largely due to aggravated glutamate excitotoxicity via both neuronal and glial mechanisms. These findings also outlined the mechanisms underlying the neuroprotection brought on by the $A_{2A}R$ antagonists in the situation of brain injury and indicate it as a potential new target for the pharmacological control of glutamate excitotoxicity.

## 4. A$_{2A}$R modulation of aberrant proteins

TBI-induced hyperphosphorylation of tau is a major pathological manifestation that contributes to cognitive impairment. Several lines of evidence reveal that tau is phosphorylated for a long period and accumulates in multiple regions of the brain after TBI. In addition, the blockage of autophagic flux aggravates the aggregation of toxic proteins, including tau protein.

## 4.1 Modulation of tau hyperphosphorylation after TBI

Recent investigations have revealed that hyperphosphorylated tau, mainly at the Ser404, but not at the Thr205 and Ser262 epitopes, increased in multiple brain regions, including the ipsilateral parietal cortex, contralateral hippocampus, and prefrontal cortex, as early as 24 h after the injury in a mouse model of TBI (Zhao Ning et al., 2017). Tau phosphorylation at Thr231 and Ser199/202 is elevated in the hippocampus and cortex 8 weeks post-injury in animal models (Sabbagh et al., 2016), and a progressive accumulation of phosphorylated tau is even observed in the cerebral cortex, mammillary bodies, and periventricular locations at 24 weeks after TBI (Kahriman et al., 2021). These findings suggest that tau is hyperphosphorylatd at different sites of the protein during different stages of TBI.

Recent investigations have shown that the inhibition or knockdown of A$_{2A}$R can attenuate hyperphosphorylated tau after TBI, and this effect is associated with the inhibition of PKA and glycogen synthase kinase-3β (GSK-3β), both of which are kinases that phosphorylate tau (Kahriman et al., 2021). Treatments with H89 (PKA antagonist), SB216763 (GSK-3 antagonist), as well as a combined treatment with H89 and SB216763, substantially reduce neuronal tau hyperphosphorylation and axonal damage (Zhao, Zhao et al., 2017). Likewise, caffeine at doses of 0.5 and 30 mg/day decreased the levels of phosphorylated GSK-3β, which were closely related to tau phosphorylation and reduced oxidative stress levels (Prasanthi et al., 2010).

Furthermore, it was confirmed in patients with frontotemporal lobar degeneration (FTLD) that the A$_{2A}$R immunoreactivity is higher in neurons with tau pathology compared to tau-negative cells when tau/A$_{2A}$R co-immunostaining was performed, demonstrating a reciprocal regulation between the emergence of tauopathy and the upregulation of neuronal A$_{2A}$R (Carvalho et al., 2019).

## 4.2 Modulation of autophagy after TBI

There are several evidences revealing that $A_{2A}R$ is involved in the regulation of autophagy via a non-canonical signaling pathway under moderate-to-severe experimental TBI in animals and cultured cells, ultimately affecting the severity of trauma and spatial cognition. Local injection of ZM241385 or $A_{2A}R$ knockdown reverses the blockage of autophagic flux and alleviates the impairment of spatial cognition by regulating the PKA/ERK2/TFEB signaling pathway (Zeng et al., 2018; Zeng et al., 2020). Another study showed that SCH58261, a selective $A_{2A}R$ antagonist, administered at the early stage of photoreceptor degeneration, protects photoreceptors through the inhibition of the inflammatory response. Interestingly, the protection of $A_{2A}R$ inhibition is prevented by the inactivation of microglia (Cheng et al., 2022).

However, the effect of $A_{2A}R$ antagonist on autophagy in peripheral tissue is opposite compared to that of in the CNS. A study showed that a high dose of LPS induced both apoptosis and autophagy of neutrophils in a mouse SIRS model and LPS-stimulated neutrophils in vitro. In this inflammation model of mice, $A_{2A}R$ activation inhibit neutrophil apoptosis by inhibiting LPS-induced autophagy though the inhibiting of the ROS-JNK pathway as well as promoting AKT signaling (Liu, Yang et al., 2016). In addition, a recent study revealed that $A_{2A}R$ regulates autophagy in myocardial ischemia–reperfusion injury. The authors confirmed that $A_{2A}R$ activation alleviates ischemia–reperfusion injury by regulating autophagy flux and apoptosis through the cAMP/PKA pathway (Xia et al., 2022).

## 5. $A_{2A}R$ modulation of cognitive dysfunction after TBI

A substantial number of patients suffering from TBI present a long-term cognitive dysfunction even after receiving both emergency and rehabilitation treatments. Therefore, TBI is regarded as an important risk factor for the early appearance of cognitive impairment (Gardner & Yaffe, 2014; Johnson, Stewart, Arena, & Smith, 2017; Pavlovic, Pekic, Stojanovic, & Popovic, 2019; Smith, Johnson, & Stewart, 2013). The hazard ratio (HR) of dementia in TBI patients increased from 2.36 (mild) to 3.77 (severe) with the severity of injury (Barnes et al., 2018).

So far, there are no specific therapeutic drugs for cognitive impairment after TBI, and the treatment of cognitive decline after TBI is similar to that of AD, including cholinesterase inhibitor (Njoku et al., 2019) and memantine hydrochloride, an NMDAR antagonist (Mokhtari et al., 2018). Other drugs under development include antibodies targeting tau protein, tau kinase inhibitors, and amyloid beta (Aβ) scavenging drugs, etc. Although these drugs have achieved good results in experimental animal models and in vitro studies, the therapeutic effect in clinical trials has not achieved as expected. A recent multi-center study of 3671 subjects showed that cholinesterase inhibitors failed to protect against TBI-induced cognitive decline (Brawman-Mintzer et al., 2021).

On the other hand, neuropathological changes in the chronic stage after TBI are different from senile neurodegeneration, with more severe neuroinflammation, axonal transport damage, synaptic dysfunction, and more significant pathological modification of tau. Thus, those used to treat AD may not be entirely appropriate, and few of them displayed effectively outcomes in clinical or pre-clinical trials. Recently, adenosine and its receptor $A_{2A}R$ have aroused wide interest from more and more investigators due to their involvement in several adverse consequences following TBI, such as cognitive dysfunction, sleep disorders, and other neuropsychiatric abnormalities. How do adenosine and $A_{2A}R$ play roles in the context of these sequelae and how to intervene?

## 5.1 Alteration of adenosine and its receptors during a chronic period of TBI

Adenosine, as an important neuromodulator, plays an effective, fine and feedback central regulatory role in maintaining the homeostasis of the nervous system through different distribution and downstream signal pathways of various receptors in the brain (Chen et al., 2007). The extracellular adenosine levels are increased rapidly after TBI by approximately 100-fold greater than baseline (Bell et al., 2001; Nilsson, Hillered, Ponten, & Ungerstedt, 1990). Do they remain elevated during the chronic phase post-injury?

Studies have revealed that extracellular adenosine is still increased in the dialysate and CSF of patients with TBI and several animal models at the chronic stage of TBI. Bell et al. found the concentration of interstitial adenosine and its metabolite xanthine were significantly higher than baseline levels in patients with severe TBI (Glasgow Coma Scale score less than 9), including samples from a patient 310 h after a car accident. This study demonstrated that interstitial adenosine concentration increased 3.1-fold during the desaturation period of jugular versus periods of normal (Bell et al., 2001), suggesting that the production of adenosine increases at the late stage of injury and is susceptible to hypoxia.

Under this condition, the increased adenosine is mostly from astrocytes. Astrocytes and neuron exert their unique roles in the CNS, thus the regulations of adenosine level in astrocytes and neurons is not the same. Astrocytes display an increase in intracellular calcium in response to synaptic activity through the activation of metabotropic or ionotropic receptors and, in turn, trigger the release of so-called gliotransmitters, such as ATP, glutamate, and D-serine. Increased levels of extracellular ATP and its degradation product adenosine at synapses thus regulate synaptic plasticity (Araque et al., 2014; Pfeiffer & Attwell, 2020). Evidence from PTSD models and photogenetic excitation of astrocytes also confirmed that enhanced astrocytes activity promoted extracellular ATP and adenosine levels (Parkinson, Xiong, & Zamzow, 2005). Increased adenosine is metabolized sequentially to inosine, hypoxanthine, xanthine, and uric acid under the actions of a series of enzymes, which promotes the production of reactive oxygen species (ROS) and leads to neuronal damage.

The influence of adenosine on post-TBI neuronal function not only depends on the level of adenosine but also on the type and abundance of local expression of various adenosine receptors. A variety of injury factors and the high level of adenosine lead to the activation of different adenosine receptors distributed in injured and uninjured brain regions, as well as the alteration of their expression. The changes in different brain regions and cell types have different effects on learning and memory, mood, and sleep. Most of the results showed that cognitive dysfunction is accompanied by increased expression of $A_{2A}R$ and decreased $A_1R$, but the changes vary with the time post-injury, site, cell type, and other factors.

In CCI-induced TBI, the level of $A_{2A}R$ remained higher than baseline until 4 weeks after injury (Zhao, Li et al., 2017). In vitro cultured glial cells, LPS stimulation increased $A_{2A}R$ expression in isolated microglia cells (Saura et al., 2005). Elevated expression of glial $A_{2A}R$ was also observed in postmortem samples from patients with AD-induced cognitive impairment (Angulo et al., 2003).

In addition, increased $A_{2A}R$ expression was also observed in the hippocampus and cortex of aging rats (Diogenes, Assaife-Lopes, Pinto-Duarte, Ribeiro, & Sebastiao, 2007; Rebola et al., 2003). Rodrigues, Canas, Lopes, Oliveira, and Cunha (2008) found that the proportion of $A_1R$ did not change in neurons, but that of $A_{2A}R$ increased to 49% at the age of 24 months. Marques, Batalha, Lopes, and Outeiro (2011) proposed in a review that the aging process is accompanied by the weakening of $A_1R$-mediated inhibition and the enhancement of $A_{2A}R$-mediated excitatory damage, thus blocking or reversing this trend is beneficial to the prevention and treatment of dementia.

Adenosine receptors regulate the downstream PKA pathway by the second messenger cAMP in opposite directions. While $A_1R$ is inhibitory G-protein (Gi/Go) coupled, thus decreasing cAMP levels, $A_{2A}R$ is excitatory G-protein (Gs) coupled, increasing cAMP levels. Understandably, $A_1R$ alteration also leads to a series of neuropathological changes, and many studies have detected changes in $A_1R$. When the injury factors persist, $A_1R$ are continuously activated by a high concentration of adenosine, which leads to decreased expression of $A_1$ receptors in the hippocampus and other brain regions. A study using autoradiography to detect $A_1R$ in the hippocampus of rats with damage in entorhinal cortex found decreased $A_1R$ expression and field potential 2 weeks after trauma, and adenosine administration aggravates the decrease in field potential, but $A_{2A}R$ expression was not detected in this work (Kahle, Ulas, & Cotman, 1993).

Postmortem examination showed that $A_1R$ in hippocampal CA1 and DG decreased by about 50% in dementia patients with AD (Jansen, Faull, Dragunow, & Synek, 1990; Kalaria, Sromek, Wilcox, & Unnerstall, 1990) and other diseases with cognitive impairment (Deckert et al., 1998). This is also confirmed by the results of positron emission tomography (PET) in AD patients (Fukumitsu et al., 2008). When the expression of $A_1R$ is decreased, $A_{2A}R$ begins to play a major role in the chronic stage of injury (Cunha, 2005). The activation of $A_{2A}R$ increases the release of excitatory amino acids and $Ca^{2+}$ influx, which further induces excitotoxicity in the brain. In addition, activated $A_{2A}R$ can aggravate the inflammatory response by activating microglia cells. Despite the differences in the subcellular distribution of $A_{2A}R$ and $A_1R$ (Pickel, Chan, Linden, & Rosin, 2006), maintaining adenosine receptor balance may be an important approach for the prevention and treatment of cognitive impairment. Further study of adenosine receptor homeostasis may contribute to a further understanding of the pathogenesis of cognitive impairment and provide new therapeutic strategies.

## 5.2 $A_{2A}R$ regulation in pathological proteins

Another important role of $A_{2A}R$ antagonists is to inhibit the production of major pathological proteins in cognitive impairment, including Aβ and tau, which has been demonstrated in various AD models (Faivre et al., 2018).

Studies have shown that Aβ may cause increased $A_{2A}R$ levels in microglia, further promoting the activation of microglia (Orr, Orr, Li, Gross, & Traynelis, 2009). Researchers reported that both the non-selective antagonist caffeine and the selective antagonist ZM241385 of $A_{2A}R$ could counteract the neuronal apoptosis caused by Aβ incubation. This in vitro study demonstrates that $A_{2A}R$ may be a target for the prevention and treatment of AD (Dall'igna, Porciuncula, Souza, Cunha, & Lara, 2003). Canas et al. (2009) found that $A_{2A}R$ antagonists can inhibit neuronal apoptosis and animal cognitive dysfunction by inhibiting the hyperphosphorylation of p38 induced by Aβ. It is suggested that there are other pathways that are independent of the cAMP/PKA pathway involved in the cognitive protection of $A_{2A}R$ inhibition. In addition, $A_{2A}R$ antagonists can also reduce the expression of β-secretase and Aβ production (Arendash et al., 2006), increase the p-glycoprotein of the blood–brain barrier, and promote Aβ clearance in the brain (Qosa, Abuznait, Hill, & Kaddoumi, 2012).

In addition, the development and exacerbation of cognitive impairment following TBI may be primarily attributed to hyperphosphorylation of tau. To address a proposed connection between tau pathology and $A_{2A}R$, Carvalho and his colleagues created a transgenic mouse with $A_{2A}R$ overexpression in THY-Tau22 mouse forebrain neurons and found spatial memory was strongly decreased accompanied by global changes of tau phosphorylation using 2D electrophoresis. Of note, $A_{2A}R$ neuronal upregulation alone in a wild-type background did not elicit basal transcriptomic changes as compared to controls, suggesting a post-transcriptional basis for the memory alteration.

Moreover, the protective effect of $A_{2A}R$ inhibition is also displayed in a mouse model of tauopathy. $A_{2A}R$ genetic or pharmacological blockade decreased neuroinflammation, tau phosphorylation and aggregation, and improved spatial learning and memory and hippocampal plasticity in the tau transgenic mouse model (Laurent et al., 2016). $A_{2A}R$ knockout can also ameliorate the abnormal polarity distribution of AQP4 caused by TBI and reduce the level of AQP4, thus promoting the clearance of extracellular phosphorylation of tau and reducing its uptake by neighboring cells (Zhao, Li et al., 2017).

## 5.3 $A_{2A}R$ and delayed cognitive dysfunction of TBI

Frontal and temporal lobes are vulnerable regions when brain suffers physical impact, which often caused a decline in cognitive function, especially in executive function, often occurs after TBI. Hippocampus is a key region for learning and memory, responsible for encoding, consolidation and retrieve of information involved in episodic and semantic memory. Executive function is a higher cognitive process involved in goal-oriented behavior, including maintaining goals, making plans, ranking, evaluating, etc. and working memory is required for goal maintenance. The prefrontal lobe is a key brain region for proper executive function. Animals suffering from TBI showed pathological and behavioral changes in PFC even if the prefrontal cortex is not directly subjected to physical impact. Differences of $A_{2A}R$ genotype can lead to changes in the connectivity of PFC-insular prominence network which is required for executive function (Geiger et al., 2016).

A study revealed that the speed of information processing in patients was significantly slower than that of the health control 6 months after injury (Kourtidou et al., 2013). Another work showed 13% of patients with mild TBI showed significant cognitive disorder 1 year after injury (Christman Schneider et al., 2022), this memory impairment was even present 11 years after mild TBI (Ahman, Saveman, Styrke, Bjornstig, & Stalnacke, 2013). $A_{2A}R$ is involved in learning and memory process under physiological conditions (Cunha, 2005), the mechanism of $A_{2A}R$ in cognitive impairment after TBI is more complex, which may be related to the heterogeneity of TBI injury and different pathological processes. Currently, there is no specific treatment for cognitive impairment after TBI. In the phase IV clinical trial of memantine, Hopkins verbal learning test revised (HVLT-R) was used to detect delayed recall after TBI and the tracking line test was used to detect executive function. Although memantine showed improvement in memory, it was discontinued due to insufficient number of cases. Another clinical trial showed memantine hydrochloride has a better effect for patients with moderate to severe TBI (Giacino et al., 2012).

In CCI-induced TBI, the level of $A_{2A}R$ expression began to increase 3 days after injury and remained a higher level until 4 weeks (Zhao, Li et al., 2017). After $A_{2A}R$ knockout, the spatial reference memory and working memory of mice were significantly improved at 1 week compared to the wild type, and the improvement of working memory was still observed at 4 weeks after the injury. In the blast injury model, $A_{2A}R$ knockout mice showed better spatial working memory and exploration ability than wild mice at 1 and 4 weeks after injury, and improvement in working memory was observed even at 8 weeks (Ning et al., 2013). In addition, local injection of ZM241385 or $A_{2A}R$ knockout after moderate to severe TBI injury can reverse the impaired autophagy flow and reduce the impairment of spatial cognitive function by regulating the PKA/ERK2/TFEB signaling pathway (Zeng et al., 2018; Zeng et al., 2020). Similarly, Pagnussat et al. (2015) found that abnormal activation of $A_{2A}R$ could lead to memory impairment in mice, and the selective antagonist SCH58261 of $A_{2A}R$ could prevent this phenomenon.

Pre-chronic consumption of caffeine (0.25 g/L for 3 weeks) can also achieve a protective effect against spatial reference memory and working memory impairment (Zhao, Zhao et al., 2017). In addition, caffeine and $A_{2A}R$ antagonists improve cognitive dysfunction in a variety of dementia animals, including APP transgenic, high-cholesterol diet, and intracranial Aβ injection (Cunha & Agostinho, 2010; Espinosa et al., 2013; Han, Jia, Li, Yang, & Min, 2013). In epidemiological studies, regular and moderate coffee consumption was found to reduce the incidence of age-related memory impairment and the risk for AD (Eskelinen, Ngandu, Tuomilehto, Soininen, & Kivipelto, 2009; Gelber, Petrovitch, Masaki, Ross, & White, 2011; Travassos et al., 2015). The mechanisms underlying posttraumatic behavioral changes are associated with neuroinflammation, impairment of synaptic function, accumulation of toxic proteins, and other pathological processes, which are described below.

## 6. $A_{2A}R$ modulation of chronic neuroinflammation in TBI

When neuroinflammation becomes chronic, it is typically associated with neurodegenerative diseases, such as Alzheimer's disease (AD), Parkinson's disease (PD), and chronic traumatic encephalopathy (CTE). Several lines of evidence have demonstrated that $A_{2A}R$ mediates neurodegeneration, and to date, it is well established that $A_{2A}R$ antagonists could be useful for the treatment of several neurodegenerative diseases, including PD, AD, multiple sclerosis (MS), and neuropsychiatric conditions.

### 6.1 Long-lasting neuroinflammation

The persistence of neuroinflammation after TBI can disrupt the homeostasis of the nervous system and lead to pathological changes. Nedeljkovic considered that sustained induction of $A_{2A}R$ and resulting perpetual $CD73/A_{2A}R$ coupling is a contributing factor for the transition from acute to chronic neuroinflammation (Nedeljkovic, 2019). Recent research has displayed the protective effect of $A_{2A}R$ is related to the inhibition of inflammasome activation. Evidence from moderate TBI showed that activation of NLRP3 inflammasome was presented in the hippocampus at 4 and 8 weeks after TBI (Tan, Zhao et al., 2021), and prolonged activation of NLRP3 inflammasome was also found in animal models with tau overexpression (Zhao et al., 2021). Likewise, recent evidence confirmed that caffeine can reduce the ischemic and hypoxic-induced brain injury in newborn rats, which is related to the decreased activity of NLRP3 inflammasome, resulting significantly decrease in iNOS, IL-1β, and TNF-α at 3 weeks after injury, while significantly increasing IL-10 and TGF-β with antiinflammatory effects. This protective effect could be reversed by CGS21680, suggesting that the protective effect of caffeine is $A_{2A}R$ dependent (Yang et al., 2022). This result was also confirmed in a LPS-induced damage model in macrophage THP-1, where caffeine significantly reduced NLRP3 expression and ASC speck formation, and consistent results were obtained with $A_{2A}R$ inhibition (Zhao et al., 2019). In this study, caffeine markedly decreased the phosphorylation levels of MAPK and NF-κB pathway members, further suppressing the translocation of NF-κB in THP-1 macrophages.

The increased inflammatory, in turn, can promote the expression of $A_{2A}R$. In lung epithelial cells, IL-1β and TNF-α induced high expression of $A_{2A}R$, which can be blocked by IkappaB kinase 2 inhibitor, indicating that NF-κB plays a major role in the increase of $A_{2A}R$ expression (Morello et al., 2006). In trimethyltin (TMT)-induced hippocampal neurodegenerative changes in rats, elevated levels of IL-1β and TNF-α remained high at 21 days after TMT administration, accompanied by increased mRNA levels of $A_{2A}R$ (Dragic, Mitrovic, Adzic, Nedeljkovic, & Grkovic, 2021). These studies suggest that there might be crosstalk between $A_{2A}R$ and the NLRP3 inflammasome, which influences the outcomes at a late stage of TBI.

Administration of the $A_{2A}R$ agonist CGS21680, but not $A_1R$ agonist DPCPX, into cultured microglia cells induced dose-dependent expression of cyclooxygenase (COX)-2 mRNA and the release of prostaglandin E2, thus exacerbated inflammatory responses (Fiebich et al., 1996). In addition, $A_{2A}R$ activation can stimulate the activation of astrocytes and increase the expression of iNOS. In a neuroinflammatory model by chronically infusion of LPS into the ventricle for 2 and 4 weeks, caffeine reduced the activation of microglia in the hippocampus in rats (Brothers, Marchalant, & Wenk, 2010).

Moreover, the BBB and glymphatic system paly significant roles in numerous neurological disorders, and increasing permeability is regarded as a major cause of increased infiltration of immune cells, production of inflammatory factors, and the dissemination of toxic proteins. During this process, $A_{2A}R$, as a receptor extensively distributed on endothelial and immune cells, is widely reported for its regulation of the BBB. In an in vitro model of the BBB constructed by primary human microvascular endothelial cells, $A_{2A}$ receptor agonists decreased the expression of tight junction protein 5 and cadherin in vascular endothelial cells, thereby increasing the permeability of the BBB (Bynoe, Viret, Yan, & Kim, 2015; Kim & Bynoe, 2015). Though this is an effective strategy for drug administration in the CNS, it leads to the persistence of neuroinflammation after TBI. Additionally, astrocytes wrap around vascular endothelial cells through their endfoot and affect the function of the BBB and glymphatic system through AQP4 distributed on one side of the blood vessel. Chronic changes of AQP4 in the glymphatic system are an important cause of toxic protein propagation (Ren et al., 2013; Ruchika et al., 2023). Glymphatic dysfunction can last for more than 1 month after TBI, and co-localization of AQP4 and $A_{2A}R$, as well as abnormal polarity of AQP4, can be detected in brain tissue from patients with brain trauma (Zhao, Li et al., 2017).

## 6.2 Impairment of synaptic structure and function

A recent study reported that $A_{2A}R$ is involved in the occurrence and development of neurodegenerative diseases by regulating synaptic function (Merighi et al., 2022). $A_{2A}R$ is enriched in synapses, including the active zone of presynaptic nerve terminals and the postsynaptic density (Rebola, Canas, Oliveira, & Cunha, 2005), and selectively regulate synaptic plasticity (Costenla et al., 2011). It has been demonstrated that memory decline is associated with impairments of synaptic structure and function in the hippocampus, which is similar to hippocampal lesions in AD patients with memory loss (Silva et al., 2018). Caffeine consumption continued for 2 months in 4-month-old mice improved the length, branching, and density of age-dependent neuronal dendrites in the hippocampal CA1 region (Vila-Luna et al., 2012). Studies on the APP/PS1 mouse model showed that $A_{2A}R$ antagonists could restore synapses and LTP in mice (Viana Da Silva et al., 2016). In addition, a gene-based association analysis of patients with mild cognitive impairment and AD revealed that $A_{2A}R$ is associated with the number of neurons, and increased expression of $A_{2A}R$ is negatively correlated with hippocampus volume (Horgusluoglu-Moloch et al., 2017).

Microglia is involved in the regulation of learning and memory and monitoring the brain parenchyma under physiological conditions. Under conditions of trauma, these cells not only release inflammatory factors but also prune synapses through their phagocytic activity, resulting synapse loss and neuronal dysfunction. ATP receptors and $A_{2A}R$ play roles in the regulation of synaptic function by being involved in microglial process extension and chemoattraction to injury (Gyoneva, Orr, & Traynelis, 2009). It has been found that the increased expression of $A_{2A}R$ in the chronic stage of injury is accompanied by decreased ATP receptor P2Y12, which is related to the amoeba morphology of microglia. Microglial deramification is a hallmark of neuroinflammation, and $A_{2A}R$ activation can lead to process retraction of microglia. In contrast, the $A_{2A}R$ antagonist SCH-58261 can reverse the amoeba-like morphology of microglia and promote branching (Orr et al., 2009).

Astrocytes serve as important supporting cells by providing energy for neurons and modulating glutamatergic synaptic transmission. After activation, astrocytes release increasing amounts of ATP, which is metabolized into extracellular adenosine and then regulates the synaptic function of neurons. Cunha et al. found that astrocyte-selective knockout of $A_{2A}R$ improved LTP in the hippocampus (Goncalves et al., 2019). Furthermore, proliferated astrocytes can envelop synapses and lead to the blocking of connections between synapses. In a blast injury model, a decrease in the density of the astrocyte marker GFAP was observed at 4 and 8 weeks after injury in $A_{2A}R$ knockout mice (Ning et al., 2013), indicating that $A_{2A}R$ is involved in reactive astrogliosis.

## 7. $A_{2A}R$ modulation of other long-lasting consequences of TBI

$A_{2A}R$ play important regulatory roles not only in cognitive dysfunction secondary to TBI but also in other sequelae and complications associated with nervous system injury, such as sleep disorders, PTSD, and epilepsy.

### 7.1 Sleep disorders

The role of adenosine in sleep has been extensively investigated. $A_1R$ and $A_{2A}R$ contribute to adenosine-mediated modulation of the sleep-wake cycle (Bjorness & Greene, 2009; Porkka-Heiskanen & Kalinchuk, 2011), and evidence for $A_{2A}R$ involvement in sleep-wake physiology mainly stems from examining the arousal effects of caffeine (Ferre, 2010; Wei, Li, & Chen, 2011).

Most patients with mild, moderate, and severe TBI experience persistent sleep-wake disturbances. A meta-analysis of more than 1700 TBI survivors revealed that approximately one-third of sleep disorders after TBI were insomnia, one-third were hypersomnia, and one-fourth were obstructive sleep apnea (Mathias & Alvaro, 2012). Another study involving 98,000 veterans also showed that about 20% had sleep disorders, and the incidence of sleep disorders in TBI soldiers was 41% higher than non-TBI soldiers (Folmer et al., 2020; Leng et al., 2021). Due to the sliding and collision of brain tissue in the cranial cavity, the frontal lobe, anterior temporal, and the base of the brain are most susceptible to damage under the action of external forces.

In a sleep-deprived mouse model, the structure and function of the BBB are damaged, and $A_{2A}R$ expression increases in the hippocampus and basal nuclei. The $A_{2A}R$ antagonist SCH58261 can up-regulate the expression of tight junction proteins (claudin-5, occluding and ZO-1) and adherens junction protein (E-cadherin) in the cortex, hippocampus and basal nuclei. These data suggest that $A_{2A}R$ may play a crucial role during sleep loss by directly modulating brain endothelial cell permeability (Hurtado-Alvarado, Dominguez-Salazar, Velazquez-Moctezuma, & Gomez-Gonzalez, 2016). In Phase III clinical trials of istradefylline (KW6002), 20–30% of PD patients were found to have improved sleep and reduced fatigue within 3 months to 1 year after orally taking 20 mg or 40 mg of istradefylline, as assessed by the PGI-I scale.

## 7.2 PTSD

PTSD is a severe and complex condition that can occur after experiencing traumatic events. The incidence of PTSD is significantly higher in individuals who have suffered from TBI (Scholten et al., 2016; Stein et al., 2023), with about 1 in 3 people developing PTSD after severe trauma. Studies have shown that adenosine derivative WS0701 could reduce hippocampal neuron apoptosis and significantly reduce anxiety symptoms in PTSD mice (Huang et al., 2014). In a mouse model of PTSD, dysfunctions in cAMP-PKA signaling pathway in the peripheral serum of PTSD patients were found to be associated with the extinction of fear memory. Activation of this pathway can promote the extinction of fear memory and reduce anxiety-like behaviors (Gao, Wang, Yang, Ji, & Zhu, 2023). Curculigoside, a compound that binds with $A_{2A}R$ and activates the PKA/CREB/BDNF/TrkB pathway, has been found to prevent PTSD-like phenotypes and synaptic deficits (Ji et al., 2023). Abnormal metabolism of ATP/ADP after trauma has also been linked to PTSD-like symptoms (Preston et al., 2021).

Anxiety is one of core symptoms of PTSD and is also a common consequence of TBI. In chronic anxiety models, $A_{2A}R$ blocking has been shown to improve cognition in dexamethasone-treated female mice by rectifying the morphology of microglia in mPFC-dHIP circuit (Duarte et al., 2019). In the amygdala, blocking $A_{2A}R$ or inhibiting CD73, an enzyme involved in adenosine production, decreased the expression of fear memory (Simoes et al., 2022). In contrast, peripheral administration of adenosine in mice increased anxiety by increasing caspase-1 activity and IL-1β levels in the amygdala, while $A_{2A}R$ knockout significantly reduced anxiety through the ATP-sensitive potassium channels and PKA pathway (Chiu et al., 2014). In vitro studies have shown that $A_{2A}R$ blockade prevents alterations in the morphology and function of astrocytes exposed to dexamethasone, as well as alterations in ATP release and basal $Ca^{2+}$ levels (Madeira et al., 2022).

## 7.3 Posttraumatic seizures and epilepsy

Epilepsy is a common consequence of TBI and increased expression of $A_{2A}R$ and CD73 has been observed in the hippocampus of patients with medial temporal lobe epilepsy (Barros-Barbosa et al., 2016). Blocking $A_{2A}R$ can prevent the transmission of epileptic discharges and contribute to synaptic remodeling in the adult hippocampus (Xu et al., 2022). $A_1R$ and its gene mutation also play a role in the development of epilepsy after TBI, as adenosine inhibits glutamate release by binding to $A_1R$ (Cotter, Kelso, & Neligan, 2017; Kochanek et al., 2006; Wagner et al., 2010). This effect is related to adenosine inhibiting glutamate release by binding to $A_1R$. Gene variants of adenosine kinase and 5'-exonuclease, a key molecule that regulates extracellular adenosine level, were associated with increased risk for post-injury epilepsy in trauma patients (Diamond et al., 2015).

In conclusion, $A_{2A}R$ plays a regulatory role in various tissues and organs after trauma. Its effects on neuroinflammation, synaptic structure and function, and the production and spread of toxic proteins contribute to the development and consequences of brain injury. Future studies should focus on the specific roles of $A_{2A}R$ in different cell types and the cross-dialogue between these cells. While $A_{2A}R$ antagonists or gene knockout have shown protective effects in brain injury, factors such as the severity, location, time post-trauma, and microenvironment changes should be carefully considered before administering drugs. Restoring the balance of adenosine receptors in the brain may be an important direction for preventing traumatic damage. The study of $A_{2A}R$ in TBI and its long-term consequences also provides insights for exploring other CNS diseases.

# References

Ahman, S., Saveman, B. I., Styrke, J., Bjornstig, U., & Stalnacke, B. M. (2013). Long-term follow-up of patients with mild traumatic brain injury: A mixed-method study. *Journal of Rehabilitation Medicine: Official Journal of the UEMS European Board of Physical and Rehabilitation Medicine, 45*, 758–764.

Aires, I. D., Madeira, M. H., Boia, R., Rodrigues-Neves, A. C., Martins, J. M., Ambrosio, A. F., & Santiago, A. R. (2019). Intravitreal injection of adenosine A(2A) receptor antagonist reduces neuroinflammation, vascular leakage and cell death in the retina of diabetic mice. *Scientific Reports, 9*, 17207.

Amato, S., Averna, M., Guidolin, D., Pedrazzi, M., Pelassa, S., Capraro, M., ... Marcoli, M. (2022). Heterodimer of A2A and oxytocin receptors regulating glutamate release in adult striatal astrocytes. *International Journal of Molecular Sciences, 23*, 2326.

Anderson, F. L., Biggs, K. E., Rankin, B. E., & Havrda, M. C. (2023). NLRP3 inflammasome in neurodegenerative disease. *Translational Research: The Journal of Laboratory and Clinical Medicine, 252*, 21–33.

Angulo, E., Casado, V., Mallol, J., Canela, E. I., Vinals, F., Ferrer, I., ... Franco, R. (2003). A1 adenosine receptors accumulate in neurodegenerative structures in Alzheimer disease and mediate both amyloid precursor protein processing and tau phosphorylation and translocation. *Brain Pathology (Zurich, Switzerland), 13*, 440–451.

Anthonymuthu, T. S., Kenny, E. M., & Bayir, H. (2016). Therapies targeting lipid peroxidation in traumatic brain injury. *Brain Research, 1640*, 57–76.

Araque, A., Carmignoto, G., Haydon, P. G., Oliet, S. H., Robitaille, R., & Volterra, A. (2014). Gliotransmitters travel in time and space. *Neuron, 81*, 728–739.

Arendash, G. W., Schleif, W., Rezai-Zadeh, K., Jackson, E. K., Zacharia, L. C., Cracchiolo, J. R., ... Tan, J. (2006). Caffeine protects Alzheimer's mice against cognitive impairment and reduces brain beta-amyloid production. *Neuroscience, 142*, 941–952.

Bai, W., Li, P., Ning, Y. L., Peng, Y., Xiong, R. P., Yang, N., ... Zhou, Y. G. (2018). Adenosine A2A receptor inhibition restores the normal transport of endothelial glutamate transporters in the brain. *Biochemical and Biophysical Research Communications, 498*, 795–802.

Bai, W., & Zhou, Y. G. (2017). Homeostasis of the intraparenchymal-blood glutamate concentration gradient: Maintenance, imbalance, and regulation. *Frontiers in Molecular Neuroscience, 10*, 400.

Ballabh, P., Braun, A., & Nedergaard, M. (2004). The blood-brain barrier: An overview: Structure, regulation, and clinical implications. *Neurobiology of Disease, 16*, 1–13.

Barnes, D. E., Byers, A. L., Gardner, R. C., Seal, K. H., Boscardin, W. J., & Yaffe, K. (2018). Association of mild traumatic brain injury with and without loss of consciousness with dementia in US Military Veterans. *JAMA Neurology, 75*, 1055–1061.

Barros-Barbosa, A. R., Ferreirinha, F., Oliveira, A., Mendes, M., Lobo, M. G., Santos, A., ... Correia-De-Sa, P. (2016). Adenosine A(2A) receptor and ecto-5′-nucleotidase/CD73 are upregulated in hippocampal astrocytes of human patients with mesial temporal lobe epilepsy (MTLE). *Purinergic Signalling, 12*, 719–734.

Bell, M. J., Robertson, C. S., Kochanek, P. M., Goodman, J. C., Gopinath, S. P., Carcillo, J. A., ... Jackson, E. K. (2001). Interstitial brain adenosine and xanthine increase during jugular venous oxygen desaturations in humans after traumatic brain injury. *Critical Care Medicine, 29*, 399–404.

Bjorness, T. E., & Greene, R. W. (2009). Adenosine and sleep. *Current Neuropharmacology, 7*, 238–245.

Boison, D., Chen, J. F., & Fredholm, B. B. (2010). Adenosine signaling and function in glial cells. *Cell Death and Differentiation, 17*, 1071–1082.

Brambilla, R., Cottini, L., Fumagalli, M., Ceruti, S., & Abbracchio, M. P. (2003). Blockade of A2A adenosine receptors prevents basic fibroblast growth factor-induced reactive astrogliosis in rat striatal primary astrocytes. *Glia, 43*, 190–194.

Brawman-Mintzer, O., Tang, X. C., Bizien, M., Harvey, P. D., Horner, M. D., Arciniegas, D. B., ... Reda, D. (2021). Rivastigmine transdermal patch treatment for moderate to severe cognitive impairment in Veterans with traumatic brain injury (RiVET Study): A randomized clinical trial. *Journal of Neurotrauma, 38*, 1943–1952.

Brodie, C., Blumberg, P. M., & Jacobson, K. A. (1998). Activation of the A2A adenosine receptor inhibits nitric oxide production in glial cells. *FEBS Letters, 429*, 139–142.

Brothers, H. M., Marchalant, Y., & Wenk, G. L. (2010). Caffeine attenuates lipopolysaccharide-induced neuroinflammation. *Neuroscience Letters, 480*, 97–100.

Bruce-Keller, A. J. (1999). Microglial-neuronal interactions in synaptic damage and recovery. *Journal of Neuroscience Research, 58*, 191–201.

Bullock, R., Zauner, A., Woodward, J. J., Myseros, J., Choi, S. C., Ward, J. D., ... Young, H. F. (1998). Factors affecting excitatory amino acid release following severe human head injury. *Journal of Neurosurgery, 89*, 507–518.

Bynoe, M. S., Viret, C., Yan, A., & Kim, D. G. (2015). Adenosine receptor signaling: A key to opening the blood-brain door. *Fluids Barriers CNS, 12*, 20.

Canas, P. M., Porciuncula, L. O., Cunha, G. M., Silva, C. G., Machado, N. J., Oliveira, J. M., ... Cunha, R. A. (2009). Adenosine A2A receptor blockade prevents synaptotoxicity and memory dysfunction caused by beta-amyloid peptides via p38 mitogen-activated protein kinase pathway. *The Journal of Neuroscience, 29*, 14741–14751.

Carvalho, K., Faivre, E., Pietrowski, M. J., Marques, X., Gomez-Murcia, V., Deleau, A., ... Blum, D. (2019). Exacerbation of C1q dysregulation, synaptic loss and memory deficits in tau pathology linked to neuronal adenosine A2A receptor. *Brain, 142*, 3636–3654.

Cassada, D. C., Tribble, C. G., Long, S. M., Laubach, V. E., Kaza, A. K., Linden, J., ... Kern, J. A. (2002). Adenosine A2A analogue ATL-146e reduces systemic tumor necrosing factor-alpha and spinal cord capillary platelet-endothelial cell adhesion molecule-1 expression after spinal cord ischemia. *Journal of Vascular Surgery: Official Publication, The Society for Vascular Surgery [and] International Society for Cardiovascular Surgery, North American Chapter, 35*, 994–998.

Cervetto, C., Venturini, A., Passalacqua, M., Guidolin, D., Genedani, S., Fuxe, K., ... Agnati, L. F. (2017). A2A-D2 receptor-receptor interaction modulates gliotransmitter release from striatal astrocyte processes. *Journal of Neurochemistry, 140*, 268–279.

Chakraborty, R., Tabassum, H., & Parvez, S. (2023). NLRP3 inflammasome in traumatic brain injury: Its implication in the disease pathophysiology and potential as a therapeutic target. *Life Sciences, 314*, 121352.

Chen, J. F., & Chern, Y. (2011). Impacts of methylxanthines and adenosine receptors on neurodegeneration: Human and experimental studies. *Handbook of Experimental Pharmacology*, 267–310.

Chen, J. F., Eltzschig, H. K., & Fredholm, B. B. (2013). Adenosine receptors as drug targets—What are the challenges? *Nature Reviews. Drug Discovery, 12*, 265–286.

Chen, J. F., Lee, C. F., & Chern, Y. (2014). Adenosine receptor neurobiology: Overview. *International Review of Neurobiology, 119*, 1–49.

Chen, J. F., & Pedata, F. (2008). Modulation of ischemic brain injury and neuroinflammation by adenosine A2A receptors. *Current Pharmaceutical Design, 14*, 1490–1499.

Chen, J. F., Sonsalla, P. K., Pedata, F., Melani, A., Domenici, M. R., Popoli, P., ... De Mendonca, A. (2007). Adenosine A2A receptors and brain injury: Broad spectrum of neuroprotection, multifaceted actions and "fine tuning" modulation. *Progress in Neurobiology, 83*, 310–331.

Chen, P. Z., He, W. J., Zhu, Z. R., E, G. J., Xu, G., Chen, D. W., & Gao, Y. Q. (2018). Adenosine A(2A) receptor involves in neuroinflammation-mediated cognitive decline through activating microglia under acute hypobaric hypoxia. *Behavioural Brain Research, 347*, 99–107.

Cheng, Y., Cao, P., Geng, C., Chu, X., Li, Y., & Cui, J. (2022). The adenosine A (2A) receptor antagonist SCH58261 protects photoreceptors by inhibiting microglial activation and the inflammatory response. *International Immunopharmacology, 112*, 109245.

Chiu, G. S., Darmody, P. T., Walsh, J. P., Moon, M. L., Kwakwa, K. A., Bray, J. K., ... Freund, G. G. (2014). Adenosine through the A2A adenosine receptor increases IL-1beta in the brain contributing to anxiety. *Brain, Behavior, and Immunity, 41*, 218–231.

Christman Schneider, A. L., Huie, J. R., Boscardin, W. J., Nelson, L., Barber, J. K., Yaffe, K., ... Investigators, T.-T. (2022). Cognitive outcome 1 year after mild traumatic brain injury: Results from the TRACK-TBI study. *Neurology, 98*, e1248–e1261.

Constantino, L. C., Pamplona, F. A., Matheus, F. C., De Carvalho, C. R., Ludka, F. K., Massari, C. M., ... Tasca, C. I. (2021). Functional interplay between adenosine A(2A) receptor and NMDA preconditioning in fear memory and glutamate uptake in the mice hippocampus. *Neurobiology of Learning and Memory, 180*, 107422.

Costenla, A. R., Diogenes, M. J., Canas, P. M., Rodrigues, R. J., Nogueira, C., Maroco, J., ... De Mendonca, A. (2011). Enhanced role of adenosine A(2A) receptors in the modulation of LTP in the rat hippocampus upon ageing. *The European Journal of Neuroscience, 34*, 12–21.

Cotter, D., Kelso, A., & Neligan, A. (2017). Genetic biomarkers of posttraumatic epilepsy: A systematic review. *Seizure: The Journal of the British Epilepsy Association, 46*, 53–58.

Cunha, R. A. (2005). Neuroprotection by adenosine in the brain: From A(1) receptor activation to A (2A) receptor blockade. *Purinergic Signalling, 1*, 111–134.

Cunha, R. A., & Agostinho, P. M. (2010). Chronic caffeine consumption prevents memory disturbance in different animal models of memory decline. *Journal of Alzheimer's Disease: JAD, 20*(Suppl 1), S95–S116.

Cusimano, M. D., Zhang, S., Mei, X. Y., Kennedy, D., Saha, A., Carpino, M., & Wolfe, D. (2021). Traumatic brain injury, abuse, and poor sustained attention in youth and young adults who previously experienced foster care. *Neurotrauma Reports, 2*, 94–102.

Dai, S. S., Li, W., An, J. H., Wang, H., Yang, N., Chen, X. Y., ... Zhou, Y. G. (2010). Adenosine A2A receptors in both bone marrow cells and non-bone marrow cells contribute to traumatic brain injury. *Journal of Neurochemistry, 113*, 1536–1544.

Dai, S. S., Xiong, R. P., Yang, N., Li, W., Zhu, P. F., & Zhou, Y. G. (2008). [Different effects of adenosine A2A receptors in the models of traumatic brain injury and peripheral tissue injury.]. *Sheng Li Xue Bao : [Acta Physiologica Sinica], 60*, 254–258.

Dai, S. S., & Zhou, Y. G. (2011). Adenosine 2A receptor: A crucial neuromodulator with bidirectional effect in neuroinflammation and brain injury. *Reviews in the Neurosciences, 22*, 231–239.

Dai, S. S., Zhou, Y. G., Li, W., An, J. H., Li, P., Yang, N., ... Chen, J. F. (2010). Local glutamate level dictates adenosine A2A receptor regulation of neuroinflammation and traumatic brain injury. *The Journal of Neuroscience, 30*, 5802–5810.

Dall'igna, O. P., Porciuncula, L. O., Souza, D. O., Cunha, R. A., & Lara, D. R. (2003). Neuroprotection by caffeine and adenosine A2A receptor blockade of beta-amyloid neurotoxicity. *British Journal of Pharmacology, 138*, 1207–1209.

Deckert, J., Abel, F., Kunig, G., Hartmann, J., Senitz, D., Maier, H., ... Riederer, P. (1998). Loss of human hippocampal adenosine A1 receptors in dementia: Evidence for lack of specificity. *Neuroscience Letters, 244*, 1–4.

Dejan, M., Snjezana, Z. M., Breyer, R. M., Aschner, M., & Montine, T. J. (2017). Chapter 55 - Neuroinflammation and oxidative injury in developmental neurotoxicity. In R. C. Gupta. (Ed.). *Reproductive and developmental toxicology* (pp. 1051–1061). (2nd ed.,). Academic Press.

Diamond, M. L., Ritter, A. C., Jackson, E. K., Conley, Y. P., Kochanek, P. M., Boison, D., & Wagner, A. K. (2015). Genetic variation in the adenosine regulatory cycle is associated with posttraumatic epilepsy development. *Epilepsia, 56*, 1198–1206.

Dias, R. B., Ribeiro, J. A., & Sebastiao, A. M. (2012). Enhancement of AMPA currents and GluR1 membrane expression through PKA-coupled adenosine A(2A) receptors. *Hippocampus, 22*, 276–291.

Diogenes, M. J., Assaife-Lopes, N., Pinto-Duarte, A., Ribeiro, J. A., & Sebastiao, A. M. (2007). Influence of age on BDNF modulation of hippocampal synaptic transmission: Interplay with adenosine A2A receptors. *Hippocampus, 17*, 577–585.

Dorsett, C. R., Mcguire, J. L., Depasquale, E. A., Gardner, A. E., Floyd, C. L., & Mccullumsmith, R. E. (2017). Glutamate neurotransmission in rodent models of traumatic brain injury. *Journal of Neurotrauma, 34*, 263–272.

Dragic, M., Mitrovic, N., Adzic, M., Nedeljkovic, N., & Grkovic, I. (2021). Microglial- and astrocyte-specific expression of purinergic signaling components and inflammatory mediators in the rat hippocampus during trimethyltin-induced neurodegeneration. *ASN Neuro, 13* 17590914211044882.

Du, H., Tan, Y., Li, C. H., Zhao, Y., Li, P., Ning, Y. L., ... Zhou, Y. G. (2022). High glutamate concentration reverses the inhibitory effect of microglial adenosine 2A receptor on NLRP3 inflammasome assembly and activation. *Neuroscience Letters, 769*, 136431.

Duan, Y., Kelley, N., & He, Y. (2020). Role of the NLRP3 inflammasome in neurodegenerative diseases and therapeutic implications. *Neural Regen Res, 15*, 1249–1250.

Duarte, J. M., Gaspar, R., Caetano, L., Patricio, P., Soares-Cunha, C., Mateus-Pinheiro, A., ... Gomes, C. A. (2019). Region-specific control of microglia by adenosine A(2A) receptors: Uncoupling anxiety and associated cognitive deficits in female rats. *Glia, 67*, 182–192.

Dunwiddie, T. V., & Masino, S. A. (2001). The role and regulation of adenosine in the central nervous system. *Annual Review of Neuroscience, 24*, 31–55.

Eskelinen, M. H., Ngandu, T., Tuomilehto, J., Soininen, H., & Kivipelto, M. (2009). Midlife coffee and tea drinking and the risk of late-life dementia: A population-based CAIDE study. *Journal of Alzheimer's Disease: JAD, 16*, 85–91.

Espinosa, J., Rocha, A., Nunes, F., Costa, M. S., Schein, V., Kazlauckas, V., ... Porciuncula, L. O. (2013). Caffeine consumption prevents memory impairment, neuronal damage, and adenosine A2A receptors upregulation in the hippocampus of a rat model of sporadic dementia. *Journal of Alzheimer's Disease: JAD, 34*, 509–518.

Faden, A. I., Demediuk, P., Panter, S. S., & Vink, R. (1989). The role of excitatory amino acids and NMDA receptors in traumatic brain injury. *Science (New York, N. Y.), 244*, 798–800.

Faivre, E., Coelho, J. E., Zornbach, K., Malik, E., Baqi, Y., Schneider, M., ... Blum, D. (2018). Beneficial effect of a selective adenosine A(2A) receptor antagonist in the APPswe/PS1dE9 mouse model of Alzheimer's disease. *Frontiers in Molecular Neuroscience, 11*, 235.

Ferre, S. (2010). Role of the central ascending neurotransmitter systems in the psychostimulant effects of caffeine. *Journal of Alzheimer's Disease: JAD, 20*(Suppl 1), S35–S49.

Ferre, S., Karcz-Kubicha, M., Hope, B. T., Popoli, P., Burgueno, J., Gutierrez, M. A., ... Ciruela, F. (2002). Synergistic interaction between adenosine A2A and glutamate mGlu5 receptors: Implications for striatal neuronal function. *Proceedings of the National Academy of Sciences of the United States of America, 99*, 11940–11945.

Ferre, S., Sarasola, L. I., Quiroz, C., & Ciruela, F. (2023). Presynaptic adenosine receptor heteromers as key modulators of glutamatergic and dopaminergic neurotransmission in the striatum. *Neuropharmacology, 223*, 109329.

Fiebich, B. L., Biber, K., Lieb, K., Van Calker, D., Berger, M., Bauer, J., & Gebicke-Haerter, P. J. (1996). Cyclooxygenase-2 expression in rat microglia is induced by adenosine A2a-receptors. *Glia, 18*, 152–160.

Floyd, C. L., Gorin, F. A., & Lyeth, B. G. (2005). Mechanical strain injury increases intracellular sodium and reverses Na+/Ca2+ exchange in cortical astrocytes. *Glia, 51*, 35–46.

Folmer, R. L., Smith, C. J., Boudreau, E. A., Hickok, A. W., Totten, A. M., Kaul, B., ... Sarmiento, K. F. (2020). Prevalence and management of sleep disorders in the Veterans Health Administration. *Sleep Medicine Reviews, 54*, 101358.

Fukumitsu, N., Ishii, K., Kimura, Y., Oda, K., Hashimoto, M., Suzuki, M., & Ishiwata, K. (2008). Adenosine A(1) receptors using 8-dicyclopropylmethyl-1-[(11)C]methyl-3-propylxanthine PET in Alzheimer's disease. *Annals of Nuclear Medicine, 22*, 841–847.

Gao, F., Wang, J., Yang, S., Ji, M., & Zhu, G. (2023). Fear extinction induced by activation of PKA ameliorates anxiety-like behavior in PTSD mice. *Neuropharmacology, 222*, 109306.

Gardner, R. C., & Yaffe, K. (2014). Traumatic brain injury may increase risk of young onset dementia. *Annals of Neurology, 75*, 339–341.

Geiger, M. J., Domschke, K., Homola, G. A., Schulz, S. M., Nowak, J., Akhrif, A., ... Neufang, S. (2016). ADORA2A genotype modulates interoceptive and exteroceptive processing in a fronto-insular network. *European Neuropsychopharmacology: The Journal of the European College of Neuropsychopharmacology, 26*, 1274–1285.

Gelber, R. P., Petrovitch, H., Masaki, K. H., Ross, G. W., & White, L. R. (2011). Coffee intake in midlife and risk of dementia and its neuropathologic correlates. *Journal of Alzheimer's Disease: JAD, 23*, 607–615.

Giacino, J. T., Whyte, J., Bagiella, E., Kalmar, K., Childs, N., Khademi, A., ... Sherer, M. (2012). Placebo-controlled trial of amantadine for severe traumatic brain injury. *The New England Journal of Medicine, 366*, 819–826.

Gomes, C., Ferreira, R., George, J., Sanches, R., Rodrigues, D. I., Goncalves, N., & Cunha, R. A. (2013). Activation of microglial cells triggers a release of brain-derived neurotrophic factor (BDNF) inducing their proliferation in an adenosine A2A receptor-dependent manner: A2A receptor blockade prevents BDNF release and proliferation of microglia. *Journal of Neuroinflammation, 10*, 16.

Goncalves, F. Q., Lopes, J. P., Silva, H. B., Lemos, C., Silva, A. C., Goncalves, N., ... Cunha, R. A. (2019). Synaptic and memory dysfunction in a beta-amyloid model of early Alzheimer's disease depends on increased formation of ATP-derived extracellular adenosine. *Neurobiology of Disease, 132*, 104570.

Gyoneva, S., Orr, A. G., & Traynelis, S. F. (2009). Differential regulation of microglial motility by ATP/ADP and adenosine. *Parkinsonism & Related Disorders, 15*(Suppl 3), S195–S199.

Hamor, P. U., Gobin, C. M., & Schwendt, M. (2020). The role of glutamate mGlu5 and adenosine A2a receptor interactions in regulating working memory performance and persistent cocaine seeking in rats. *Progress in Neuro-psychopharmacology & Biological Psychiatry, 103*, 109979.

Han, K., Jia, N., Li, J., Yang, L., & Min, L. Q. (2013). Chronic caffeine treatment reverses memory impairment and the expression of brain BNDF and TrkB in the PS1/APP double transgenic mouse model of Alzheimer's disease. *Molecular Medicine Reports, 8*, 737–740.

Hasko, G., Pacher, P., Vizi, E. S., & Illes, P. (2005). Adenosine receptor signaling in the brain immune system. *Trends in Pharmacological Sciences, 26*, 511–516.

Higley, M. J., & Sabatini, B. L. (2010). Competitive regulation of synaptic Ca2+ influx by D2 dopamine and A2A adenosine receptors. *Nature Neuroscience, 13*, 958–966.

Hindley, S., Herman, M. A., & Rathbone, M. P. (1994). Stimulation of reactive astrogliosis in vivo by extracellular adenosine diphosphate or an adenosine A2 receptor agonist. *Journal of Neuroscience Research, 38*, 399–406.

Horgusluoglu-Moloch, E., Nho, K., Risacher, S. L., Kim, S., Foroud, T., Shaw, L. M., ... ... Alzheimer's Disease Neuroimaging, I. (2017). Targeted neurogenesis pathway-based gene analysis identifies ADORA2A associated with hippocampal volume in mild cognitive impairment and Alzheimer's disease. *Neurobiology of Aging, 60*, 92–103.

Hou, X., Li, Y., Huang, Y., Zhao, H., & Gui, L. (2020). Adenosine receptor A1-A2a heteromers regulate EAAT2 expression and glutamate uptake via YY1-induced repression of PPARgamma transcription. *PPAR Research, 2020*, 2410264.

Howlett, J. R., Nelson, L. D., & Stein, M. B. (2022). Mental health consequences of traumatic brain injury. *Biological Psychiatry, 91*, 413–420.

Huang, Z. L., Liu, R., Bai, X. Y., Zhao, G., Song, J. K., Wu, S., & Du, G. H. (2014). Protective effects of the novel adenosine derivative WS0701 in a mouse model of posttraumatic stress disorder. *Acta Pharmacologica Sinica, 35*, 24–32.

Hurtado-Alvarado, G., Dominguez-Salazar, E., Velazquez-Moctezuma, J., & Gomez-Gonzalez, B. (2016). A2A adenosine receptor antagonism reverts the blood-brain barrier dysfunction induced by sleep restriction. *PLoS One, 11*, e0167236.

Jansen, K. L., Faull, R. L., Dragunow, M., & Synek, B. L. (1990). Alzheimer's disease: Changes in hippocampal N-methyl-D-aspartate, quisqualate, neurotensin, adenosine, benzodiazepine, serotonin and opioid receptors—An autoradiographic study. *Neuroscience, 39*, 613–627.

Jassam, Y. N., Izzy, S., Whalen, M., Mcgavern, D. B., & El Khoury, J. (2017). Neuroimmunology of traumatic brain injury: Time for a paradigm shift. *Neuron, 95*, 1246–1265.

Ji, M., Zhang, Z., Gao, F., Yang, S., Wang, J., Wang, X., & Zhu, G. (2023). Curculigoside rescues hippocampal synaptic deficits elicited by PTSD through activating cAMP-PKA signaling. *Phytotherapy Research: PTR, 37*, 759–773.

Johnson, V. E., Stewart, W., Arena, J. D., & Smith, D. H. (2017). Traumatic brain injury as a trigger of neurodegeneration. *Advances in Neurobiology, 15*, 383–400.

Jones, P. A., Smith, R. A., & Stone, T. W. (1998a). Protection against hippocampal kainate excitotoxicity by intracerebral administration of an adenosine A2A receptor antagonist. *Brain Research, 800*, 328–335.

Jones, P. A., Smith, R. A., & Stone, T. W. (1998b). Protection against kainate-induced excitotoxicity by adenosine A2A receptor agonists and antagonists. *Neuroscience, 85*, 229–237.

Kahle, J. S., Ulas, J., & Cotman, C. W. (1993). Increased sensitivity to adenosine in the rat dentate gyrus molecular layer two weeks after partial entorhinal lesions. *Brain Research, 609*, 201–210.

Kahriman, A., Bouley, J., Smith, T. W., Bosco, D. A., Woerman, A. L., & Henninger, N. (2021). Mouse closed head traumatic brain injury replicates the histological tau pathology pattern of human disease: Characterization of a novel model and systematic review of the literature. *Acta Neuropathologica Communications, 9*, 118.

Kalaria, R. N., Sromek, S., Wilcox, B. J., & Unnerstall, J. R. (1990). Hippocampal adenosine A1 receptors are decreased in Alzheimer's disease. *Neuroscience Letters, 118*, 257–260.

Kaur, P., & Sharma, S. (2018). Recent advances in pathophysiology of traumatic brain injury. *Current Neuropharmacology, 16*, 1224–1238.

Kim, D. G., & Bynoe, M. S. (2015). A2A adenosine receptor regulates the human blood-brain barrier permeability. *Molecular Neurobiology, 52*, 664–678.

Kochanek, P. M., Vagni, V. A., Janesko, K. L., Washington, C. B., Crumrine, P. K., Garman, R. H., ... Jackson, E. K. (2006). Adenosine A1 receptor knockout mice develop lethal status epilepticus after experimental traumatic brain injury. *Journal of Cerebral Blood Flow and Metabolism: Official Journal of the International Society of Cerebral Blood Flow and Metabolism, 26*, 565–575.

Kofalvi, A., Moreno, E., Cordomi, A., Cai, N. S., Fernandez-Duenas, V., Ferreira, S. G., ... Ferre, S. (2020). Control of glutamate release by complexes of adenosine and cannabinoid receptors. *BMC Biology, 18*, 9.

Kourtidou, P., Mccauley, S. R., Bigler, E. D., Traipe, E., Wu, T. C., Chu, Z. D., ... Wilde, E. A. (2013). Centrum semiovale and corpus callosum integrity in relation to information processing speed in patients with severe traumatic brain injury. *The Journal of Head Trauma Rehabilitation, 28*, 433–441.

Krawczyk, D. C., Han, K., Martinez, D., Rakic, J., Kmiecik, M. J., Chang, Z., ... Didehbani, N. (2019). Executive function training in chronic traumatic brain injury patients: Study protocol. *Trials, 20*, 435.

Kumar, A., & Loane, D. J. (2012). Neuroinflammation after traumatic brain injury: Opportunities for therapeutic intervention. *Brain, Behavior, and Immunity, 26*, 1191–1201.

Lambertucci, C., Marucci, G., Catarzi, D., Colotta, V., Francucci, B., Spinaci, A., ... Volpini, R. (2022). A(2A) adenosine receptor antagonists and their potential in neurological disorders. *Current Medicinal Chemistry, 29*, 4780–4795.

Latini, S., & Pedata, F. (2001). Adenosine in the central nervous system: Release mechanisms and extracellular concentrations. *Journal of Neurochemistry, 79*, 463–484.

Laurent, C., Burnouf, S., Ferry, B., Batalha, V. L., Coelho, J. E., Baqi, Y., ... Blum, D. (2016). A2A adenosine receptor deletion is protective in a mouse model of tauopathy. *Molecular Psychiatry, 21*, 97–107.

Lea, P. M., Custer, S. J., Vicini, S., & Faden, A. I. (2002). Neuronal and glial mGluR5 modulation prevents stretch-induced enhancement of NMDA receptor current. *Pharmacology, Biochemistry, and Behavior, 73*, 287–298.

Leng, Y., Byers, A. L., Barnes, D. E., Peltz, C. B., Li, Y., & Yaffe, K. (2021). Traumatic brain injury and incidence risk of sleep disorders in nearly 200,000 US Veterans. *Neurology, 96*, e1792–e1799.

Lewen, A., Fujimura, M., Sugawara, T., Matz, P., Copin, J. C., & Chan, P. H. (2001). Oxidative stress-dependent release of mitochondrial cytochrome c after traumatic brain injury. *Journal of Cerebral Blood Flow and Metabolism: Official Journal of the International Society of Cerebral Blood Flow and Metabolism, 21*, 914–920.

Li, W., Dai, S., An, J., Li, P., Chen, X., Xiong, R., ... Zhou, Y. (2008). Chronic but not acute treatment with caffeine attenuates traumatic brain injury in the mouse cortical impact model. *Neuroscience, 151*, 1198–1207.

Li, W., Dai, S., An, J., Xiong, R., Li, P., Chen, X., ... Zhou, Y. (2009). Genetic inactivation of adenosine A2A receptors attenuates acute traumatic brain injury in the mouse cortical impact model. *Experimental Neurology, 215*, 69–76.

Li, Y., Oskouian, R. J., Day, Y. J., Rieger, J. M., Liu, L., Kern, J. A., & Linden, J. (2006). Mouse spinal cord compression injury is reduced by either activation of the adenosine A2A receptor on bone marrow-derived cells or deletion of the A2A receptor on non-bone marrow-derived cells. *Neuroscience, 141*, 2029–2039.

Liu, X., Huang, P., Wang, J., Yang, Z., Huang, S., Luo, X., ... Zhong, Y. (2016). The effect of A2A receptor antagonist on microglial activation in experimental glaucoma. *Investigative Ophthalmology & Visual Science, 57*, 776–786.

Liu, Y., Chu, S., Hu, Y., Yang, S., Li, X., Zheng, Q., ... Chen, N. H. (2021). Exogenous adenosine antagonizes excitatory amino acid toxicity in primary astrocytes. *Cellular and Molecular Neurobiology, 41*, 687–704.

Liu, Y. W., Yang, T., Zhao, L., Ni, Z., Yang, N., He, F., & Dai, S. S. (2016). Activation of adenosine 2A receptor inhibits neutrophil apoptosis in an autophagy-dependent manner in mice with systemic inflammatory response syndrome. *Scientific Reports, 6*, 33614.

Lopes, C. R., Cunha, R. A., & Agostinho, P. (2021). Astrocytes and adenosine A(2A) receptors: Active players in Alzheimer's disease. *Frontiers in Neuroscience, 15*, 666710.

Madeira, D., Dias, L., Santos, P., Cunha, R. A., Agostinho, P., & Canas, P. M. (2022). Adenosine A(2A) receptors blockade attenuates dexamethasone-induced alterations in cultured astrocytes. *Purinergic Signalling, 18*, 199–204.

Mallah, K., Couch, C., Alshareef, M., Borucki, D., Yang, X., Alawieh, A., & Tomlinson, S. (2021). Complement mediates neuroinflammation and cognitive decline at extended chronic time points after traumatic brain injury. *Acta Neuropathologica Communications, 9*, 72.

Marcoli, M., Bonfanti, A., Roccatagliata, P., Chiaramonte, G., Ongini, E., Raiteri, M., & Maura, G. (2004). Glutamate efflux from human cerebrocortical slices during ischemia: Vesicular-like mode of glutamate release and sensitivity to A(2A) adenosine receptor blockade. *Neuropharmacology, 47*, 884–891.

Marcoli, M., Raiteri, L., Bonfanti, A., Monopoli, A., Ongini, E., Raiteri, M., & Maura, G. (2003). Sensitivity to selective adenosine A1 and A2A receptor antagonists of the release of glutamate induced by ischemia in rat cerebrocortical slices. *Neuropharmacology, 45*, 201–210.

Marques, S., Batalha, V. L., Lopes, L. V., & Outeiro, T. F. (2011). Modulating Alzheimer's disease through caffeine: A putative link to epigenetics. *Journal of Alzheimer's Disease: JAD, 24*(Suppl 2), 161–171.

Marti Navia, A., Dal Ben, D., Lambertucci, C., Spinaci, A., Volpini, R., Marques-Morgado, I., ... Buccioni, M. (2020). Adenosine receptors as neuroinflammation modulators: Role of A(1) agonists and A(2A) antagonists. *Cells, 9*, 1739.

Marucci, G., Ben, D. D., Lambertucci, C., Navia, A. M., Spinaci, A., Volpini, R., & Buccioni, M. (2021). Combined therapy of A(1)AR agonists and A(2A)AR antagonists in neuroinflammation. *Molecules (Basel, Switzerland), 26*, 1188.

Mathias, J. L., & Alvaro, P. K. (2012). Prevalence of sleep disturbances, disorders, and problems following traumatic brain injury: A meta-analysis. *Sleep Medicine, 13*, 898–905.

Matos, M., Augusto, E., Agostinho, P., Cunha, R. A., & Chen, J. F. (2013). Antagonistic interaction between adenosine A2A receptors and Na+/K+-ATPase-alpha2 controlling glutamate uptake in astrocytes. *The Journal of Neuroscience, 33*, 18492–18502.

Mayne, M., Fotheringham, J., Yan, H. J., Power, C., Del Bigio, M. R., Peeling, J., & Geiger, J. D. (2001). Adenosine A2A receptor activation reduces proinflammatory events and decreases cell death following intracerebral hemorrhage. *Annals of Neurology, 49*, 727–735.

Mcdonald, S., Dalton, K. I., Rushby, J. A., & Landin-Romero, R. (2019). Loss of white matter connections after severe traumatic brain injury (TBI) and its relationship to social cognition. *Brain Imaging and Behavior, 13*, 819–829.

Medeiros, G. C., Twose, C., Weller, A., Dougherty, J. W., 3rd, Goes, F. S., Sair, H. I., ... Roy, D. (2022). Neuroimaging correlates of depression after traumatic brain injury: A systematic review. *Journal of Neurotrauma, 39*, 755–772.

Merighi, S., Borea, P. A., Varani, K., Vincenzi, F., Jacobson, K. A., & Gessi, S. (2022). A(2A) adenosine receptor antagonists in neurodegenerative diseases. *Current Medicinal Chemistry, 29*, 4138–4151.

Michinaga, S., & Koyama, Y. (2021). Pathophysiological responses and roles of astrocytes in traumatic brain injury. *International Journal of Molecular Sciences, 22*, 6418.

Mokhtari, M., Nayeb-Aghaei, H., Kouchek, M., Miri, M. M., Goharani, R., Amoozandeh, A., ... Sistanizad, M. (2018). Effect of memantine on serum levels of neuron-specific enolase and on the Glasgow Coma Scale in patients with moderate traumatic brain injury. *Journal of Clinical Pharmacology, 58*, 42–47.

Morello, S., Ito, K., Yamamura, S., Lee, K. Y., Jazrawi, E., Desouza, P., ... Adcock, I. M. (2006). IL-1 beta and TNF-alpha regulation of the adenosine receptor (A2A) expression: Differential requirement for NF-kappa B binding to the proximal promoter. *Journal of Immunology, 177*, 7173–7183.

Morganti-Kossmann, M. C., Satgunaseelan, L., Bye, N., & Kossmann, T. (2007). Modulation of immune response by head injury. *Injury, 38*, 1392–1400.

Mouro, F. M., Rombo, D. M., Dias, R. B., Ribeiro, J. A., & Sebastiao, A. M. (2018). Adenosine A(2A) receptors facilitate synaptic NMDA currents in CA1 pyramidal neurons. *British Journal of Pharmacology, 175*, 4386–4397.

Mullah, S. H., Inaji, M., Nariai, T., Ishibashi, S., & Ohno, K. (2013). A selective adenosine A2A receptor antagonist ameliorated hyperlocomotion in an animal model of lateral fluid percussion brain injury. *Acta Neurochirurgica. Supplement, 118*, 89–92.

Nedeljkovic, N. (2019). Complex regulation of ecto-5′-nucleotidase/CD73 and A(2A)R-mediated adenosine signaling at neurovascular unit: A link between acute and chronic neuroinflammation. *Pharmacological Research: The Official Journal of the Italian Pharmacological Society, 144*, 99–115.

Nicholls, D., & Attwell, D. (1990). The release and uptake of excitatory amino acids. *Trends in Pharmacological Sciences, 11*, 462–468.

Nilsson, P., Hillered, L., Ponten, U., & Ungerstedt, U. (1990). Changes in cortical extracellular levels of energy-related metabolites and amino acids following concussive brain injury in rats. *Journal of Cerebral Blood Flow and Metabolism: Official Journal of the International Society of Cerebral Blood Flow and Metabolism, 10*, 631–637.

Ning, Y. L., Yang, N., Chen, X., Xiong, R. P., Zhang, X. Z., Li, P., ... Zhou, Y. G. (2013). Adenosine A2A receptor deficiency alleviates blast-induced cognitive dysfunction. *Journal of Cerebral Blood Flow and Metabolism: Official Journal of the International Society of Cerebral Blood Flow and Metabolism, 33*, 1789–1798.

Nishizaki, T., Nagai, K., Nomura, T., Tada, H., Kanno, T., Tozaki, H., ... Saito, N. (2002). A new neuromodulatory pathway with a glial contribution mediated via A(2a) adenosine receptors. *Glia, 39*, 133–147.

Njoku, I., Radabaugh, H. L., Nicholas, M. A., Kutash, L. A., O'neil, D. A., Marshall, I. P., ... Bondi, C. O. (2019). Chronic treatment with galantamine rescues reversal learning in an attentional set-shifting test after experimental brain trauma. *Experimental Neurology, 315*, 32–41.

O'brien, W. T., Pham, L., Symons, G. F., Monif, M., Shultz, S. R., & Mcdonald, S. J. (2020). The NLRP3 inflammasome in traumatic brain injury: Potential as a biomarker and therapeutic target. *Journal of Neuroinflammation, 17*, 104.

Orr, A. G., Orr, A. L., Li, X. J., Gross, R. E., & Traynelis, S. F. (2009). Adenosine A(2A) receptor mediates microglial process retraction. *Nature Neuroscience, 12*, 872–878.

Ouyang, X., Ghani, A., Malik, A., Wilder, T., Colegio, O. R., Flavell, R. A., ... Mehal, W. Z. (2013). Adenosine is required for sustained inflammasome activation via the A(2)A receptor and the HIF-1alpha pathway. *Nature Communications, 4*, 2909.

Pagnussat, N., Almeida, A. S., Marques, D. M., Nunes, F., Chenet, G. C., Botton, P. H., ... Porciuncula, L. O. (2015). Adenosine A(2A) receptors are necessary and sufficient to trigger memory impairment in adult mice. *British Journal of Pharmacology, 172*, 3831–3845.

Paiva, I., Carvalho, K., Santos, P., Cellai, L., Pavlou, M. A. S., Jain, G., ... Blum, D. (2019). A(2A) R-induced transcriptional deregulation in astrocytes: An in vitro study. *Glia, 67*, 2329–2342.

Parkinson, F. E., Xiong, W., & Zamzow, C. R. (2005). Astrocytes and neurons: Different roles in regulating adenosine levels. *Neurological Research, 27*, 153–160.

Patraca, I., Martinez, N., Busquets, O., Marti, A., Pedros, I., Beas-Zarate, C., ... Folch, J. (2017). Anti-inflammatory role of Leptin in glial cells through p38 MAPK pathway inhibition. *Pharmacological Reports: PR, 69*, 409–418.

Paudel, Y. N., Shaikh, M. F., Chakraborti, A., Kumari, Y., Aledo-Serrano, A., Aleksovska, K., ... Othman, I. (2018). HMGB1: A common biomarker and potential target for TBI, neuroinflammation, epilepsy, and cognitive dysfunction. *Frontiers in Neuroscience, 12*, 628.

Pavlovic, D., Pekic, S., Stojanovic, M., & Popovic, V. (2019). Traumatic brain injury: Neuropathological, neurocognitive and neurobehavioral sequelae. *Pituitary, 22*, 270–282.

Pfeiffer, T., & Attwell, D. (2020). Brain's immune cells put the brakes on neurons. *Nature, 586*, 366–367.

Pickel, V. M., Chan, J., Linden, J., & Rosin, D. L. (2006). Subcellular distributions of adenosine A1 and A2A receptors in the rat dorsomedial nucleus of the solitary tract at the level of the area postrema. *Synapse (New York, N. Y.), 60*, 496–509.

Pinna, A., Serra, M., Marongiu, J., & Morelli, M. (2020). Pharmacological interactions between adenosine A(2A) receptor antagonists and different neurotransmitter systems. *Parkinsonism & Related Disorders, 80*(Suppl 1), S37–S44.

Pintor, A., Galluzzo, M., Grieco, R., Pezzola, A., Reggio, R., & Popoli, P. (2004). Adenosine A 2A receptor antagonists prevent the increase in striatal glutamate levels induced by glutamate uptake inhibitors. *Journal of Neurochemistry, 89*, 152–156.

Porkka-Heiskanen, T., & Kalinchuk, A. V. (2011). Adenosine, energy metabolism and sleep homeostasis. *Sleep Medicine Reviews, 15*, 123–135.

Prasanthi, J. R., Dasari, B., Marwarha, G., Larson, T., Chen, X., Geiger, J. D., & Ghribi, O. (2010). Caffeine protects against oxidative stress and Alzheimer's disease-like pathology in rabbit hippocampus induced by cholesterol-enriched diet. *Free Radical Biology & Medicine, 49,* 1212–1220.

Preston, G., Emmerzaal, T., Radenkovic, S., Lanza, I. R., Oglesbee, D., Morava, E., & Kozicz, T. (2021). Cerebellar and multi-system metabolic reprogramming associated with trauma exposure and post-traumatic stress disorder (PTSD)-like behavior in mice. *Neurobiol Stress, 14,* 100300.

Qosa, H., Abuznait, A. H., Hill, R. A., & Kaddoumi, A. (2012). Enhanced brain amyloid-beta clearance by rifampicin and caffeine as a possible protective mechanism against Alzheimer's disease. *Journal of Alzheimer's Disease: JAD, 31,* 151–165.

Rebola, N., Canas, P. M., Oliveira, C. R., & Cunha, R. A. (2005). Different synaptic and subsynaptic localization of adenosine A2A receptors in the hippocampus and striatum of the rat. *Neuroscience, 132,* 893–903.

Rebola, N., Lujan, R., Cunha, R. A., & Mulle, C. (2008). Adenosine A2A receptors are essential for long-term potentiation of NMDA-EPSCs at hippocampal mossy fiber synapses. *Neuron, 57,* 121–134.

Rebola, N., Sebastiao, A. M., De Mendonca, A., Oliveira, C. R., Ribeiro, J. A., & Cunha, R. A. (2003). Enhanced adenosine A2A receptor facilitation of synaptic transmission in the hippocampus of aged rats. *Journal of Neurophysiology, 90,* 1295–1303.

Rebola, N., Simoes, A. P., Canas, P. M., Tome, A. R., Andrade, G. M., Barry, C. E., ... Cunha, R. A. (2011). Adenosine A2A receptors control neuroinflammation and consequent hippocampal neuronal dysfunction. *Journal of Neurochemistry, 117,* 100–111.

Ren, Z., Iliff, J. J., Yang, L., Yang, J., Chen, X., Chen, M. J., ... Nedergaard, M. (2013). Hit & Run' model of closed-skull traumatic brain injury (TBI) reveals complex patterns of post-traumatic AQP4 dysregulation. *Journal of Cerebral Blood Flow and Metabolism: Official Journal of the International Society of Cerebral Blood Flow and Metabolism, 33,* 834–845.

Ribeiro, J. A. (2005). What can adenosine neuromodulation do for neuroprotection? *Current Drug Targets. CNS and Neurological Disorders, 4,* 325–329.

Robertson, C. L., Bell, M. J., Kochanek, P. M., Adelson, P. D., Ruppel, R. A., Carcillo, J. A., ... Jackson, E. K. (2001). Increased adenosine in cerebrospinal fluid after severe traumatic brain injury in infants and children: Association with severity of injury and excitotoxicity. *Critical Care Medicine, 29,* 2287–2293.

Rodrigues, R. J., Canas, P. M., Lopes, L. V., Oliveira, C. R., & Cunha, R. A. (2008). Modification of adenosine modulation of acetylcholine release in the hippocampus of aged rats. *Neurobiology of Aging, 29,* 1597–1601.

Rossi, D. J., Oshima, T., & Attwell, D. (2000). Glutamate release in severe brain ischaemia is mainly by reversed uptake. *Nature, 403,* 316–321.

Roth, T. L., Nayak, D., Atanasijevic, T., Koretsky, A. P., Latour, L. L., & Mcgavern, D. B. (2014). Transcranial amelioration of inflammation and cell death after brain injury. *Nature, 505,* 223–228.

Ruchika, F., Shah, S., Neupane, D., Vijay, R., Mehkri, Y., & Lucke-Wold, B. (2023). Understanding the molecular progression of chronic traumatic encephalopathy in traumatic brain injury, aging and neurodegenerative disease. *International Journal of Molecular Sciences, 24,* 1847.

Sabbagh, J. J., Fontaine, S. N., Shelton, L. B., Blair, L. J., Hunt, J. B., Jr., Zhang, B., ... Dickey, C. A. (2016). Noncontact rotational head injury produces transient cognitive deficits but lasting neuropathological changes. *Journal of Neurotrauma, 33,* 1751–1760.

Sachse, K. T., Jackson, E. K., Wisniewski, S. R., Gillespie, D. G., Puccio, A. M., Clark, R. S., ... Kochanek, P. M. (2008). Increases in cerebrospinal fluid caffeine concentration are associated with favorable outcome after severe traumatic brain injury in humans. *Journal of Cerebral Blood Flow and Metabolism: Official Journal of the International Society of Cerebral Blood Flow and Metabolism, 28,* 395–401.

Saura, J., Angulo, E., Ejarque, A., Casado, V., Tusell, J. M., Moratalla, R., ... Serratosa, J. (2005). Adenosine A2A receptor stimulation potentiates nitric oxide release by activated microglia. *Journal of Neurochemistry, 95*, 919–929.

Schiffmann, S. N., Jacobs, O., & Vanderhaeghen, J. J. (1991). Striatal restricted adenosine A2 receptor (RDC8) is expressed by enkephalin but not by substance P neurons: An in situ hybridization histochemistry study. *Journal of Neurochemistry, 57*, 1062–1067.

Scholten, A. C., Haagsma, J. A., Cnossen, M. C., Olff, M., Van Beeck, E. F., & Polinder, S. (2016). Prevalence of and risk factors for anxiety and depressive disorders after traumatic brain injury: A systematic review. *Journal of Neurotrauma, 33*, 1969–1994.

Shi, J., Zhao, Y., Wang, K., Shi, X., Wang, Y., Huang, H., ... Shao, F. (2015). Cleavage of GSDMD by inflammatory caspases determines pyroptotic cell death. *Nature, 526*, 660–665.

Shi, K., Zhang, J., Dong, J. F., & Shi, F. D. (2019). Dissemination of brain inflammation in traumatic brain injury. *Cellular & Molecular Immunology, 16*, 523–530.

Shinozaki, Y., Shibata, K., Yoshida, K., Shigetomi, E., Gachet, C., Ikenaka, K., ... Koizumi, S. (2017). Transformation of astrocytes to a neuroprotective phenotype by microglia via P2Y(1) receptor downregulation. *Cell Reports, 19*, 1151–1164.

Silva, A. C., Lemos, C., Goncalves, F. Q., Pliassova, A. V., Machado, N. J., Silva, H. B., ... Agostinho, P. (2018). Blockade of adenosine A(2A) receptors recovers early deficits of memory and plasticity in the triple transgenic mouse model of Alzheimer's disease. *Neurobiology of Disease, 117*, 72–81.

Simoes, A. P., Goncalves, F. Q., Rial, D., Ferreira, S. G., Lopes, J. P., Canas, P. M., & Cunha, R. A. (2022). CD73-mediated formation of extracellular adenosine is responsible for adenosine A(2A) receptor-mediated control of fear memory and amygdala plasticity. *International Journal of Molecular Sciences, 23*, 12826.

Skeberdis, V. A., Lan, J., Opitz, T., Zheng, X., Bennett, M. V., & Zukin, R. S. (2001). mGluR1-mediated potentiation of NMDA receptors involves a rise in intracellular calcium and activation of protein kinase C. *Neuropharmacology, 40*, 856–865.

Smith, D. H., Johnson, V. E., & Stewart, W. (2013). Chronic neuropathologies of single and repetitive TBI: Substrates of dementia? *Nature Reviews Neurology, 9*, 211–221.

Stein, M. B., Jain, S., Parodi, L., Choi, K. W., Maihofer, A. X., Nelson, L. D., ... Investigators, T.-T. (2023). Polygenic risk for mental disorders as predictors of posttraumatic stress disorder after mild traumatic brain injury. *Translational Psychiatry, 13*, 13.

Swanson, K. V., Deng, M., & Ting, J. P. (2019). The NLRP3 inflammasome: Molecular activation and regulation to therapeutics. *Nature Reviews. Immunology, 19*, 477–489.

Tan, S. W., Li, P., Ning, Y. L., Zhao, Y., Yang, N., & Zhou, Y. G. (2021). A novel method for detecting heteromeric complexes at synaptic level by combining a modified method of proximity ligation assay with transmission electron microscopy. *Neurochemistry International, 149*, 105145.

Tan, S. W., Zhao, Y., Li, P., Ning, Y. L., Huang, Z. Z., Yang, N., ... Zhou, Y. G. (2021). HMGB1 mediates cognitive impairment caused by the NLRP3 inflammasome in the late stage of traumatic brain injury. *Journal of Neuroinflammation, 18*, 241.

Travassos, M., Santana, I., Baldeiras, I., Tsolaki, M., Gkatzima, O., Sermin, G., ... De Mendon, A. (2015). Does caffeine consumption modify cerebrospinal fluid amyloid-beta levels in patients with Alzheimer's disease? *Journal of Alzheimer's Disease: JAD, 47*, 1069–1078.

Van Vliet, E. A., Ndode-Ekane, X. E., Lehto, L. J., Gorter, J. A., Andrade, P., Aronica, E., ... Pitkanen, A. (2020). Long-lasting blood-brain barrier dysfunction and neuroinflammation after traumatic brain injury. *Neurobiology of Disease, 145*, 105080.

Velayutham, P., Murthy, M., & Babu, K. S. (2019). Time-dependent bidirectional neuroprotection by adenosine 2A receptor in experimental traumatic brain injury. *World Neurosurgery, 125*, e743–e753.

Viana Da Silva, S., Haberl, M. G., Zhang, P., Bethge, P., Lemos, C., Goncalves, N., ... Mulle, C. (2016). Early synaptic deficits in the APP/PS1 mouse model of Alzheimer's disease involve neuronal adenosine A2A receptors. *Nature Communications, 7*, 11915.

Vila-Luna, S., Cabrera-Isidoro, S., Vila-Luna, L., Juarez-Diaz, I., Bata-Garcia, J. L., Alvarez-Cervera, F. J., ... Gongora-Alfaro, J. L. (2012). Chronic caffeine consumption prevents cognitive decline from young to middle age in rats, and is associated with increased length, branching, and spine density of basal dendrites in CA1 hippocampal neurons. *Neuroscience, 202*, 384–395.

Vlajkovic, S. M., Housley, G. D., & Thorne, P. R. (2009). Adenosine and the auditory system. *Current Neuropharmacology, 7*, 246–256.

Wagner, A. K., Miller, M. A., Scanlon, J., Ren, D., Kochanek, P. M., & Conley, Y. P. (2010). Adenosine A1 receptor gene variants associated with post-traumatic seizures after severe TBI. *Epilepsy Research, 90*, 259–272.

Wei, C. J., Li, W., & Chen, J. F. (2011). Normal and abnormal functions of adenosine receptors in the central nervous system revealed by genetic knockout studies. *Biochimica et Biophysica Acta, 1808*, 1358–1379.

Wicher, G., Wallenquist, U., Lei, Y., Enoksson, M., Li, X., Fuchs, B., ... Forsberg-Nilsson, K. (2017). Interleukin-33 promotes recruitment of microglia/macrophages in response to traumatic brain injury. *Journal of Neurotrauma, 34*, 3173–3182.

Wu, A. G., Zhou, X. G., Qiao, G., Yu, L., Tang, Y., Yan, L., ... Wu, J. M. (2021). Targeting microglial autophagic degradation in NLRP3 inflammasome-mediated neurodegenerative diseases. *Ageing Research Reviews, 65*, 101202.

Xia, Y., He, F., Moukeila Yacouba, M. B., Zhou, H., Li, J., Xiong, Y., ... Ke, J. (2022). Adenosine A2a receptor regulates autophagy flux and apoptosis to alleviate ischemia-reperfusion injury via the cAMP/PKA signaling pathway. *Frontiers in Cardiovascular Medicine, 9*, 755619.

Xu, X., Beleza, R. O., Goncalves, F. Q., Valbuena, S., Alcada-Morais, S., Goncalves, N., ... Marques, J. M. (2022). Adenosine A(2A) receptors control synaptic remodeling in the adult brain. *Scientific Reports, 12*, 14690.

Yang, L., Yu, X., Zhang, Y., Liu, N., Xue, X., & Fu, J. (2022). Caffeine treatment started before injury reduces hypoxic-ischemic white-matter damage in neonatal rats by regulating phenotypic microglia polarization. *Pediatric Research, 92*, 1543–1554.

Yu, J., Zhong, Y., Shen, X., Cheng, Y., Qi, J., & Wang, J. (2012). In vitro effect of adenosine A2A receptor antagonist SCH 442416 on the expression of glutamine synthetase and glutamate aspartate transporter in rat retinal Muller cells at elevated hydrostatic pressure. *Oncology Reports, 27*, 748–752.

Yu, L., Huang, Z., Mariani, J., Wang, Y., Moskowitz, M., & Chen, J. F. (2004). Selective inactivation or reconstitution of adenosine A2A receptors in bone marrow cells reveals their significant contribution to the development of ischemic brain injury. *Nature Medicine, 10*, 1081–1087.

Yu, L., Shen, H. Y., Coelho, J. E., Araujo, I. M., Huang, Q. Y., Day, Y. J., ... Chen, J. F. (2008). Adenosine A2A receptor antagonists exert motor and neuroprotective effects by distinct cellular mechanisms. *Annals of Neurology, 63*, 338–346.

Zeng, X. J., Li, P., Ning, Y. L., Zhao, Y., Peng, Y., Yang, N., ... Zhou, Y. G. (2020). A2A R inhibition in alleviating spatial recognition memory impairment after TBI is associated with improvement in autophagic flux in RSC. *Journal of Cellular and Molecular Medicine, 24*, 7000–7014.

Zeng, X. J., Li, P., Ning, Y. L., Zhao, Y., Peng, Y., Yang, N., ... Zhou, Y. G. (2018). Impaired autophagic flux is associated with the severity of trauma and the role of A2AR in brain cells after traumatic brain injury. *Cell Death and Disease, 9*, 252.

Zhao, W., Ma, L., Cai, C., & Gong, X. (2019). Caffeine inhibits NLRP3 inflammasome activation by suppressing MAPK/NF-kappaB and A2aR signaling in LPS-induced THP-1 macrophages. *International Journal of Biological Sciences, 15*, 1571–1581.

Zhao, Y., Tan, S. W., Huang, Z. Z., Shan, F. B., Li, P., Ning, Y. L., ... Zhou, Y. G. (2021). NLRP3 inflammasome-dependent increases in high mobility group box 1 involved in the cognitive dysfunction caused by tau-overexpression. *Frontiers in Aging Neuroscience, 13*, 721474.

Zhao, Z., Nelson, A. R., Betsholtz, C., & Zlokovic, B. V. (2015). Establishment and dysfunction of the blood-brain barrier. *Cell, 163*, 1064–1078.

Zhao, Z. A., Li, P., Ye, S. Y., Ning, Y. L., Wang, H., Peng, Y., ... Zhou, Y. G. (2017). Perivascular AQP4 dysregulation in the hippocampal CA1 area after traumatic brain injury is alleviated by adenosine A2A receptor inactivation. *Scientific Reports, 7*, 2254.

Zhao, Z. A., Ning, Y. L., Li, P., Yang, N., Peng, Y., Xiong, R. P., ... Zhou, Y. G. (2017). Widespread hyperphosphorylated tau in the working memory circuit early after cortical impact injury of brain (Original study). *Behavioural Brain Research, 323*, 146–153.

Zhao, Z. A., Zhao, Y., Ning, Y. L., Yang, N., Peng, Y., Li, P., ... Zhou, Y. G. (2017). Adenosine A2A receptor inactivation alleviates early-onset cognitive dysfunction after traumatic brain injury involving an inhibition of tau hyperphosphorylation. *Translational Psychiatry, 7*, e1123.

# CHAPTER ELEVEN

# Chemobrain: An accelerated aging process linking adenosine $A_{2A}$ receptor signaling in cancer survivors

Alfredo Oliveros[a], Michael Poleschuk[a], Peter D. Cole[b], Detlev Boison[a,*], and Mi-Hyeon Jang[a,*]

[a]Department of Neurosurgery, Robert Wood Johnson Medical School, Rutgers, The State University of New Jersey, Piscataway, NJ, United States
[b]Division of Pediatric Hematology/Oncology, Rutgers Cancer Institute of New Jersey, New Brunswick, NJ, USA
*Corresponding authors. e-mail address: detlev.boison@rutgers.edu; db1114@rwjms.rutgers.edu; mihyeon.jang@rutgers.edu

## Contents

| | |
|---|---|
| 1. Introduction | 268 |
| 2. Adenosine $A_{2A}R$ in cognitive function | 271 |
|    2.1 Physiological role of $A_{2A}R$ and cognitive improvement | 271 |
|    2.2 $A_{2A}R$ and cognitive improvement in pathological conditions | 272 |
|    2.3 Convergence of the adenosine $A_{2A}R$, aging and AD | 272 |
| 3. Findings of cognitive dysfunction in chemobrain | 274 |
|    3.1 Clinical imaging studies | 274 |
| 4. Chemobrain as an accelerated aging phenotype in adult and pediatric cancer | 275 |
|    4.1 Evidence of accelerated aging in adult cancer patients | 275 |
|    4.2 Evidence of accelerated aging phenotypes in pediatric cancer patients | 278 |
|    4.3 Accelerated aging mitochondrial defects associated with cancer and chemobrain | 279 |
|    4.4 Accelerated aging, telomeres, and chemobrain | 281 |
|    4.5 Epigenetic related alterations in chemobrain and accelerated aging | 283 |
| 5. The adenosine $A_{2A}$ receptor as a potential mechanism of accelerated aging in chemobrain and cancer | 285 |
|    5.1 Anticancer effects and immune cell modulation of the adenosine $A_{2A}R$ in malignancies | 285 |
|    5.2 Interaction between chemotherapy, adenosine $A_{2A}R$ and chemobrain | 288 |
|    5.3 Adenosine $A_{2A}R$ and chemobrain | 289 |
| 6. Concluding remarks | 291 |
| Acknowledgments | 292 |
| Conflict of interest statement | 292 |
| References | 292 |

## Abstract

Chemotherapy has a significant positive impact in cancer treatment outcomes, reducing recurrence and mortality. However, many cancer surviving children and adults suffer from aberrant chemotherapy neurotoxic effects on learning, memory, attention, executive functioning, and processing speed. This chemotherapy-induced cognitive impairment (CICI) is referred to as "chemobrain" or "chemofog". While the underlying mechanisms mediating CICI are still unclear, there is strong evidence that chemotherapy accelerates the biological aging process, manifesting as effects which include telomere shortening, epigenetic dysregulation, oxidative stress, mitochondrial defects, impaired neurogenesis, and neuroinflammation, all of which are known to contribute to increased anxiety and neurocognitive decline. Despite the increased prevalence of CICI, there exists a lack of mechanistic understanding by which chemotherapy detrimentally affects cognition in cancer survivors. Moreover, there are no approved therapeutic interventions for this condition. To address this gap in knowledge, this review attempts to identify how adenosine signaling, particularly through the adenosine $A_{2A}$ receptor, can be an essential tool to attenuate accelerated aging phenotypes. Importantly, the adenosine $A_{2A}$ receptor uniquely stands at the crossroads of cancer treatment and improved cognition, given that it is widely known to control tumor induced immunosuppression in the tumor microenvironment, while also posited to be an essential regulator of cognition in neurodegenerative disease. Consequently, we propose that the adenosine $A_{2A}$ receptor may provide a multifaceted therapeutic strategy to enhance anticancer activity, while combating chemotherapy induced cognitive deficits, both which are essential to provide novel therapeutic interventions against accelerated aging in cancer survivors.

## 1. Introduction

The discovery and development of a wide range of chemotherapeutic drugs has a significantly positive impact in cancer treatment. However, while chemotherapy reduced recurrence and mortality and increased survival rate in a variety of cancers, particularly in breast cancer, many cancer survivors also suffer from aberrant neurotoxic effects of chemotherapy on multiple cognitive domains including learning, memory, attention, and executive functioning. This chemotherapy-induced cognitive impairment (CICI) is often referred to as "chemobrain" or "chemofog" (Falleti, Sanfilippo, Maruff, Weih, & Phillips, 2005). While it is possible that cancer can cause cognitive decline due to tumorigenic inflammation (Ahles, Saykin et al., 2008; Wefel, Lenzi et al., 2004), CICI may occur both in the acute period of treatment and chronic period following treatment, and has been reported in up to 75% of patients (Koppelmans, Breteler et al., 2012; Wefel & Schagen, 2012; Weiss, 2008).

A subset of survivors (15–20%) experience measurably persistent cognitive decline even after the termination of a chemotherapy regimen (de Ruiter, Reneman et al., 2011; Heflin, Meyerowitz et al., 2005; Lange, Joly et al., 2019; Lange, Licaj et al., 2019). This can affect both adult and pediatric cancer populations, and unfortunately, since there are no current FDA approved clinical treatments, CICI represents a significant public health concern (ACS, 2019). Therefore, understanding the neurobiological underpinnings of CICI will foster development of strategies to alleviate these neurotoxic effects which will ultimately improve the quality of life for cancer survivors.

While the underlying mechanisms mediating CICI are largely unknown, both clinical and preclinical evidence suggests that chemotherapy adversely alters brain function similar to the aging process, manifesting as effects which include increased anxiety, neurocognitive decline, impaired neurogenesis, and increased gliosis. In addition, chemotherapy accelerates biological and molecular aging (Demaria et al., 2017; Sanoff, Deal et al., 2014; Yoo, Tang et al., 2021) and increases the risk for the development of Alzheimer's disease (Kesler, Rao, Ray, Rao, & I. Alzheimer's Disease Neuroimaging, 2017), all of which have been observed in breast cancer patients. These observations suggest that CICI and brain aging may share common pathogenic mechanisms mediating cognitive impairment. However, how chemotherapy accelerates brain aging and whether this detrimental effect can be prevented remains unclear. In this regard, adenosine $A_{2A}$ receptor ($A_{2A}R$) signaling has emerged as a hallmark of age-related diseases which encompass neurodegeneration as well as cancer. Notably, the adenosine $A_{2A}R$ has recently garnered attention as a potential therapeutic target for chemotherapy-induced adverse effects ranging from cognitive impairment, peripheral neuropathy, to nephrotoxicity (Dewaeles, Carvalho et al., 2022; Oliveros, Yoo et al., 2022). Throughout this chapter we will take into consideration how adenosine $A_{2A}R$ receptor function is involved in cognition from a physiological perspective, as well as its involvement in pathological conditions such as accelerated aging and CICI (Fig. 1). To facilitate this discussion, we briefly introduce the phenomenon of chemobrain from a clinical perspective in adult and pediatric cancer survivors, followed by a description of studies associating accelerated aging mechanisms with the chemotherapy treatment related toxicities that are known to bestow neurocellular dysfunctions and cognitive impairments. We then examine adenosine $A_{2A}R$ involvement in cancer-related immunosuppression and posit that the

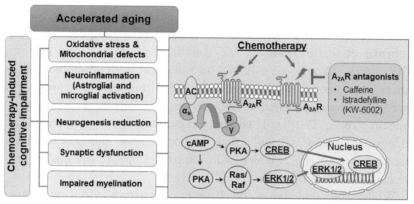

**Fig. 1** Potential cellular and molecular mechanisms mediating the convergence of chemotherapy-induced cognitive impairment and accelerated aging phenotypes. Chemotherapy induced cognitive impairments (CICI) can be experienced by cancer patients with varying genotoxic mechanisms of action, including antifolate antimetabolites (methotrexate), platinum-based compounds (cisplatin), as well as anthracyclines (doxorubicin), amongst others. Notably, mounting evidence suggests that generally, chemotherapies target multiple cellular and molecular mechanisms that, when disrupted, are widely known to negatively affect hallmarks of cognitive function, such as reductions in the neurogenic potential of the hippocampus, deficient functional neuronal architecture of dendrites yielding stunted dendrite spine densities, and impaired oligodendrocyte progenitor development which hinders myelination capacity. Several lines of evidence suggest that oxidative stress generated by mitochondrial defects, as well as neuroinflammatory sequelae resulting from proinflammatory cytokine penetration of the brain parenchyma (stemming from chemotherapy induced weakening of the blood brain barrier), are candidate hypotheses that may explain how CICI develops. Interestingly, the phenotypes that typify CICI also share a commonality with cellular and molecular detriments observed in cancer-related accelerated aging and neurodegenerative-related accelerated aging. These shared common pathological mechanisms include mitochondrial oxidative stress, and an increased proinflammatory cytokine secretory response engaged by chemotherapy's apoptotic and necrotic effect on malignancies. Similarly, CICI and neurodegeneration associated accelerated aging share impairments in neurogenesis, synaptic function, and myelination. One pathway that may elucidate the commonality between CICI, cancer-related accelerated aging, and neurodegenerative associated accelerated aging, is through inhibition of adenosine $A_{2A}R$ signaling. Inhibition of $A_{2A}R$ has been demonstrated to sensitize chemotherapy against malignancies as well as prevent tumor induced immune evasion in the tumor microenvironment. Importantly, increased adenosine $A_{2A}R$ is implicated with cognitive deficits in Alzheimer's disease. $A_{2A}R$ inhibition is associated with in improved cognition in preclinical models of neurodegenerative disease, as well as preservation of neuronal architecture, neurogenesis, and cognitive function in a preclinical model of cisplatin induced CICI. Taken together, we propose that CICI may be classified as a form of accelerated aging, and that the adenosine $A_{2A}R$ could serve as a unique nexus that can provide therapeutic efficacy in cancer survivors and accelerated aging phenotypes.

adenosine $A_{2A}R$ is uniquely positioned to be a therapeutic target in neurocellular dysfunctions and cognitive impairments stemming from chemotherapy related accelerated brain aging (Fig. 1).

## 2. Adenosine $A_{2A}R$ in cognitive function
### 2.1 Physiological role of $A_{2A}R$ and cognitive improvement

Adenosine is an endogenous neurotransmitter that plays an essential role in synaptic plasticity and maintenance of brain homeostasis through four distinct adenosine receptors: $A_1R$, $A_{2A}R$, $A_{2B}R$ and $A_3R$ (Chen, 2014). Through adenylate cyclase activation, the Gs-protein coupled adenosine $A_{2A}R$ engages cyclic adenosine monophosphate (cAMP) resulting in subsequent protein kinase-A mediated (PKA) phosphorylation of downstream CREB and ERK resulting transcriptional activity (Chen Choi & Cunha, 2023; Nam, Bruner, & Choi, 2013). Known to have high abundance in the striatum, and to a lesser degree, in the hippocampus and cortex (Liu, Chen et al., 2019), the $A_{2A}R$ is increasingly a notable regulator of neuro-glial interactions (Orr, Hsiao et al., 2015; Orr, Orr, Li, Gross, & Traynelis, 2009), neurodevelopment (Rodrigues, Marques, & Cunha, 2019), corticolimbic synaptic plasticity (Reis, Silva et al., 2019), and memory function (Li et al., 2018; Temido-Ferreira, Coelho, Pousinha, & Lopes, 2019). In regards to physiological expression and functionality, the $A_{2A}R$ can be found at presynaptic and postsynaptic terminals of neurons, where through receptor-receptor interactions with the adenosine $A_1R$, or the dopamine $D_2$ receptor, for example, it can fine tune synaptic long-term potentiation (LTP) and long-term depression (LTD), the putative hallmark mechanisms by which memory and emotive behavior are controlled (Chen, Lee, & Chern, 2014; Rebola, Lujan, Cunha, & Mulle, 2008; Rebola, Rodrigues et al., 2005). Importantly, postsynaptic activation/inhibition of the $A_{2A}R$ is effective in the control of aberrant synaptic plasticity, thus making this receptor important for improvement of memory function in normal aging physiology, as well as in pathophysiological conditions that impair learning and memory function (Chen, 2014). The physiological role of $A_{2A}R$ in cognitive improvement is exemplified by the global use of the cognitive enhancer caffeine, which is a partial $A_{2A}R$ antagonist and known to improve cognitive function in normal aging and neurodegenerative conditions (e.g., Alzheimer's disease), as well as in preclinical rodent models of learning and memory (Canas, Porciuncula et al., 2009; Chen, 2014; Kaster, Machado et al., 2015; Laurent, Burnouf et al., 2016; Merighi, Borea et al., 2022).

## 2.2 A$_{2A}$R and cognitive improvement in pathological conditions

Interestingly, a high level of A$_{2A}$R expression is observed in the aged brain, as well as in brains from Alzheimer's disease and Parkinson's disease (PD) patients and mouse models of these conditions (Calon, Dridi et al., 2004; Chen, 2014; Orr, Hsiao et al., 2015; Temido-Ferreira, Ferreira et al., 2020). Importantly, A$_{2A}$R antagonists are known to improve deficits in synaptic plasticity as well as learning and memory function in mouse models of AD and PD (Da Silva et al., 2016; Ferreira, Temido-Ferreira et al., 2017; Li, Chen et al., 2018; Li, Silva et al., 2015; Orr, Lo et al., 2018; Silva, Lemos et al., 2018; Viana Da Silva Haberl et al., 2016), whereas overactivation of neuronal A$_{2A}$R impairs memory function (Carvalho, Faivre et al., 2019), suggesting a causal relationship between A$_{2A}$R in neurodegeneration and therapy. More recently, the A$_{2A}$R has also been demonstrated to be involved in neurotoxicity and cognitive impairments associated with chemotherapy (Gyau & Deaglio, 2023; Oliveros, Yoo et al., 2022). Given adenosine's capacity to affect normal and pathological conditions, and in particular those pertaining to brain function through adenosine receptors (Kaur, Weadick et al., 2022), we propose that the adenosine A$_{2A}$ receptor uniquely stands at the crossroads of cancer treatment and cognition, as it may represent a novel therapeutic target for cancer survivors.

## 2.3 Convergence of the adenosine A$_{2A}$R, aging and AD

The proposed existence of two distinct subtypes of purinergic receptors by Geoffrey Burnstock revealed a distinct difference in affinity for either adenosine (P1 receptors) or conversely, P2 receptors, which have a higher affinity for ATP (Burnstock 1978). Further investigation revealed that P1 receptors could be classified by adenosine's inherent ability to either inhibit (A$_1$R) or stimulate (A$_{2A}$R) downstream cAMP signaling in cultured mouse perinatal brain cells (Van Calker, Muller, & Hamprecht, 1979). Similarly, adenosine receptor function was identified by their ability to modulate acetylcholine and glutamatergic neurotransmitter release in the striatum and hippocampus (Ferre, Sarasola, Quiroz, & Ciruela, 2023; Sperlagh & Vizi, 2011; Spignoli, Pedata, & Pepeu, 1984). In terms of AD related neurodegeneration, increased hippocampal expression of astroglial $A_{2A}R$ has been reported in postmortem AD brains, as well as in preclinical AD-related human amyloid precursor protein (hAPP) mouse models (Orr, Hsiao et al., 2015). Dias and colleagues also identified that functional inhibition of astroglial P2 receptors by the specific $A_{2A}R$ antagonist SCH58261 occurred in a PKA dependent manner, whereas the presence of amyloid-ß (1–42) peptides was sufficient to disrupt this interaction (Dias, Madeira et al., 2022). In support of the unique relationship between aging and $A_{2A}R$ expression, $A_{2A}R$ densities and downstream cAMP signaling is increased in 2-year-old rat limbic cortex when compared to 6-week old rats (Lopes, Cunha, & Ribeiro, 1999). Similarly, the frontal cortex of postmortem AD brains have been shown to exhibit increased $A_{2A}R$ expression (Albasanz, Perez, Barrachina, Ferrer, & Martin, 2008), thus emphasizing the adenosine $A_{2A}R$ as a potential regulator of accelerated aging (Dias, Madeira et al., 2022; Lopes, Cunha, & Agostinho, 2021). In support of this hypothesis, a significant body of preclinical studies have demonstrated that $A_{2A}R$ inhibition offers cognitive improvements (Liu, Chen et al., 2019). Of note, activation of the hypothalamic-pituitary-adrenal (HPA) axis can result in excessive cortisol release during advanced aging, and this HPA axis engagement can be further exacerbated by forebrain $A_{2A}R$ activation, thus resulting in synaptic plasticity deficits and memory impairments (Batalha, Ferreira et al., 2016). Conversely, impairments in synaptic plasticity and memory function can be restored by global $A_{2A}R$ deletion, selective deletion of $A_{2A}R$ in forebrain neurons, the non-specific $A_{2A}R$ antagonist caffeine, and the specific $A_{2A}R$ antagonists istradefylline (KW 6002) and SCH-58261 (Batalha, Ferreira et al., 2016; Kaster, Machado et al., 2015). Notably, $A_{2A}R$ elevations have been associated with dysfunctional hippocampal regulation of the HPA axis, resulting in accelerated clinical AD progression, memory impairments, and hippocampal ß-amyloid and Tau accumulation in AD mouse models (Csernansky, Dong et al., 2006; Green, Billings, Roozendaal, McGaugh, & LaFerla, 2006; Temido-Ferreira, Ferreira et al., 2020). Conversely, genetic $A_{2A}R$ deletion or pharmacological inhibition of $A_{2A}R$ that target hippocampal neural cells are reported to improve accelerated aging memory, synaptic plasticity and neurocellular dysfunctions caused by amyloid-ß (1–42) peptide induced neurotoxicity, and Tau accumulation in preclinical mouse models of AD (Canas, Porciuncula et al., 2009; Carvalho, Faivre et al., 2019; Goncalves, Lopes et al., 2019; Kaster, Machado et al., 2015; Laurent, Burnouf et al., 2016; Temido-Ferreira, Coelho et al., 2019).

Taken together, these reports suggest that the $A_{2A}R$ plays a significant role in regulation of aging, and neurodegenerative disease. Further investigations are warranted to determine whether the adenosine $A_{2A}R$ inhibition could play a mechanistic role in attenuation of accelerated aging phenotypes.

## 3. Findings of cognitive dysfunction in chemobrain
### 3.1 Clinical imaging studies

Due to advanced imaging technology, the impact of chemotherapy on brain functionality has been widely investigated, with magnetic resonance imaging (MRI) playing a pivotal role in identifying which specific brain regions are negatively affected, often typified by reductions in brain volumes. For example, Mc Donald and colleagues demonstrated that breast cancer patients exhibited decreased gray matter in several brain regions, including the frontal and temporal cortices, the cerebellum, and the right thalamus, immediately after chemotherapy (McDonald, Conroy, Ahles, West, & Saykin, 2010). Importantly, frontal gray matter reduction observed in breast cancer patients after chemotherapy was associated with deficits in cortical executive function and memory processes (McDonald, Conroy et al., 2010). Functional MRI (fMRI) studies in chemotherapy-treated women also found reduced activation during cognitive tasks in multiple brain regions including the left caudal lateral prefrontal cortex, the dorsolateral prefrontal cortex, and the parahippocampal gyrus, all of which were significantly correlated with impairments in executive functioning, planning performance and recognition memory (De Ruiter, Reneman et al., 2011; Kesler, Kent, & O'Hara, 2011). In contrast to studies showing hypoactivation, several groups also report increased activation in multiple brain regions. For instance, an fMRI study conducted by Cimprich et al. (2010) investigated verbal working memory in patients with breast cancer both prior and post chemotherapy treatment and found greater levels of activation in the right inferior frontal gyrus and the frontoparietal attentional network (FPN), including bilateral frontal and parietal regions (Cimprich, Reuter-Lorenz et al., 2010). Another study conducted by Scherling and colleagues (2011) examined neurofunctional differences in working memory in breast cancer patients compared to controls prior to chemotherapy treatment (Scherling, Collins, Mackenzie, Bielajew, & Smith, 2011). This investigation revealed that relative to controls, breast cancer patients made fewer commissioned errors and were slower than controls, and these effects were associated with increased activation of the left inferior frontal gyrus, left insula, bilateral thalamus and right midbrain during the working memory tasks. This led the authors to conclude that breast cancer patients found the task more challenging than the control group, thus suggesting that breast cancer patients, to maintain a baseline level of performance, necessitate a greater recruitment of brain regions in order to maintain accuracy in memory tasks (Simo, Rifa-Ros, Rodriguez-Fornells, & Bruna, 2013). Nevertheless, these studies provide evidence that the symptomatology of chemobrain has a solid biological basis, rather than being purely psychological.

## 4. Chemobrain as an accelerated aging phenotype in adult and pediatric cancer

### 4.1 Evidence of accelerated aging in adult cancer patients

Aging is the time dependent progression of dynamic cellular, molecular, metabolic, biochemical, and cognitive changes across the lifespan of an organism. Inescapably, aging is typified by pathologies that include accumulation of cellular damage, increased oxidative stress, epigenetic and genomic instability, less-than adequate cellular repair mechanisms, telomere attrition, ineffective signaling cascades, detrimental transcriptional activity, increased peripheral inflammation, blood brain barrier disruption and exacerbated neuroinflammation (Armstrong & Boonekamp, 2023; Chiang, Huo, Kavelaars, & Heijnen, 2019; Knox, Aburto, Clarke, Cryan, & O'Driscoll, 2022; Lopez-Otin, Blasco, Partridge, Serrano, & Kroemer, 2022; Ness, Armstrong et al., 2015; Schmauck-Medina, Moliere et al., 2022; Squassina, Manchia et al., 2020). Despite healthy aging inevitably increasing these pathological processes, and often in the absence of apparent clinical syndromes (Banks, Reed, Logsdon, Rhea, & Erickson, 2021), a biological system's inherent inability to prevent aging-related processes will eventually manifest into overt clinical disease. The resulting age-related complications include (but are not limited to) cardiovascular and cardiopulmonary pathologies, neurodegenerative related motor and memory dysfunctions, and for the purposes of this review, cancerous malignancies and the accompanying neuropathological and cognitive disturbances associated with chemotherapy (Carroll, Bower, & Ganz, 2022; Lopez-Otin, Blasco et al., 2022).

Not surprisingly, the resulting off-target effects of chemotherapy described from studies involving patients with breast cancer, are hypothesized to resemble pathologies akin to accelerated aging, given that both chemotherapy and accelerated aging are accompanied by dysfunctionality in several cellular hallmarks of cognitive function (Nguyen & Ehrlich, 2020). These include a compromised ability for healthy neural stem cell (NSC) maturation (i.e. adult neurogenesis), a loss of dendritic spines and dendrite arborization complexity, as well as inadequate neurotransmitter release (Onzi et al., 2022). In addition, chemotherapy induced accelerated aging can lead to negative impacts on cellular mechanisms of neuronal support that control adequate cognition, such as inoperative oligodendrocyte maturation leading to myelination deficits (Cardoso et al., 2020; Geraghty, Gibson et al., 2019; Gibson & Monje, 2019; Matsos, Loomes et al., 2017). It has been well established that cancer development, as well as chemotherapy's apoptotic and necrotic effect on tumor cells, is associated with increased systemic inflammation and an immune response near the tumor site (Stagg & Smyth, 2010). Consequently, chemotherapy's ability to damage DNA and increase oxidative stress is accompanied by blood brain barrier (BBB) breakdown (Wardill, Mander et al., 2016). Accordingly, chemotherapy induced BBB breakdown increases the propensity for peripheral permeation of the brain parenchyma by proinflammatory cytokines that increase the likelihood of reactive astrogliosis taking precedence over astroglial neuronal support, and in addition, can shift microglia behavior from surveillance to reactivity (Gibson, Nagaraja et al., 2019; Jia, Zhou et al., 2023; Monje, Vogel et al., 2007). Not surprisingly it is posited that the increase in inflammatory cytokines by chemotherapy is an aspect of accelerated aging, termed senescence associated secretory phenotype (SASP), which has also been described to contribute to cognitive dysfunction Alzheimer's disease (Guerrero, De Strooper, & Arancibia-Carcamo, 2021; Wang, Prizment, Thyagarajan, & Blaes, 2021). Simply stated, cellular senescence is a natural, age-related process by which cells decline in their ability to divide, thus entering into a terminally quiescent, albeit functional state. Cellular senescence can also be engaged by oncogenic signaling, nutrient deprivation, mitochondrial dysfunction, and genotoxic chemotherapeutics, all of which activate accelerated aging sequelae (Gorgoulis, Adams et al., 2019). Consequently, SASP can increase secretion of growth factors, monocyte chemoattractant proteins, receptor proteins, chemokines and proinflammatory cytokines such as IL-6, IL-8, TGF-ß, TNF-α, amongst others (Gorgoulis, Adams et al., 2019; Wang, Prizment et al., 2021). Overlapping with SASP released proinflammatory markers, chemotherapy treated cancer patients also exhibit increased levels of peripheral and brain inflammatory cytokines (e.g. IL-6, IL-8, TNF-α) that are associated with decreased cognitive function (Carroll, Bower et al., 2022; Hoogland, Nelson et al., 2019; Laird, McMillan et al., 2013; Rummel, Chaiswing, Bondada, St Clair, & Butterfield, 2021). Preclinical animal models of methotrexate, doxorubicin and cisplatin CICI also support these findings, as increased BBB breakdown fosters increased entry of peripheral cytokines, such as IL-6, IL-8, IL-1ß and TNF-α which promote both astroglial and microglial reactivity (Alexander, Mahalingam et al., 2022; Chiang, Huo et al., 2019; Gibson, Nagaraja et al., 2019; Savchuk & Monje, 2022), causing further proinflammatory sequelae culminating in neurotoxicity that damages white matter, neurons, and consequently, disrupts cognitive function (Cardoso et al., 2020; George, Semendric, Hutchinson, & Whittaker, 2021; Lacourt & Heijnen, 2017; Ren, Clair, & Butterfield, 2017; Ren, Keeney et al., 2019). Clearly, the complexity of detriments exerted by cancer and cancer chemotherapy that result in cancer related cognitive dysfunction are multifaceted, and can be attributed to several factors, including high stress from cancer diagnosis, peripheral proinflammatory sequelae derived from the innate and/or adaptive immune responses to malignancies, or from chemotherapy and/or radiotherapy treatment devised to eradicate tumors (Argyriou, Assimakopoulos, Iconomou, Giannakopoulou, & Kalofonos, 2011; Wefel, Kesler, Noll, & Schagen, 2015). In regards to the accelerated aging similarities between CICI and neurodegenerative disease, recent studies demonstrate that chemotherapy increases the risk for development of AD. To that effect, the apolipoprotein E4 allele (APOE4), one of the strongest genetic risk factors for AD, links chemotherapy treatment and APOE4 as a potential risk factor in cognitive impairments associated with anthracycline chemotherapy (Fernandez, Varma, Flowers, & Rebeck, 2020). For instance, breast and testicular cancer patients receiving doxorubicin chemotherapy that are genetic carriers of APOE4, are also more likely to exhibit worse attentional processing speed and working memory impairments (Ahles, Li et al., 2014; Ahles, Saykin et al., 2003; Amidi, Agerbaek et al., 2017). Supporting this hypothesis, doxorubicin treatment in preclinical mouse models of familial Alzheimer's disease mutations (5XFAD), which recapitulate AD cellular pathogenic and behavioral impairments, had increased reactive astrogliosis surrounding cortical amyloid beta deposits (Ng, Biran et al., 2022). CICI is also shown to be associated with accelerated development of tau (a hallmark of AD) clustering in the brain, as a sign of accelerated aging (Chiang, Huo et al., 2019). Furthermore, transgenic APOE4 knock in mice administered doxorubicin displayed reduced cortical and hippocampal volumes alongside impairments in spatial memory (Demby, Rodriguez et al., 2020; Speidell, Demby et al., 2019). Another potentially viable senescence associated target that was recently identified to be increased in cortical and hippocampal tissues of aged rodent models is the cyclin dependent kinase inhibitor and tumor suppressor $p16^{INK4A}$ (Wang Muthu Karuppan et al., 2021; Xie, Zhi, & Meng, 2021). Notably, $p16^{INK4A}$ is also routinely identified as a biomarker of chemotherapy induced senescent tissues in accelerated aging models (Baker, Wijshake et al., 2011; Sanoff, Deal et al., 2014; Uziel, Lahav et al., 2020). Hence, future investigations into the potential functional role that $p16^{INK4A}$ can exert in CICI may provide novel pharmacological or genetic therapeutic strategies to combat accelerated aging in CICI. Collectively, the convergence of these observations suggests that cancer survivors experiencing CICI display various phenotypes that are hallmarks of the accelerated aging process. These include increases in peripheral proinflammatory mediators that penetrate the brain parenchyma due to BBB dysfunction, elevated reactive astrogliosis and microgliosis stemming from increased neuroinflammation, and defective myelination capacity that hinders effective corticolimbic function. Together, these phenotypes foment cognitive deficits that can be exacerbated by genetic risk factors (i.e. APOE4), thus necessitating rapid development of treatment strategies to improve quality of life in cancer survivors.

## 4.2 Evidence of accelerated aging phenotypes in pediatric cancer patients

Similar as what was observed in adult cancer patients, approximately 500,000 survivors from childhood cancer live in the United States and up to 35% will experience persistent neurocognitive impairment months or years after treatment (Williams & Cole, 2021). Given that chemotherapies target proliferating cells, it is not surprising that neural stem cells and oligodendrocyte progenitors undergoing neurogenic development are particularly vulnerable (Mogavero, Bruni, DelRosso, & Ferri, 2020; Sekeres, Bradley-Garcia, Martinez-Canabal, & Winocur, 2021). Therefore, CICI in pediatric and adolescent cancer survivors is particularly egregious and damaging to brain development in this population. Studies utilizing fMRI in ALL pediatric patients have identified reduced volumes in multiple brain structures, including the cortex, hippocampus, white matter structures (e.g. corpus callosum), amygdala, thalamus and striatum (van der Plas, Schachar et al., 2017; Stefancin, Cahaney et al., 2020). More importantly, these brain volume deficits are significantly associated with deficiencies in working memory, attentional processing, sleep disruptions, behavioral response inhibition, and poor planning, which are confirmed by studies showing that 67% of pediatric cancer survivors report attention deficits, and up to 35% of survivors report memory dysfunction (Brace, Lee, Cole, & Sussman, 2019; Cole, Finkelstein et al., 2015; Mogavero, Bruni et al., 2020). Confirming these findings, preclinical adolescent rodent studies that employ methotrexate, a frontline chemotherapy utilized to treat pediatric cancer patients, similarly show disruptions in cognition, synaptic plasticity, myelination and neurogenesis (Geraghty, Gibson et al., 2019; Gibson, Nagaraja et al., 2019; Wen, Maxwell et al., 2018; Wen, Patel et al., 2022). Several neurobiological processes are disrupted in chemotherapy induced accelerated aging which are hypothesized to contribute to the frail physicality and cognitive impairments in young cancer survivors, which strikingly resemble similar patterns observed in frail advanced aged older populations. These include increased oxidative stress, telomere shortening, and inflammatory sequelae which can explain the aggravated physical frailty, and increased morbidity from secondary cardiovascular and cardiopulmonary pathologies, as well as attention, memory, and learning dysfunction, in addition to psychological distress experienced by pediatric cancer survivors (Dewar, Ahn, Eraj, Mahal, & Sanford, 2021; Williams & Cole, 2021; Williams, Krull et al., 2021).

## 4.3 Accelerated aging mitochondrial defects associated with cancer and chemobrain

Mitochondria are essential cellular organelles that provide energy for biochemical processes and metabolic function and dictate successful longitudinal survivorship across the lifespan. However, despite the essentiality of this organelle for survival, mitochondrial function deteriorates with normal aging (Tang, Oliveros, & Jang, 2019). Aging-related impairments in mitochondria include decreased function of respiratory chain protein complexes, reduced oxidative phosphorylation capacity, deficits in adenosine triphosphate (ATP) generation, and enhanced cytochrome-C release which engage the mitochondrial apoptotic cascade (Tang, Oliveros et al., 2019). Moreover, mitochondrial impairments elevate generation of reactive oxygen species (ROS) which promote downstream inflammasome activation of caspases, and eventually cell death (Lopez-Otin, Blasco et al., 2022). Not surprisingly, exacerbated deterioration of mitochondria is exhibited by diseases that typify accelerated aging, including neurodegenerative conditions and cancer treatment (Amorim, Coppotelli et al., 2022). Consequently, many cancer chemotherapies, including doxorubicin, cyclophosphamide, cisplatin, fluorouracil (5-FU), methotrexate, paclitaxel and others, in their efforts to induce death of cellular malignancies, can generate mitochondrial ROS that negligently affect non-cancerous cells near the tumor site, and permeate the BBB to negatively affect the function of distal corticolimbic brain structures (Rummel, Chaiswing et al., 2021). Chemotherapy treatment depletes the abundance of free-radical scavenging antioxidant biomolecules (e.g., glutathione, catalase, superoxide dismutase, and others), resulting in heightened oxidative stress that is posited to be a main contributing mechanism in central nervous system (CNS) toxicity in CICI (Joshi, Aluise et al., 2010; Sahu, Langeh, Singh, & Singh, 2021; Welbat, Naewla et al., 2020). In terms of aging related diseases, metformin, an antihyperglycemic compound that lowers hepatic glucose production to treat type-2 diabetes (Bailey, 2017), has recently garnered attention for its promising ability to reduce hippocampal-related oxidative stress and cognitive deficits through its effects on the mitochondrial fission/fusion protein Drp1 in a mouse model of diabetes (Hu, Zhou et al., 2022). As an antiaging compound, metformin is also purported to be effective in attenuating cognitive decline in diabetic patients (Samaras, Makkar et al., 2020), whereas metformin's action on cortical mitochondrial complex I protein NDUFA2 was recently linked to reductions in Alzheimer's disease (Zheng, Xu et al., 2022). Metformin has also been demonstrated to be effective in reducing cancer related oxidative stress (Kulkarni, Gubbi, & Barzilai, 2020) and more importantly, ameliorating cisplatin-induced cognitive dysfunctions, dendrite spine density impairments and neuronal myelination deficits in a mouse model of CICI (Zhou, Kavelaars, & Heijnen, 2016). Notably, it is hypothesized that cisplatin-induced CICI fosters an accelerated aging phenotype, as cisplatin stimulates the accumulation of Tau aggregates which are accompanied by decreased expression of the postsynaptic marker PSD-95 and adjacent astrogliosis (Chiang, Huo et al., 2019). Further supporting the role of mitochondria in chemotherapy induced accelerated aging, a significant body of work by Cobi Heijen's group posits that accumulation of the tumor suppressor protein p53 compromises synaptic mitochondrial integrity, which culminates in cisplatin-induced neurotoxicity, cortical myelin impairments, and hippocampal memory dysfunction (Chiu, Boukelmoune et al., 2018; Chiu, Maj et al., 2017). Naturally, implementation of compounds that promote metabolic stabilization of mitochondrial dynamics and reduction of oxidative stress can be indispensable strategies to restore mitochondrial function in CICI. For example, cisplatin administration to cortical neurons derived from human induced pluripotent stem cells (iPSCs), as well as studies from rat hippocampus, exhibited significant swollen vacuolizations in mitochondria, increased ROS levels, and $\gamma H_2AX$ DNA damage, coupled with disruptions in membrane potential and deficits in ATP generation (Lomeli, Di, Czerniawski, Guzowski, & Bota, 2017; Rashid, Oliveros, Kim, & Jang, 2022). This is reminiscent of the oxidative stress and neurotoxicity observed in patients treated with chemotherapy (Torre, Dey, Woods, & Feany, 2021). However, interventions with compounds that increase antioxidant activity and improve metabolic function in mitochondria, including nicotinamide mononucleotide (NMN) and N-acetylcisteine (NAC), have been effective in restoring mitochondrial respiration, and increasing ATP production, while preventing mitochondrial vacuolization, DNA damage and cognitive dysfunction (Lomeli, Di et al., 2017; Rashid, Oliveros et al., 2022). In support, restoration of healthy mitochondria via non-invasive nasal administration shows significant promise in attenuation of cisplatin-induced CICI (Alexander, Mahalingam et al., 2022; Alexander, Seua et al., 2021). Given that nasal administration of mitochondria is a simple, non-invasive route of administration that can be applied in the clinic (Lofts, Abu-Hijleh, Rigg, Mishra, & Hoare, 2022), further investigations are needed to ascertain whether mitochondrial replacement therapy in CICI is a general mechanism that can be implemented only with cisplatin or whether this strategy can also be applied to chemotherapies with different mechanisms of action (e.g. methotrexate, cyclophosphamide, 5-FU, etc.) to ameliorate the neurocellular and cognitive deficits typified by this condition.

### 4.4 Accelerated aging, telomeres, and chemobrain

Across the lifespan of organisms, including humans, the ribonucleoprotein enzyme telomerase provides chromosomal endpoint stabilization and telomere lengthening to protect genomic DNA integrity upon repeated cell divisions, thereby avoiding telomere attrition

(Blackburn, Epel, & Lin, 2015). Interestingly, telomere attrition has been described as a risk factor in neuropsychiatric conditions (e.g. depression, psychosis, post-traumatic stress disorder), and recently, it was revealed as a mechanism in preclinical Alzheimer's disease models (Lindqvist, Epel et al., 2015; Shim, Horner et al., 2021). Exemplifying this, a recent study examining telomere reverse transcriptase (TERT) haploinsufficiency in Alzheimer's disease mouse models found accumulation of amyloid precursor protein and amyloid-ß, resulting in decreased dendrite spine densities and cognitive deficits, whereas overexpression of TERT effectively reversed these aging pathologies (Shim, Horner et al., 2021). Interestingly, there is some evidence for the involvement of adenosine and its receptors, in regulation of telomere function. For example, a study conducted by Fishman and colleagues showed that exposure of adenosine suppressed telomere signals in Nb2-11C rat lymphoma cells (Fishman, Bar-Yehuda et al., 2000), whereas the adenosine $A_{2A}R$ antagonist SCH-442416 was demonstrated to stabilize the telomeric G-quadruplex DNA complex, suggesting that the $A_{2A}R$ may play a notable function in telomerase dynamics (Salem, Haty, & Ghattas, 2022). In spite of the necessity for possessing active telomerase for healthy aging, the telomerase-telomere complex decreases in expression and function with age, consequently making telomere shortening a key hallmark in aging-related disease (Lopez-Otin, Blasco et al., 2022). Paradoxically, cancer cells undergoing unchecked cell division beyond the Hayflick limit (Romaniuk, Paszel-Jaworska et al., 2019), avoid senescence or apoptosis by overactivation of telomerase, thus necessitating genotoxic chemotherapy treatment to treat malignancies (Blackburn, Epel et al., 2015). Over the last several decades, tactful utilization of chemotherapy has garnered positive outcomes in terminating malignancies at the cost of inducing telomere attrition, which is posited to be a potential mechanism that effectuates CICI and is akin to accelerated aging (Wang, Prizment et al., 2021). Recent studies in breast cancer survivors treated with chemotherapy and longitudinally assessed up to 6 years following treatment, exhibited higher levels of DNA damage and lower peripheral blood mononuclear cells (PBMCs) telomerase activity, which was associated with impairments in executive function and attention (Carroll, Van Dyk et al., 2019; Scuric, Carroll et al., 2017). Supporting this hypothesis, breast cancer patients treated with combinations of either docetaxel, doxorubicin, cyclophosphamide, carboplatin and trastuzumab showed decreased telomere lengths associated with older age, while significant associations were also found between shorter chromosomal telomere lengths and disruptions in cognitive domains spanning visual memory, psychomotor processing speed, reaction time, and executive functioning (Alhareeri, Archer et al., 2020). Further evidence implicating adenosine regulation of telomeres was described in a study where increased adenosine levels via the loss of adenosine deaminase (ADA), which converts adenosine to inosine, reduced telomerase activity and accelerated senescence in CD8+ T-lymphocytes (Parish, Kim et al., 2010). Conversely, decreased production of adenosine by mesenchymal stromal cells (MSC) immortalized via retroviral transfection of the telomerase reverse transcriptase (TERT), were less immunosuppressive towards lymphocytes, hence denoting the bi-directional relationship between telomerase function and adenosine (Beckenkamp, da Fontoura et al., 2020). Given that chemotherapies can detrimentally reduce telomerase activity and prevent accelerated aging mediated telomere lengthening (Cupit-Link, Kirkland et al., 2017), future studies are required to determine whether adenosine $A_{2A}R$ antagonists can modify telomerase activity or telomere lengthening in the context of chemobrain. Investigation of this concept can engender novel mechanisms for attenuation of CICI in cancer-related accelerated aging. Addressing these critical issues may help devise effective therapeutic strategies that eradicate tumors, while preserving cognitive function in cancer survivors.

## 4.5 Epigenetic related alterations in chemobrain and accelerated aging

Dysfunctional histone acetylation or methylation of the chromatin structures encapsulating DNA in the epigenome can promote aberrant transcriptional modifications that contribute to the aging process (Lopez-Otin, Blasco et al., 2022). Epigenetically, decreases in the function of the sirtuin (SIRT) family of protein deacetylases is associated with impaired nicotinamide adenine dinucleotide (NAD+) metabolic regulation, resulting in accelerated aging (SIRT7 deletion) and degenerative (SIRT6 deletion) phenotypes in mouse models similar to BubR1 progeroid mice, a well-known genotype that exhibits accelerated aging defects (Cho, Yoo et al., 2019; Corujo-Ramirez, Dua, Yoo, Oliveros, & Jang, 2020; Lopez-Otin, Blasco et al., 2022). Conversely, a mouse model of SIRT6 overexpression that increases de novo NAD+ synthesis, has been shown to delay age-related frailty and preserve healthy aging (Roichman, Elhanati et al., 2021).

Interestingly, inhibition of the histone deacetylase HDAC6 has been associated with pathogenic Tau accumulation in preclinical studies of Alzheimer's disease (Trzeciakiewicz, Ajit et al., 2020; Tseng, Xie et al., 2017), whereas selective HDAC6 inhibition with Ricolinostat (ACY-1215) showed significant antitumor activity in multiple myeloma clinical studies (Vogl, Raje et al., 2017). Given the importance of histone modification in conveying antitumor efficacy cancer, while preserving healthy aging related phenotypes as discussed above, it is conceivable that control of histone modification may confer some neuroprotective effects in chemobrain. Accordingly, administration of the selective HDAC6 inhibitor Ricolinostat in young mice (8–10 weeks of age) showed effective attenuation of cisplatin-induced cognitive deficits, in conjunction with reversal of hippocampal Tau phosphorylation, restoration of mitochondrial function, and synaptic integrity (Ma, Huo, Jarpe, Kavelaars, & Heijnen, 2018). In aged mice (7–8 months of age), cisplatin chemotherapy decreased PSD-95 expression and significantly accumulated Tau cluster formations in the hippocampus, suggesting that cisplatin accelerates aging akin to what is observed in rodent models of Alzheimer's disease (Chiang, Huo et al., 2019; Tseng, Xie et al., 2017). Indeed, our own recent study demonstrated how cisplatin chemotherapy decreased functional activity of nicotinamide phosphorybosyltransferase (Nampt) and SIRT2, thus disrupting NAD+ metabolic activity, which contributes to hippocampal neural stem cell impairments, neuronal maturation deficits, as well as spatial and recall memory dysfunction (Yoo, Tang et al., 2021), akin to neurodegenerative conditions and accelerated aging. Intriguingly, North and colleagues (2014) performed an investigation using the BubR1 accelerated aging mouse model which revealed that administration of the NAD+ precursor nicotine mononucleotide (NMN), as well as SIRT2 overexpression, causally increased BubR1 protein levels to promote life extension (North, Rosenberg et al., 2014). Importantly, the work of North and colleagues (2014) is uniquely in line with our findings where we demonstrate that cisplatin chemotherapy foments accelerated aging, given that NMN supplementation or NAMPT overexpression, can efficaciously attenuate cisplatin-induced chemobrain by regulating expression levels of SIRT2 (Yoo, Tang et al., 2021). Overall, targeting the NAD+ metabolic pathway may provide future effective treatment strategies that can provide clinical benefits to cancer survivors experiencing accelerated aging sequelae related to chemotherapy.

## 5. The adenosine $A_{2A}$ receptor as a potential mechanism of accelerated aging in chemobrain and cancer

### 5.1 Anticancer effects and immune cell modulation of the adenosine $A_{2A}R$ in malignancies

Adenosine can exert tight control of adenosine receptor function to control cellular proliferation and exert immunosuppression during cancer and chemotherapy treatment, through (1) the extracellular enzymatic conversion of existing ATP, ADP or AMP by the ectonucleotidases CD39 and CD73, or (2) through the release of ATP from mitochondrial intracellular stores, which again, utilize CD39 and CD73 to generate adenosine (Boison & Yegutkin, 2019; Congreve, Brown, Borodovsky, & Lamb, 2018). Indeed, in the CNS, astrocyte mediated ATP release and subsequent $Ca^{2+}$ influx in neural cells triggers mitochondrial oxidative stress culminating in neuronal damage and cell death, which are neurotoxic biological consequences associated with accelerated aging and neurodegeneration (Rodrigues, Tome, & Cunha, 2015). However, the above mechanisms are by no means the only pathways that generate adenosine, and a detailed description of these processes have been extensively reviewed (Boison & Yegutkin, 2019; Ohta, 2016). The net effect of CD39 and CD73 enzymatic activity is to elevate extracellular levels of adenosine in the tumor microenvironment, thereby naturally activating adenosine $A_{2A}R$ on lymphocytes in an immunosuppressive capacity, thus protecting tumors from destruction (Ohta, Gorelik et al., 2006; Sitkovsky, Kjaergaard, Lukashev, & Ohta, 2008).

In chronic hypoxic microenvironments typified by extensive cellular proliferation, such as B-lymphocyte immune memory development in germinal centers, or the tumor microenvironment, cellular stress can induce persistent adenosine release and subsequent activation of $A_{2A}R$, which plays an important role in conducting immune cell development, and tumor-controlled immunosuppression of T-lymphocytes (Abbott, Silva et al., 2017; Blay, White, & Hoskin, 1997; Ohta, 2016; Ohta, Gorelik et al., 2006; Young, Mittal et al., 2014). In the tumor microenvironment, adenosine achieves this by accumulating in the extracellular space surrounding solid tumors (Blay, White et al., 1997). Extracellular adenosine accumulation, in addition to the degradation of ATP by the ectonucleotidases CD39 and CD73 (released by apoptotic or necrotic cells exposed to chemotherapy), can also be released by activated monocytes, neutrophils, and dendritic cells, all of which are chemotactically summoned in response to ATP "danger signals" emanating from the tumor microenvironment (Stagg & Smyth, 2010). ATP mediated "danger signal" adenosine accumulation can result from mechanical injury to epithelium (Mikolajewicz, Mohammed, Morris, & Komarova, 2018; Yin, Xu, Zhang, Kumar, & Yu, 2007), or immunogenic cell death started by cytotoxic chemotherapy and subsequent responses by tumor targeting lymphocytes (Martins, Tesniere et al., 2009; Martins, Wang et al., 2014). Mechanistically, overexpression of cell surface ectonucleotidase CD39 promotes ATP conversion towards adenosine, thus attenuating ATP mediated "danger signaling" (Imai, Goepfert, Kaczmarek, & Robson, 2000). Indeed, Ghiringhelli and collaborators further demonstrated that oxaliplatin and doxorubicin chemotherapy causes release of ATP derived from dying tumor cells that subsequently activate dendritic cell NLRP1 inflammasome to release proinflammatory IL-1β (Ghiringhelli, Apetoh et al., 2009). Therefore, adenosine acting through the $A_{2A}R$ can tightly regulate proinflammatory mediators to eradicate malignancies, and anti-inflammatory mediators to engage in immunosuppression, to protect against general tissue destruction by injury, genotoxicity or immune cell responses (Stagg & Smyth, 2010).

Interestingly, $A_{2A}R$ activation can have significant regulatory effects on natural killer (NK) and lymphokine activated killer (LAK) cell function. $A_{2A}R$ activation can potentiate immunosuppression by limiting proinflammatory cytokine release (e.g., INF-γ, TNF-α) and effective recognition and elimination of malignancies by NK (Raskovalova, Huang et al., 2005) and LAK (Lokshin, Raskovalova et al., 2006) cells in the tumor microenvironment. Conversely, Young and colleagues found that $A_{2A}R$ deficiency in mice allows full maturation and proliferation of NK cells, resulting in decreased tumor initiation (Young, Ngiow et al., 2018). Not surprisingly, during LPS induced inflammation, adenosine levels can also have an anti-inflammatory effect by acting on $A_{2A}R$ function in antigen presenting dendritic cells in a cAMP-PKA dependent manner, so that these cells preferentially release immunosuppressive IL-10 (Kayhan, Koyas, Akdemir, Savas, & Cekic, 2019), and limit proinflammatory IL-2 release, thus extending immune tolerance (Challier, Bruniquel, Sewell, & Laugel, 2013).

Immunogenic cell death notwithstanding, tumor cells have cleverly circumvented ATP signaled immune cell destruction through upregulation of cell surface expression of CD39 and CD73 ectonucleotidases, both of which respectively convert ATP/ADP and AMP into immunosuppressive adenosine in the tumor microenvironment (Kaur, Weadick et al., 2022).

In the tumor microenvironment, a mechanism involving adenosine receptors on dendritic cells can foster release of immunosuppressive cytokines to prevent tumor destruction by NK cells or lymphocytes as has been reported to occur through the ectonucleotidases CD39 and CD73 on tumors, tumor associated macrophages (TAMs), and T-lymphocytes to promote immunosuppressive adenosine circulation and generation (Jacoberger-Foissac et al., 2023; Montalban de Barrio et al., 2016). Nonetheless, enhancing chemotherapy effectiveness can be achieved by several strategies that reduce adenosine generation, including monoclonal antibody mediated blockade of CD39 and CD73 (Perrot, Michaud et al., 2019), by the antidiabetic medication metformin inhibiting CD39/CD73 function to increase CD8+ T-lymphocytes antitumor activity which prevens adenosine elevation-mediated immunosuppression in the ovarian cancer tumor microenvironment (Li, Wang et al., 2018). Similarly, the CD73 inhibitor AB680 (Jacoberger-Foissac et al., 2023; Lawson, Kalisiak et al., 2020), the CD39 inhibitor POM-1, as well as adenosine deaminase (d'Almeida et al., 2016) are also known to enhance chemotherapy effectiveness.

Despite the fact that anti-inflammatory mechanisms via adenosine at the $A_{2A}R$ may be at play, inhibition of this receptor has shown promise in preventing immunosuppression and increasing effective anticancer therapy in immuno-oncology (Congreve, Brown et al., 2018; Leone, Lo, & Powell, 2015). Indeed, preclinical studies have demonstrated that $A_{2A}R$ antagonists markedly enhance antitumor immunity, tumor vaccines, checkpoint blockade and adoptive T cell therapy (Mediavilla-Varela, Luddy et al., 2013; Ohta, Gorelik et al., 2006; Willingham, Ho et al., 2018). For example, global $A_{2A}R$ deletion ($A_{2A}R^{-/-}$ mice) fostered 60% tumor rejection, and increased survival, effects that were not detected in control mice (Ohta, Gorelik et al., 2006). Accordingly, specific $A_{2A}R$ antagonism with ZM-241385, SCH-58261 or the non-specific antagonist caffeine enhanced the antitumor effects of CD8+ T-lymphocytes in tumor xenograft models (Mediavilla-Varela, Luddy et al., 2013; Ohta, Gorelik et al., 2006). Similarly, potent $A_{2A}R$ blockade with CPI-444 increased IL-2 and INF-γ production in vitro, while blunting tumor growth and extending survival, either alone or in combination with anti-PD-L1 or anti-CTLA-4 antibodies in syngeneic mouse tumor models (Willingham, Ho et al., 2018). Moreover, adoptive immunotherapy transplantation of $A_{2A}R^{-/-}$ T-lymphocytes in tumor bearing mice yielded increased INF-γ secretion, while specific antagonism with the FDA-approved drug Istradefylline (KW-6002) improved adoptive T-lymphocyte immunotherapy, likely in a CREB mediated fashion (Bai, Zhang et al., 2022; Kjaergaard, Hatfield, Jones, Ohta, & Sitkovsky, 2018). Collectively, $A_{2A}R$ antagonists potentially represent the next generation of immune checkpoint inhibition in cancer immunotherapy, which may have far-reaching therapeutic effects on both cancer and CICI (Ohta, 2016). This is important, given that istradefylline was recently demonstrated to prevent cisplatin induced cognitive dysfunctions without affecting antitumor efficacy in a mouse model of chemobrain (Oliveros, Yoo et al., 2022).

## 5.2 Interaction between chemotherapy, adenosine $A_{2A}R$ and chemobrain

The exact mechanism by which chemotherapy acts on adenosine receptors to potentiate chemobrain is not well understood. To this effect, Bednarska-Szczepaniak and colleagues (2019) provide a clue for how $A_{2A}R$ inhibition may prevent aberrant cisplatin uptake by mature and/or developing neurons in the CNS. In their study, the adenosine A1 receptor antagonist PSB-36 was recently shown to sensitize cisplatin's cytotoxic effects in an ovarian cancer cell line known to be resistant to the platinum-based compound, while conversely, application of the specific adenosine $A_{2A}R$ antagonist ZM241385 to this cisplatin-resistant ovarian cancer cell line reduced cellular uptake of cisplatin and attenuated cisplatin-induced cell cycle arrest (Bednarska-Szczepaniak, Krzyzanowski, Klink, & Nowak, 2019). Echoing these findings, increased uptake of adenosine via the equilibrative nucleoside transporter-1 (ENT1) in an ovarian cancer cell line was reported to enhance chemosensitivity to cisplatin in an AMPK mediated fashion, whereas this effect was abrogated by the phosphodiesterase inhibitor dipyridamole (Sureechatchaiyan, Hamacher, Brockmann, Stork, & Kassack, 2018). These studies suggest that increased adenosine presence enhances cisplatin's apoptotic effects, which is good for malignancy eradication outside the CNS, but detrimental to neural stem cells in the brain, hence fostering induction of chemobrain. Further support highlighting the regulatory control that $A_{2A}R$ exerts in preventing chemotherapy induced mitochondrial dysfunction was described by Silva and colleagues, where selective $A_{2A}R$ antagonism by SCH58261 and ZM241385 prevented mitochondrial cytochrome-C release and caspase-3 activation from hippocampal neurons exposed to the antitumor compound Staurosporine (Silva, Porciuncula, Canas, Oliveira, & Cunha, 2007). Other adenosine receptors may also be involved in chemobrain, as one study demonstrated that the selective adenosine $A_3$ receptor ($A_3R$) agonist MRS5980 ameliorates cisplatin induced neurotoxicities, including increases in oxidative stress, mitochondrial dysfunction, cognitive deficits, and neuropathy (Singh, Mahalingam et al., 2022). Therefore, targeting adenosine receptors ($A_{2A}R$, $A_3R$), and nucleoside transporters (ENT1) can similarly provide novel therapeutic strategies to protect neuronal function during chemotherapy.

## 5.3 Adenosine $A_{2A}R$ and chemobrain

Cisplatin is an efficacious platinum-based compound widely used as a frontline therapy to treat cancers of the prostate, ovaries and breast (Balmana, Tung et al., 2014; Dasari & Tchounwou, 2014). Recent evidence shows that this chemotherapy affects cognitive function and neurogenic development in preclinical models of chemobrain (Matsos & Johnston, 2019; Zhou, Kavelaars et al., 2016). In agreement, our group recently revealed that one mechanism by which cisplatin detrimentally affects functional neuronal morphology and cognition is through disruption of nicotinamide adenine dinucleotide (NAD+) metabolic pathways (Yoo, Tang et al., 2021), which importantly, can be ameliorated by nicotinamide mononucleotide (NMN) supplementation. While cisplatin can accumulate in the cortex to exert cognitive impairments (Alexander, Seua et al., 2021; Huo, Reyes, Heijnen, & Kavelaars, 2018; Yoo, Tang et al., 2021), we and others have found particular vulnerability to neuronal cells in the hippocampus, a brain structure known for its control of learning and memory. In chemobrain, the molecular pathways by which cognitive dysfunction occurs is largely unexplored. Therefore, given the prominence for the hippocampus to replenish neuronal populations in its function as a neurogenic niche (Ming & Song, 2011), and in light of the fact that impairments in neurogenesis and neuronal dendrite architecture are pathogenic hallmarks of accelerated aging and cognitive disorders (Hering & Sheng, 2001), shedding light on novel mechanisms by which chemotherapy can impair cognition is of paramount importance. Hence, we posit that activation of cellular stress response mediators, such as adenosine, frequently occurs at the convergence of cancer and chemotherapy.

In the brain, the functional expression of $A_{2A}R$ in the striatum and hippocampus uniquely places this receptor as an important investigative target in the study of how chemobrain detrimentally affects synaptic plasticity, neural stem cell development, memory, and emotive behavior

(Horgusluoglu-Moloch, Nho et al., 2017; Peyton, Oliveros, Choi, & Jang, 2021; Ribeiro, Glaser, Oliveira-Giacomelli, & Ulrich, 2019). Interestingly, preclinical animal models and clinical studies examining the effects of chemotherapy on brain function have reported a stark similarity in expression of molecular markers that underlie accelerated brain aging and chemobrain (Carroll, Van Dyk et al., 2019; Chiang, Huo et al., 2019). Cisplatin targets several corticolimbic structures involved in cognitive function, including the cortex (Alexander, Seua et al., 2021; Huo, Reyes et al., 2018). Our recent study identified disruptions in the adenosine $A_{2A}R$ and its downstream signaling effectors, cAMP, and CREB in the hippocampus, as key contributors in cisplatin-induced chemobrain (Oliveros, Yoo et al., 2022). Notably, we found $A_{2A}R$ in hippocampal excitatory neurons to be uniquely vulnerable to cisplatin insult since we detected specific elevations in hippocampal $A_{2A}R$ concomitant with impaired neurogenesis, increased anxiety-like behavior, as well as dysfunctions in recall memory and spatial memory. Furthermore, specific $A_{2A}R$ inhibition with istradefylline attenuated these detrimental effects (Oliveros, Yoo et al., 2022). Our findings thus prompts the question: can robust elevations in hippocampal $A_{2A}R$ by cisplatin, which are associated with cognitive dysfunctions, be akin to accelerated aging? Indeed, the evidence proposed in this review certainly agrees with this notion, given how increased $A_{2A}R$ is closely associated with accelerated aging in neurodegenerative conditions, such as Alzheimer's disease, and cancer (Carvalho, Faivre et al., 2019; Orr, Hsiao et al., 2015; Temido-Ferreira, Ferreira et al., 2020). In support, selective $A_{2A}R$ inhibition through other antagonists such as SCH58261 or through the nonselective $A_{2A}R$ antagonist caffeine, also enhances synaptic plasticity and spatial working memory in aging and Alzheimer's disease conditions in mice (Kaster, Machado et al., 2015; Reis, Silva et al., 2019; Viana Da Silva Haberl et al., 2016).

Currently, there is a dearth of therapeutics available to treat chemobrain. Given adenosine's increasing prominence in emerging cancer therapies (Boison & Yegutkin, 2019), in conjunction with chemotherapy's inherent ability to potentiate neuronal impairments and cognitive dysfunction resembling accelerated brain aging (Yoo, Tang et al., 2021), it stands to reason that $A_{2A}R$ inhibition may ameliorate cisplatin induced CICI, thus offering a novel therapeutic intervention against this chemobrain-associated condition. Our group's recent findings demonstrate that istradefylline, an FDA approved drug for the treatment of Parkinson's disease (Chen & Cunha, 2020), can be efficacious in cisplatin chemobrain

(Oliveros, Yoo et al., 2022). Istradefylline has been shown to rectify synaptic plasticity, dendritic spine densities and improve cognitive function in a fragile-x-syndrome mouse model (Ferrante, Boussadia et al., 2021). However, other studies also show that $A_{2A}R$ activation can increase neurogenic potential (Ribeiro, Ferreira et al., 2021) while loss of $A_{2A}R$ function impairs cognition (Moscoso-Castro, Lopez-Cano, Gracia-Rubio, Ciruela, & Valverde, 2017), suggesting that pharmacological or genetic neuromodulation of the $A_{2A}R$ may be dependent on downstream transcriptional activities which can alter $A_{2A}R$ abundance to regain homeostatic functionality. Indeed, our results show the neuromodulatory nature of $A_{2A}R$, as istradefylline attenuated cisplatin-induced increases in cAMP, CREB and ERK (unpublished observations) in conjunction with hippocampal cellular and neurogenic improvements (Oliveros, Yoo et al., 2022). This is interesting since increases in CREB and ERK are generally associated with increased neurogenesis, whereas our results show an inverse relationship. However, in support of our findings, pathogenic conditions that generate injury elevate Ras/Raf/ERK with impaired neurogenesis and dendrite spine development, while conversely, inhibition of Ras/Raf/ERK restores these impairments (Chen, Rusnak, Lombroso, & Sidhu, 2009; Xu, Cao, Sun, Liu, & Feng, 2016; Yang, Cao et al., 2013). Nonetheless, it is possible that cisplatin-mediated activation of $A_{2A}R$ and cAMP, promotes a B-RAF and Rap1 mediated compensatory increase in CREB and ERK (Stork & Schmitt, 2002; Takahashi, Li, Dillon, & Stork, 2017). Further studies are required to ascertain the relationship of diminished neurogenesis and increased neuronal CREB or ERK activity that may occur due to cisplatin induced neurotoxic injury. Finally, $A_{2A}R$ antagonism via istradefylline may have broad neuroprotective effects against chemotherapies with non-platinum mechanisms of action (Chang, Chung et al., 2020; Gibson, Nagaraja et al., 2019). However, further studies are warranted to test this hypothesis.

## 6. Concluding remarks

In conclusion, this review attempts to bridge the current gap in knowledge between how adenosine, through $A_{2A}R$ activation, may be a novel constituent of accelerated aging caused by cancer and chemotherapy, thus resulting in cognitive dysfunction. More importantly, we also provide information on the potential therapeutic potential that

$A_{2A}R$ inhibition may confer, not only in chemobrain, but also in cancer treatment. Given that $A_{2A}R$ antagonists are proven to enhance antitumor activity (Congreve, Brown et al., 2018), as well as safely provide neuroprotection in neurodegenerative conditions (Chen & Cunha, 2020), inhibiting $A_{2A}R$ may have far-reaching synergistic effects on cancer treatment and prevention of the development of chemobrain, as well as providing a novel therapeutic strategy in accelerated aging phenotypes.

## Acknowledgments

This work was supported by the NIH (R01CA242158, R01AG058560), Regenerative Medicine Minnesota (RMM091718DS005), and Rutgers Cancer Institute of New Jersey (CINJ) survivorship award to M.H.J., and the NIH (NS103740, NS065957, NS127846) and the US Department of the Army through contract W81XWH2210638 to D.B. Support to AO was provided by the American Association for Cancer Research-Bosarge Family Foundation-Waun Ki Hong Scholar Regenerative Cancer Medicine Award (19-40-60-OLIV) and the Rutgers CINJ Pediatric Cancer and Blood Disorders Research Center.

## Conflict of interest statement

Dr. Detlev Boison is cofounder and CDO of PrevEp Inc. Other authors report no conflicts of interest.

## References

Abbott, R. K., Silva, M., Labuda, J., Thayer, M., Cain, D. W., Philbrook, P., ... Sitkovsky, M. (2017). The GS protein-coupled A2a adenosine receptor controls T cell help in the germinal center. *The Journal of Biological Chemistry, 292*(4), 1211–1217.

ACS. (2019). Cancer treatment and survivorship: Facts and figures 2019-2021. *Atlanta: American Cancer Society, 2019*, 1–48.

Ahles, T. A., Li, Y., McDonald, B. C., Schwartz, G. N., Kaufman, P. A., Tsongalis, G. J., ... Saykin, A. J. (2014). Longitudinal assessment of cognitive changes associated with adjuvant treatment for breast cancer: the impact of APOE and smoking. *Psycho-oncology, 23*(12), 1382–1390.

Ahles, T. A., Saykin, A. J., McDonald, B. C., Furstenberg, C. T., Cole, B. F., Hanscom, B. S., ... Kaufman, P. A. (2008). Cognitive function in breast cancer patients prior to adjuvant treatment. *Breast Cancer Research and Treatment, 110*(1), 143–152.

Ahles, T. A., Saykin, A. J., Noll, W. W., Furstenberg, C. T., Guerin, S., Cole, B., & Mott, L. A. (2003). The relationship of APOE genotype to neuropsychological performance in long-term cancer survivors treated with standard dose chemotherapy. *Psycho-oncology, 12*(6), 612–619.

Albasanz, J. L., Perez, S., Barrachina, M., Ferrer, I., & Martin, M. (2008). Up-regulation of adenosine receptors in the frontal cortex in Alzheimer's disease. *Brain Pathology (Zurich, Switzerland), 18*(2), 211–219.

Alexander, J. F., Mahalingam, R., Seua, A. V., Wu, S., Arroyo, L. D., Horbelt, T., ... Heijnen, C. J. (2022). Targeting the meningeal compartment to resolve chemobrain and neuropathy via nasal delivery of functionalized mitochondria. *Advanced Healthcare Materials, 11*(8) e2102153.

Alexander, J. F., Seua, A. V., Arroyo, L. D., Ray, P. R., Wangzhou, A., Heibeta-Luckemann, L., ... Heijnen, C. J. (2021). Nasal administration of mitochondria reverses chemotherapy-induced cognitive deficits. *Theranostics, 11*(7), 3109–3130.

Alhareeri, A. A., Archer, K. J., Fu, H., Lyon, D. E., Elswick, R. K., Kelly, J. R. D. L., ... Jackson-Cook, C. K. (2020). Telomere lengths in women treated for breast cancer show associations with chemotherapy, pain symptoms, and cognitive domain measures: A longitudinal study. *Breast Cancer Research: BCR, 22*(1), 137.

Amidi, A., Agerbaek, M., Wu, L. M., Pedersen, A. D., Mehlsen, M., Clausen, C. R., ... Zachariae, R. (2017). Changes in cognitive functions and cerebral grey matter and their associations with inflammatory markers, endocrine markers, and APOE genotypes in testicular cancer patients undergoing treatment. *Brain Imaging and Behavior, 11*(3), 769–783.

Amorim, J. A., Coppotelli, G., Rolo, A. P., Palmeira, C. M., Ross, J. M., & Sinclair, D. A. (2022). Mitochondrial and metabolic dysfunction in ageing and age-related diseases. *Nature Reviews Endocrinology, 18*(4), 243–258.

Argyriou, A. A., Assimakopoulos, K., Iconomou, G., Giannakopoulou, F., & Kalofonos, H. P. (2011). Either called "chemobrain" or "chemofog," the long-term chemotherapy-induced cognitive decline in cancer survivors is real. *Journal of Pain and Symptom Management, 41*(1), 126–139.

Armstrong, E., & Boonekamp, J. (2023). Does oxidative stress shorten telomeres in vivo? A meta-analysis. *Ageing Research Reviews, 85*, 101854.

Bai, Y., Zhang, X., Zheng, J., Liu, Z., Yang, Z., & Zhang, X. (2022). Overcoming high level adenosine-mediated immunosuppression by DZD2269, a potent and selective A2aR antagonist. *Journal of Experimental & Clinical Cancer Research: CR, 41*(1), 302.

Bailey, C. J. (2017). Metformin: Historical overview. *Diabetologia, 60*(9), 1566–1576.

Baker, D. J., Wijshake, T., Tchkonia, T., LeBrasseur, N. K., Childs, B. G., Van de Sluis, B., ... Van Deursen, J. M. (2011). Clearance of p16Ink4a-positive senescent cells delays ageing-associated disorders. *Nature, 479*(7372), 232–236.

Balmana, J., Tung, N. M., Isakoff, S. J., Grana, B., Ryan, P. D., Saura, C., ... Garber, J. E. (2014). Phase I trial of olaparib in combination with cisplatin for the treatment of patients with advanced breast, ovarian and other solid tumors. *Annals of Oncology: Official Journal of the European Society for Medical Oncology/ESMO, 25*(8), 1656–1663.

Banks, W. A., Reed, M. J., Logsdon, A. F., Rhea, E. M., & Erickson, M. A. (2021). Healthy aging and the blood-brain barrier. *Nature Aging, 1*(3), 243–254.

Batalha, V. L., Ferreira, D. G., Coelho, J. E., Valadas, J. S., Gomes, R., Temido-Ferreira, M., ... Lopes, L. V. (2016). The caffeine-binding adenosine A2a receptor induces age-like HPA-axis dysfunction by targeting glucocorticoid receptor function. *Scientific Reports, 6*, 31493.

Beckenkamp, L. R., Da Fontoura, D. M. S., Korb, V. G., De Campos, R. P., Onzi, G. R., Iser, I. C., ... Wink, M. R. (2020). Immortalization of mesenchymal stromal cells by TERT affects adenosine metabolism and impairs their immunosuppressive capacity. *Stem Cell Reviews and Reports, 16*(4), 776–791.

Bednarska-Szczepaniak, K., Krzyzanowski, D., Klink, M., & Nowak, M. (2019). Adenosine analogues as opposite modulators of the cisplatin resistance of ovarian cancer cells. *Anti-cancer Agents in Medicinal Chemistry, 19*(4), 473–486.

Blackburn, E. H., Epel, E. S., & Lin, J. (2015). Human telomere biology: A contributory and interactive factor in aging, disease risks, and protection. *Science (New York, N. Y.), 350*(6265), 1193–1198.

Blay, J., White, T. D., & Hoskin, D. W. (1997). The extracellular fluid of solid carcinomas contains immunosuppressive concentrations of adenosine. *Cancer Research, 57*(13), 2602–2605.

Boison, D., & Yegutkin, G. G. (2019). Adenosine metabolism: Emerging concepts for cancer therapy. *Cancer Cell, 36*(6), 582–596.

Brace, K. M., Lee, W. W., Cole, P. D., & Sussman, E. S. (2019). Childhood leukemia survivors exhibit deficiencies in sensory and cognitive processes, as reflected by event-related brain potentials after completion of curative chemotherapy: A preliminary investigation. *Journal of Clinical and Experimental Neuropsychology, 41*(8), 814–831.

Calon, F., Dridi, M., Hornykiewicz, O., Bedard, P. J., Rajput, A. H., & Di Paolo, T. (2004). Increased adenosine A2A receptors in the brain of Parkinson's disease patients with dyskinesias. *Brain, 127*(Pt 5), 1075–1084.

Canas, P. M., Porciuncula, L. O., Cunha, G. M., Silva, C. G., Machado, N. J., Oliveira, J. M., ... Cunha, R. A. (2009). Adenosine A2A receptor blockade prevents synaptotoxicity and memory dysfunction caused by beta-amyloid peptides via p38 mitogen-activated protein kinase pathway. *The Journal of Neuroscience, 29*(47), 14741–14751.

Cardoso, C. V., De Barros, M. P., Bachi, A. L. L., Bernardi, M. M., Kirsten, T. B., De Fatima Monteiro Martins, M., ... Bondan, E. F. (2020). Chemobrain in rats: Behavioral, morphological, oxidative and inflammatory effects of doxorubicin administration. *Behavioural Brain Research, 378*, 112233.

Carroll, J. E., Bower, J. E., & Ganz, P. A. (2022). Cancer-related accelerated ageing and biobehavioural modifiers: A framework for research and clinical care. *Nature Reviews Clinical Oncology, 19*(3), 173–187.

Carroll, J. E., Van Dyk, K., Bower, J. E., Scuric, Z., Petersen, L., Schiestl, R., ... Ganz, P. A. (2019). Cognitive performance in survivors of breast cancer and markers of biological aging. *Cancer, 125*(2), 298–306.

Carvalho, K., Faivre, E., Pietrowski, M. J., Marques, X., Gomez-Murcia, V., Deleau, A., ... Blum, D. (2019). Exacerbation of C1q dysregulation, synaptic loss and memory deficits in tau pathology linked to neuronal adenosine A2A receptor. *Brain, 142*(11), 3636–3654.

Challier, J., Bruniquel, D., Sewell, A. K., & Laugel, B. (2013). Adenosine and cAMP signalling skew human dendritic cell differentiation towards a tolerogenic phenotype with defective CD8(+) T-cell priming capacity. *Immunology, 138*(4), 402–410.

Chang, A., Chung, N. C., Lawther, A. J., Ziegler, A. I., Shackleford, D. M., Sloan, E. K., & Walker, A. K. (2020). The anti-inflammatory drug aspirin does not protect against chemotherapy-induced memory impairment by paclitaxel in mice. *Frontiers in Oncology, 10*, 564965.

Chen, J., Rusnak, M., Lombroso, P. J., & Sidhu, A. (2009). Dopamine promotes striatal neuronal apoptotic death via ERK signaling cascades. *The European Journal of Neuroscience, 29*(2), 287–306.

Chen, J. F. (2014). Adenosine receptor control of cognition in normal and disease. *International Review of Neurobiology, 119*, 257–307.

Chen, J. F., Choi, D. S., & Cunha, R. A. (2023). Striatopallidal adenosine A(2A) receptor modulation of goal-directed behavior: Homeostatic control with cognitive flexibility. *Neuropharmacology, 226*, 109421.

Chen, J. F., & Cunha, R. A. (2020). The belated US FDA approval of the adenosine A2A receptor antagonist istradefylline for treatment of Parkinson's disease. *Purinergic Signalling, 16*(2), 167–174.

Chen, J. F., Lee, C. F., & Chern, Y. (2014). Adenosine receptor neurobiology: Overview. *International Review of Neurobiology, 119*, 1–49.

Chiang, A. C. A., Huo, X., Kavelaars, A., & Heijnen, C. J. (2019). Chemotherapy accelerates age-related development of tauopathy and results in loss of synaptic integrity and cognitive impairment. *Brain, Behavior, and Immunity, 79*, 319–325.

Chiu, G. S., Boukelmoune, N., Chiang, A. C. A., Peng, B., Rao, V., Kingsley, C., ... Heijnen, C. J. (2018). Nasal administration of mesenchymal stem cells restores cisplatin-induced cognitive impairment and brain damage in mice. *Oncotarget, 9*(85), 35581–35597.

Chiu, G. S., Maj, M. A., Rizvi, S., Dantzer, R., Vichaya, E. G., Laumet, G., ... Heijnen, C. J. (2017). Pifithrin-mu prevents cisplatin-induced chemobrain by preserving neuronal mitochondrial function. *Cancer Research, 77*(3), 742–752.

Cho, C. H., Yoo, K. H., Oliveros, A., Paulson, S., Hussaini, S. M. Q., Van Deursen, J. M., & Jang, M. H. (2019). sFRP3 inhibition improves age-related cellular changes in BubR1 progeroid mice. *Aging Cell, 18*(2) e12899.

Cimprich, B., Reuter-Lorenz, P., Nelson, J., Clark, P. M., Therrien, B., Normolle, D., ... Welsh, R. C. (2010). Prechemotherapy alterations in brain function in women with breast cancer. *Journal of Clinical and Experimental Neuropsychology, 32*(3), 324–331.

Cole, P. D., Finkelstein, Y., Stevenson, K. E., Blonquist, T. M., Vijayanathan, V., Silverman, L. B., ... Waber, D. P. (2015). Polymorphisms in genes related to oxidative stress are associated with inferior cognitive function after therapy for childhood acute lymphoblastic leukemia. *Journal of Clinical Oncology: Official Journal of the American Society of Clinical Oncology, 33*(19), 2205–2211.

Congreve, M., Brown, G. A., Borodovsky, A., & Lamb, M. L. (2018). Targeting adenosine A2A receptor antagonism for treatment of cancer. *Expert Opinion on Drug Discovery, 13*(11), 997–1003.

Corujo-Ramirez, A. M., Dua, M., Yoo, K. H., Oliveros, A., & Jang, M. H. (2020). Genetic inhibition of sFRP3 prevents glial reactivity in a mouse model of accelerated aging. *International Neurourology Journal, 24*(Suppl 2), 72–78.

Csernansky, J. G., Dong, H., Fagan, A. M., Wang, L., Xiong, C., Holtzman, D. M., & Morris, J. C. (2006). Plasma cortisol and progression of dementia in subjects with Alzheimer-type dementia. *The American Journal of Psychiatry, 163*(12), 2164–2169.

Cupit-Link, M. C., Kirkland, J. L., Ness, K. K., Armstrong, G. T., Tchkonia, T., LeBrasseur, N. K., ... Hashmi, S. K. (2017). Biology of premature ageing in survivors of cancer. *ESMO Open, 2*(5), e000250.

d'Almeida, S. M., Kauffenstein, G., Roy, C., Bassett, L., Papargyris, L., Henrion, D., ... Tabiasco, J. (2016). The ecto-ATPDase CD39 is involved in the acquisition of the immunoregulatory phenotype by M-CSF-macrophages and ovarian cancer tumor-associated macrophages: Regulatory role of IL-27. *Oncoimmunology, 5*(7), e1178025.

Da Silva, S. V., Haberl, M. G., Zhang, P., Bethge, P., Lemos, C., Goncalves, N., ... Mulle, C. (2016). Early synaptic deficits in the APP/PS1 mouse model of Alzheimer's disease involve neuronal adenosine A(2A) receptors. *Nature Communications, 7*, 11915.

Dasari, S., & Tchounwou, P. B. (2014). Cisplatin in cancer therapy: Molecular mechanisms of action. *European Journal of Pharmacology, 740*, 364–378.

De Ruiter, M. B., Reneman, L., Boogerd, W., Veltman, D. J., van Dam, F. S., Nederveen, A. J., ... Schagen, S. B. (2011). Cerebral hyporesponsiveness and cognitive impairment 10 years after chemotherapy for breast cancer. *Human Brain Mapping, 32*(8), 1206–1219.

Demaria, M., O'Leary, M. N., Chang, J., Shao, L., Liu, S., Alimirah, F., ... Campisi, J. (2017). Cellular senescence promotes adverse effects of chemotherapy and cancer relapse. *Cancer Discovery, 7*(2), 165–176.

Demby, T. C., Rodriguez, O., McCarthy, C. W., Lee, Y. C., Albanese, C., Mandelblatt, J., & Rebeck, G. W. (2020). A mouse model of chemotherapy-related cognitive impairments integrating the risk factors of aging and APOE4 genotype. *Behavioural Brain Research, 384*, 112534.

Dewaeles, E., Carvalho, K., Fellah, S., Sim, J., Boukrout, N., Caillierez, R., ... Cauffiez, C. (2022). Istradefylline protects from cisplatin-induced nephrotoxicity and peripheral neuropathy while preserving cisplatin antitumor effects. *The Journal of Clinical Investigation, 132*(22), e152924.

Dewar, E. O., Ahn, C., Eraj, S., Mahal, B. A., & Sanford, N. N. (2021). Psychological distress and cognition among long-term survivors of adolescent and young adult cancer in the USA. *Journal of Cancer Survivorship, 15*(5), 776–784.

Dias, L., Madeira, D., Dias, R., Tome, A. R., Cunha, R. A., & Agostinho, P. (2022). Abeta (1-42) peptides blunt the adenosine A(2A) receptor-mediated control of the interplay between P(2)X(7) and P(2)Y(1) receptors mediated calcium responses in astrocytes. *Cellular and Molecular Life Sciences: CMLS, 79*(8), 457.

Falleti, M. G., Sanfilippo, A., Maruff, P., Weih, L., & Phillips, K. A. (2005). The nature and severity of cognitive impairment associated with adjuvant chemotherapy in women with breast cancer: A meta-analysis of the current literature. *Brain and Cognition, 59*(1), 60–70.

Fernandez, H. R., Varma, A., Flowers, S. A., & Rebeck, G. W. (2020). Cancer chemotherapy related cognitive impairment and the impact of the Alzheimer's disease risk factor APOE. *Cancers (Basel), 12*(12), 3842.

Ferrante, A., Boussadia, Z., Borreca, A., Mallozzi, C., Pedini, G., Pacini, L., ... Martire, A. (2021). Adenosine A2A receptor inhibition reduces synaptic and cognitive hippocampal alterations in Fmr1 KO mice. *Translational Psychiatry, 11*(1), 112.

Ferre, S., Sarasola, L. I., Quiroz, C., & Ciruela, F. (2023). Presynaptic adenosine receptor heteromers as key modulators of glutamatergic and dopaminergic neurotransmission in the striatum. *Neuropharmacology, 223*, 109329.

Ferreira, D. G., Temido-Ferreira, M., Vicente Miranda, H., Batalha, V. L., Coelho, J. E., Szego, E. M., ... Outeiro, T. F. (2017). Alpha-synuclein interacts with PrP(C) to induce cognitive impairment through mGluR5 and NMDAR2B. *Nature Neuroscience, 20*(11), 1569–1579.

Fishman, P., Bar-Yehuda, S., Ohana, G., Pathak, S., Wasserman, L., Barer, F., & Multani, A. S. (2000). Adenosine acts as an inhibitor of lymphoma cell growth: A major role for the A3 adenosine receptor. *European Journal of Cancer, 36*(11), 1452–1458.

George, R. P., Semendric, I., Hutchinson, M. R., & Whittaker, A. L. (2021). Neuroimmune reactivity marker expression in rodent models of chemotherapy-induced cognitive impairment: A systematic scoping review. *Brain, Behavior, and Immunity, 94*, 392–409.

Geraghty, A. C., Gibson, E. M., Ghanem, R. A., Greene, J. J., Ocampo, A., Goldstein, A. K., ... Monje, M. (2019). Loss of adaptive myelination contributes to methotrexate chemotherapy-related cognitive impairment. *Neuron, 103*(2), 250–265.e258.

Ghiringhelli, F., Apetoh, L., Tesniere, A., Aymeric, L., Ma, Y., Ortiz, C., ... Zitvogel, L. (2009). Activation of the NLRP3 inflammasome in dendritic cells induces IL-1beta-dependent adaptive immunity against tumors. *Nature Medicine, 15*(10), 1170–1178.

Gibson, E. M., & Monje, M. (2019). Emerging mechanistic underpinnings and therapeutic targets for chemotherapy-related cognitive impairment. *Current Opinion in Oncology, 31*(6), 531–539.

Gibson, E. M., Nagaraja, S., Ocampo, A., Tam, L. T., Wood, L. S., Pallegar, P. N., ... Monje, M. (2019). Methotrexate chemotherapy induces persistent tri-glial dysregulation that underlies chemotherapy-related cognitive impairment. *Cell, 176*(1–2), 43–55.e13.

Goncalves, F. Q., Lopes, J. P., Silva, H. B., Lemos, C., Silva, A. C., Goncalves, N., ... Cunha, R. A. (2019). Synaptic and memory dysfunction in a beta-amyloid model of early Alzheimer's disease depends on increased formation of ATP-derived extracellular adenosine. *Neurobiology of Disease, 132*, 104570.

Gorgoulis, V., Adams, P. D., Alimonti, A., Bennett, D. C., Bischof, O., Bishop, C., ... Demaria, M. (2019). Cellular senescence: Defining a path forward. *Cell, 179*(4), 813–827.

Green, K. N., Billings, L. M., Roozendaal, B., McGaugh, J. L., & LaFerla, F. M. (2006). Glucocorticoids increase amyloid-beta and tau pathology in a mouse model of Alzheimer's disease. *The Journal of Neuroscience, 26*(35), 9047–9056.

Guerrero, A., De Strooper, B., & Arancibia-Carcamo, I. L. (2021). Cellular senescence at the crossroads of inflammation and Alzheimer's disease. *Trends in Neurosciences, 44*(9), 714–727.

Gyau, B. B., & Deaglio, S. (2023). A(2A) receptor signaling drives cisplatin-mediated hippocampal neurotoxicity and cognitive defects in mice. *Purinergic Signalling.* https://doi.org/10.1007/s11302-023-09919-0.

Heflin, L. H., Meyerowitz, B. E., Hall, P., Lichtenstein, P., Johansson, B., Pedersen, N. L., & Gatz, M. (2005). Cancer as a risk factor for long-term cognitive deficits and dementia. *Journal of the National Cancer Institute, 97*(11), 854–856.

Hering, H., & Sheng, M. (2001). Dendritic spines: Structure, dynamics and regulation. *Nature Reviews. Neuroscience, 2*(12), 880–888.

Hoogland, A. I., Nelson, A. M., Gonzalez, B. D., Small, B. J., Breen, E. C., Sutton, S. K., ... Jim, H. S. L. (2019). Worsening cognitive performance is associated with increases in systemic inflammation following hematopoietic cell transplantation. *Brain, Behavior, and Immunity, 80,* 308–314.

Horgusluoglu-Moloch, E., Nho, K., Risacher, S. L., Kim, S., Foroud, T., Shaw, L. M., ... Saykin, A. J. & I. Alzheimer's Disease Neuroimaging. (2017). Targeted neurogenesis pathway-based gene analysis identifies ADORA2A associated with hippocampal volume in mild cognitive impairment and Alzheimer's disease. *Neurobiology of Aging, 60,* 92–103.

Hu, Y., Zhou, Y., Yang, Y., Tang, H., Si, Y., Chen, Z., ... Fang, H. (2022). Metformin protects against diabetes-induced cognitive dysfunction by inhibiting mitochondrial fission protein DRP1. *Frontiers in Pharmacology, 13,* 832707.

Huo, X., Reyes, T. M., Heijnen, C. J., & Kavelaars, A. (2018). Cisplatin treatment induces attention deficits and impairs synaptic integrity in the prefrontal cortex in mice. *Scientific Reports, 8*(1), 17400.

Imai, M., Goepfert, C., Kaczmarek, E., & Robson, S. C. (2000). CD39 modulates IL-1 release from activated endothelial cells. *Biochemical and Biophysical Research Communications, 270*(1), 272–278.

Jacoberger-Foissac, C., Cousineau, I., Bareche, Y., Allard, D., Chrobak, P., Allard, B., ... Stagg, J. (2023). CD73 inhibits cGAS-STING and cooperates with CD39 to promote pancreatic cancer. *Cancer Immunology Research, 11*(1), 56–71.

Jia, L., Zhou, Y., Ma, L., Li, W., Chan, C., Zhang, S., & Zhao, Y. (2023). Inhibition of NLRP3 alleviated chemotherapy-induced cognitive impairment in rats. *Neuroscience Letters, 793,* 136975.

Joshi, G., Aluise, C. D., Cole, M. P., Sultana, R., Pierce, W. M., Vore, M., ... Butterfield, D. A. (2010). Alterations in brain antioxidant enzymes and redox proteomic identification of oxidized brain proteins induced by the anti-cancer drug adriamycin: Implications for oxidative stress-mediated chemobrain. *Neuroscience, 166*(3), 796–807.

Kaster, M. P., Machado, N. J., Silva, H. B., Nunes, A., Ardais, A. P., Santana, M., ... Cunha, R. A. (2015). Caffeine acts through neuronal adenosine A2A receptors to prevent mood and memory dysfunction triggered by chronic stress. *Proceedings of the National Academy of Sciences of the United States of America, 112*(25), 7833–7838.

Kaur, T., Weadick, B., Mace, T. A., Desai, K., Odom, H., & Govindarajan, R. (2022). Nucleoside transporters and immunosuppressive adenosine signaling in the tumor microenvironment: Potential therapeutic opportunities. *Pharmacology & Therapeutics, 240,* 108300.

Kayhan, M., Koyas, A., Akdemir, I., Savas, A. C., & Cekic, C. (2019). Adenosine receptor signaling targets both PKA and Epac pathways to polarize dendritic cells to a suppressive phenotype. *Journal of Immunology, 203*(12), 3247–3255.

Kesler, S. R., Kent, J. S., & O'Hara, R. (2011). Prefrontal cortex and executive function impairments in primary breast cancer. *Archives of Neurology, 68*(11), 1447–1453.

Kesler, S. R., Rao, V., Ray, W. J., & Rao, A. I. Alzheimer's Disease Neuroimaging. (2017). Probability of Alzheimer's disease in breast cancer survivors based on gray-matter structural network efficiency. *Alzheimer's & Dementia (Amst), 9,* 67–75.

Kjaergaard, J., Hatfield, S., Jones, G., Ohta, A., & Sitkovsky, M. (2018). A2A adenosine receptor gene deletion or synthetic A2A antagonist liberate tumor-reactive CD8(+) T cells from tumor-induced immunosuppression. *Journal of Immunology, 201*(2), 782–791.

Knox, E. G., Aburto, M. R., Clarke, G., Cryan, J. F., & O'Driscoll, C. M. (2022). The blood-brain barrier in aging and neurodegeneration. *Molecular Psychiatry, 27*(6), 2659–2673.

Koppelmans, V., Breteler, M. M., Boogerd, W., Seynaeve, C., Gundy, C., & Schagen, S. B. (2012). Neuropsychological performance in survivors of breast cancer more than 20 years after adjuvant chemotherapy. *Journal of Clinical Oncology: Official Journal of the American Society of Clinical Oncology, 30*(10), 1080–1086.

Kulkarni, A. S., Gubbi, S., & Barzilai, N. (2020). Benefits of metformin in attenuating the hallmarks of aging. *Cell Metabolism, 32*(1), 15–30.

Lacourt, T. E., & Heijnen, C. J. (2017). Mechanisms of neurotoxic symptoms as a result of breast cancer and its treatment: Considerations on the contribution of stress, inflammation, and cellular bioenergetics. *Current Breast Cancer Reports, 9*(2), 70–81.

Laird, B. J., McMillan, D. C., Fayers, P., Fearon, K., Kaasa, S., Fallon, M. T., & Klepstad, P. (2013). The systemic inflammatory response and its relationship to pain and other symptoms in advanced cancer. *The Oncologist, 18*(9), 1050–1055.

Lange, M., Joly, F., Vardy, J., Ahles, T., Dubois, M., Tron, L., ... Castel, H. (2019). Cancer-related cognitive impairment: An update on state of the art, detection, and management strategies in cancer survivors. *Annals of Oncology: Official Journal of the European Society for Medical Oncology / ESMO, 30*(12), 1925–1940.

Lange, M., Licaj, I., Clarisse, B., Humbert, X., Grellard, J. M., Tron, L., & Joly, F. (2019). Cognitive complaints in cancer survivors and expectations for support: Results from a web-based survey. *Cancer Medicine, 8*(5), 2654–2663.

Laurent, C., Burnouf, S., Ferry, B., Batalha, V. L., Coelho, J. E., Baqi, Y., ... Blum, D. (2016). A2A adenosine receptor deletion is protective in a mouse model of tauopathy. *Molecular Psychiatry, 21*(1), 97–107.

Lawson, K. V., Kalisiak, J., Lindsey, E. A., Newcomb, E. T., Leleti, M. R., Debien, L., ... Powers, J. P. (2020). Discovery of AB680: A potent and selective inhibitor of CD73. *Journal of Medicinal Chemistry, 63*(20), 11448–11468.

Leone, R. D., Lo, Y. C., & Powell, J. D. (2015). A2aR antagonists: Next generation checkpoint blockade for cancer immunotherapy. *Computational and Structural Biotechnology Journal, 13*, 265–272.

Li, L., Wang, L., Li, J., Fan, Z., Yang, L., Zhang, Z., ... Zhang, Y. (2018). Metformin-induced reduction of CD39 and CD73 blocks myeloid-derived suppressor cell activity in patients with ovarian cancer. *Cancer Research, 78*(7), 1779–1791.

Li, W., Silva, H. B., Real, J., Wang, Y. M., Rial, D., Li, P., ... Chen, J. F. (2015). Inactivation of adenosine A2A receptors reverses working memory deficits at early stages of Huntington's disease models. *Neurobiology of Disease, 79*, 70–80.

Li, Z., Chen, X., Wang, T., Gao, Y., Li, F., Chen, L., ... Chen, J. F. (2018). The corticostriatal adenosine A2A receptor controls maintenance and retrieval of spatial working memory. *Biological Psychiatry, 83*(6), 530–541.

Lindqvist, D., Epel, E. S., Mellon, S. H., Penninx, B. W., Revesz, D., Verhoeven, J. E., ... Wolkowitz, O. M. (2015). Psychiatric disorders and leukocyte telomere length: Underlying mechanisms linking mental illness with cellular aging. *Neuroscience and Biobehavioral Reviews, 55*, 333–364.

Liu, Y. J., Chen, J., Li, X., Zhou, X., Hu, Y. M., Chu, S. F., ... Chen, N. H. (2019). Research progress on adenosine in central nervous system diseases. *CNS Neuroscience & Therapeutics, 25*(9), 899–910.

Lofts, A., Abu-Hijleh, F., Rigg, N., Mishra, R. K., & Hoare, T. (2022). Using the intranasal route to administer drugs to treat neurological and psychiatric illnesses: Rationale, successes, and future needs. *CNS Drugs, 36*(7), 739–770.

Lokshin, A., Raskovalova, T., Huang, X., Zacharia, L. C., Jackson, E. K., & Gorelik, E. (2006). Adenosine-mediated inhibition of the cytotoxic activity and cytokine production by activated natural killer cells. *Cancer Research, 66*(15), 7758–7765.

Lomeli, N., Di, K., Czerniawski, J., Guzowski, J. F., & Bota, D. A. (2017). Cisplatin-induced mitochondrial dysfunction is associated with impaired cognitive function in rats. *Free Radical Biology & Medicine, 102*, 274–286.

Lopes, C. R., Cunha, R. A., & Agostinho, P. (2021). Astrocytes and adenosine A(2A) receptors: Active players in Alzheimer's disease. *Frontiers in Neuroscience, 15*, 666710.

Lopes, L. V., Cunha, R. A., & Ribeiro, J. A. (1999). Increase in the number, G protein coupling, and efficiency of facilitatory adenosine A2A receptors in the limbic cortex, but not striatum, of aged rats. *Journal of Neurochemistry, 73*(4), 1733–1738.

Lopez-Otin, C., Blasco, M. A., Partridge, L., Serrano, M., & Kroemer, G. (2022). Hallmarks of aging: An expanding universe. *Cell, 186*, 243–278.

Ma, J., Huo, X., Jarpe, M. B., Kavelaars, A., & Heijnen, C. J. (2018). Pharmacological inhibition of HDAC6 reverses cognitive impairment and tau pathology as a result of cisplatin treatment. *Acta Neuropathologica Communications, 6*(1), 103.

Martins, I., Tesniere, A., Kepp, O., Michaud, M., Schlemmer, F., Senovilla, L., ... Kroemer, G. (2009). Chemotherapy induces ATP release from tumor cells. *Cell Cycle (Georgetown, Tex.), 8*(22), 3723–3728.

Martins, I., Wang, Y., Michaud, M., Ma, Y., Sukkurwala, A. Q., Shen, S., ... Kroemer, G. (2014). Molecular mechanisms of ATP secretion during immunogenic cell death. *Cell Death and Differentiation, 21*(1), 79–91.

Matsos, A., & Johnston, I. N. (2019). Chemotherapy-induced cognitive impairments: A systematic review of the animal literature. *Neuroscience and Biobehavioral Reviews, 102*, 382–399.

Matsos, A., Loomes, M., Zhou, I., Macmillan, E., Sabel, I., Rotziokos, E., ... Johnston, I. N. (2017). Chemotherapy-induced cognitive impairments: White matter pathologies. *Cancer Treatment Reviews, 61*, 6–14.

McDonald, B. C., Conroy, S. K., Ahles, T. A., West, J. D., & Saykin, A. J. (2010). Gray matter reduction associated with systemic chemotherapy for breast cancer: A prospective MRI study. *Breast Cancer Research and Treatment, 123*(3), 819–828.

Mediavilla-Varela, M., Luddy, K., Noyes, D., Khalil, F. K., Neuger, A. M., Soliman, H., & Antonia, S. J. (2013). Antagonism of adenosine A2A receptor expressed by lung adenocarcinoma tumor cells and cancer associated fibroblasts inhibits their growth. *Cancer Biology & Therapy, 14*(9), 860–868.

Merighi, S., Borea, P. A., Varani, K., Vincenzi, F., Jacobson, K. A., & Gessi, S. (2022). A(2A) adenosine receptor antagonists in neurodegenerative diseases. *Current Medicinal Chemistry, 29*(24), 4138–4151.

Mikolajewicz, N., Mohammed, A., Morris, M., & Komarova, S. V. (2018). Mechanically stimulated ATP release from mammalian cells: Systematic review and meta-analysis. *Journal of Cell Science, 131*(22), jcs223354.

Ming, G. L., & Song, H. (2011). Adult neurogenesis in the mammalian brain: Significant answers and significant questions. *Neuron, 70*(4), 687–702.

Mogavero, M. P., Bruni, O., DelRosso, L. M., & Ferri, R. (2020). Neurodevelopmental consequences of pediatric cancer and its treatment: The role of sleep. *Brain Sciences, 10*(7), 411.

Monje, M. L., Vogel, H., Masek, M., Ligon, K. L., Fisher, P. G., & Palmer, T. D. (2007). Impaired human hippocampal neurogenesis after treatment for central nervous system malignancies. *Annals of Neurology, 62*(5), 515–520.

Montalban de Barrio, I., Penski, C., Schlahsa, L., Stein, R. G., Diessner, J., Wöchel, A., ... Häusler, S. F. M. (2016). Adenosine-generating ovarian cancer cells attract myeloid cells which differentiate into adenosine-generating tumor associated macrophages — a self amplifying, CD39- and CD73-dependent mechanism for tumor immune escape. *Journal for ImmunoTherapy of Cancer, 4*(49).

Moscoso-Castro, M., Lopez-Cano, M., Gracia-Rubio, I., Ciruela, F., & Valverde, O. (2017). Cognitive impairments associated with alterations in synaptic proteins induced by the genetic loss of adenosine A2A receptors in mice. *Neuropharmacology, 126*, 48–57.

Nam, H. W., Bruner, R. C., & Choi, D. S. (2013). Adenosine signaling in striatal circuits and alcohol use disorders. *Molecules and Cells, 36*(3), 195–202.

Ness, K. K., Armstrong, G. T., Kundu, M., Wilson, C. L., Tchkonia, T., & Kirkland, J. L. (2015). Frailty in childhood cancer survivors. *Cancer, 121*(10), 1540–1547.

Ng, C. A. S., Biran, L. P., Galvano, E., Mandelblatt, J., Vicini, S., & Rebeck, G. W. (2022). Chemotherapy promotes astrocytic response to Abeta deposition, but not Abeta levels, in a mouse model of amyloid and APOE. *Neurobiology of Disease, 175*, 105915.

Nguyen, L. D., & Ehrlich, B. E. (2020). Cellular mechanisms and treatments for chemobrain: Insight from aging and neurodegenerative diseases. *EMBO Molecular Medicine, 12*(6), e12075.

North, B. J., Rosenberg, M. A., Jeganathan, K. B., Hafner, A. V., Michan, S., Dai, J., ... Sinclair, D. A. (2014). SIRT2 induces the checkpoint kinase BubR1 to increase lifespan. *The EMBO Journal, 33*(13), 1438–1453.

Ohta, A. (2016). A metabolic immune checkpoint: Adenosine in tumor microenvironment. *Frontiers in Immunology, 7*, 109.

Ohta, A., Gorelik, E., Prasad, S. J., Ronchese, F., Lukashev, D., Wong, M. K., ... Sitkovsky, M. (2006). A2A adenosine receptor protects tumors from antitumor T cells. *Proceedings of the National Academy of Sciences of the United States of America, 103*(35), 13132–13137.

Oliveros, A., Yoo, K. H., Rashid, M. A., Corujo-Ramirez, A., Hur, B., Sung, J., ... Jang, M. H. (2022). Adenosine A2A receptor blockade prevents cisplatin-induced impairments in neurogenesis and cognitive function. *Proceedings of the National Academy of Sciences of the United States of America, 119*(28) e2206415119.

Onzi, G. R., D'Agustini, N., Garcia, S. C., Guterres, S. S., Pohlmann, P. R., Rosa, D. D., & Pohlmann, A. R. (2022). Chemobrain in breast cancer: Mechanisms, clinical manifestations, and potential interventions. *Drug Safety: An International Journal of Medical Toxicology and Drug Experience, 45*(6), 601–621.

Orr, A. G., Hsiao, E. C., Wang, M. M., Ho, K., Kim, D. H., Wang, X., ... Mucke, L. (2015). Astrocytic adenosine receptor A2A and Gs-coupled signaling regulate memory. *Nature Neuroscience, 18*(3), 423–434.

Orr, A. G., Lo, I., Schumacher, H., Ho, K., Gill, M., Guo, W., ... Mucke, L. (2018). Istradefylline reduces memory deficits in aging mice with amyloid pathology. *Neurobiology of Disease, 110*, 29–36.

Orr, A. G., Orr, A. L., Li, X. J., Gross, R. E., & Traynelis, S. F. (2009). Adenosine A(2A) receptor mediates microglial process retraction. *Nature Neuroscience, 12*(7), 872–878.

Parish, S. T., Kim, S., Sekhon, R. K., Wu, J. E., Kawakatsu, Y., & Effros, R. B. (2010). Adenosine deaminase modulation of telomerase activity and replicative senescence in human CD8 T lymphocytes. *Journal of Immunology, 184*(6), 2847–2854.

Perrot, I., Michaud, H. A., Giraudon-Paoli, M., Augier, S., Docquier, A., Gros, L., ... Bonnefoy, N. (2019). Blocking antibodies targeting the CD39/CD73 immunosuppressive pathway unleash immune responses in combination cancer therapies. *Cell Reports, 27*(8), 2411–2425 e2419.

Peyton, L., Oliveros, A., Choi, D. S., & Jang, M. H. (2021). Hippocampal regenerative medicine: Neurogenic implications for addiction and mental disorders. *Experimental & Molecular Medicine, 53*(3), 358–368.

Rashid, M. A., Oliveros, A., Kim, Y. S., & Jang, M. H. (2022). Nicotinamide mononucleotide prevents cisplatin-induced mitochondrial defects in cortical neurons derived from human induced pluripotent stem cells. *Brain Plasticity, 8*(2), 143–152.

Raskovalova, T., Huang, X., Sitkovsky, M., Zacharia, L. C., Jackson, E. K., & Gorelik, E. (2005). Gs protein-coupled adenosine receptor signaling and lytic function of activated NK cells. *Journal of Immunology, 175*(7), 4383–4391.

Rebola, N., Lujan, R., Cunha, R. A., & Mulle, C. (2008). Adenosine A2A receptors are essential for long-term potentiation of NMDA-EPSCs at hippocampal mossy fiber synapses. *Neuron, 57*(1), 121–134.

Rebola, N., Rodrigues, R. J., Lopes, L. V., Richardson, P. J., Oliveira, C. R., & Cunha, R. A. (2005). Adenosine A1 and A2A receptors are co-expressed in pyramidal neurons and co-localized in glutamatergic nerve terminals of the rat hippocampus. *Neuroscience, 133*(1), 79–83.

Reis, S. L., Silva, H. B., Almeida, M., Cunha, R. A., Simoes, A. P., & Canas, P. M. (2019). Adenosine A1 and A2A receptors differently control synaptic plasticity in the mouse dorsal and ventral hippocampus. *Journal of Neurochemistry, 151*(2), 227–237.

Ren, X., Keeney, J. T. R., Miriyala, S., Noel, T., Powell, D. K., Chaiswing, L., ... Butterfield, D. A. (2019). The triangle of death of neurons: Oxidative damage, mitochondrial dysfunction, and loss of choline-containing biomolecules in brains of mice treated with doxorubicin. Advanced insights into mechanisms of chemotherapy induced cognitive impairment ("chemobrain") involving TNF-alpha. *Free Radical Biology & Medicine, 134*, 1–8.

Ren, X., Clair, D. K. S. T., & Butterfield, D. A. (2017). Dysregulation of cytokine mediated chemotherapy induced cognitive impairment. *Pharmacological Research: The Official Journal of the Italian Pharmacological Society, 117*, 267–273.

Ribeiro, D. E., Glaser, T., Oliveira-Giacomelli, A., & Ulrich, H. (2019). Purinergic receptors in neurogenic processes. *Brain Research Bulletin, 151*, 3–11.

Ribeiro, F. F., Ferreira, F., Rodrigues, R. S., Soares, R., Pedro, D. M., Duarte-Samartinho, M., ... Xapelli, S. (2021). Regulation of hippocampal postnatal and adult neurogenesis by adenosine A2A receptor: Interaction with brain-derived neurotrophic factor. *Stem Cells, 39*, 1362–1381.

Rodrigues, R. J., Marques, J. M., & Cunha, R. A. (2019). Purinergic signalling and brain development. *Seminars in Cell & Developmental Biology, 95*, 34–41.

Rodrigues, R. J., Tome, A. R., & Cunha, R. A. (2015). ATP as a multi-target danger signal in the brain. *Frontiers in Neuroscience, 9*, 148.

Roichman, A., Elhanati, S., Aon, M. A., Abramovich, I., Di Francesco, A., Shahar, Y., ... Cohen, H. Y. (2021). Restoration of energy homeostasis by SIRT6 extends healthy lifespan. *Nature Communications, 12*(1), 3208.

Romaniuk, A., Paszel-Jaworska, A., Toton, E., Lisiak, N., Holysz, H., Krolak, A., ... Rubis, B. (2019). The non-canonical functions of telomerase: To turn off or not to turn off. *Molecular Biology Reports, 46*(1), 1401–1411.

Rummel, N. G., Chaiswing, L., Bondada, S., St Clair, D. K., & Butterfield, D. A. (2021). Chemotherapy-induced cognitive impairment: Focus on the intersection of oxidative stress and TNFalpha. *Cellular and Molecular Life Sciences: CMLS, 78*(19–20), 6533–6540.

Sahu, K., Langeh, U., Singh, C., & Singh, A. (2021). Crosstalk between anticancer drugs and mitochondrial functions. *Current Research in Pharmacology and Drug Discovery, 2*, 100047.

Salem, A. A., Haty, I. A. E. L., & Ghattas, M. A. (2022). GW-2974 and SCH-442416 modulators of tyrosine kinase and adenosine receptors can also stabilize human telomeric G-quadruplex DNA. *PLoS One, 17*(12) e0277963.

Samaras, K., Makkar, S., Crawford, J. D., Kochan, N. A., Wen, W., Draper, B., ... Sachdev, P. S. (2020). Metformin use is associated with slowed cognitive decline and reduced incident dementia in older adults with type 2 diabetes: The Sydney memory and ageing study. *Diabetes Care, 43*(11), 2691–2701.

Sanoff, H. K., Deal, A. M., Krishnamurthy, J., Torrice, C., Dillon, P., Sorrentino, J., ... Muss, H. B. (2014). Effect of cytotoxic chemotherapy on markers of molecular age in patients with breast cancer. *Journal of the National Cancer Institute, 106*(4) dju057.

Savchuk, S., & Monje, M. (2022). Mini-review: Aplastic myelin following chemotherapy. *Neuroscience Letters, 790,* 136861.

Scherling, C., Collins, B., Mackenzie, J., Bielajew, C., & Smith, A. (2011). Pre-chemotherapy differences in visuospatial working memory in breast cancer patients compared to controls: An FMRI study. *Frontiers in Human Neuroscience, 5,* 122.

Schmauck-Medina, T., Moliere, A., Lautrup, S., Zhang, J., Chlopicki, S., Madsen, H. B., ... Fang, E. F. (2022). New hallmarks of ageing: A 2022 Copenhagen ageing meeting summary. *Aging (Albany NY), 14*(16), 6829–6839.

Scuric, Z., Carroll, J. E., Bower, J. E., Ramos-Perlberg, S., Petersen, L., Esquivel, S., ... Schiestl, R. (2017). Biomarkers of aging associated with past treatments in breast cancer survivors. *NPJ Breast Cancer, 3,* 50.

Sekeres, M. J., Bradley-Garcia, M., Martinez-Canabal, A., & Winocur, G. (2021). Chemotherapy-induced cognitive impairment and hippocampal neurogenesis: A review of physiological mechanisms and interventions. *International Journal of Molecular Sciences, 22*(23), 12697.

Shim, H. S., Horner, J. W., Wu, C. J., Li, J., Lan, Z. D., Jiang, S., ... DePinho, R. A. (2021). Telomerase reverse transcriptase preserves neuron survival and cognition in Alzheimer's disease models. *Nature Aging, 1*(12), 1162–1174.

Silva, A. C., Lemos, C., Goncalves, F. Q., Pliassova, A. V., Machado, N. J., Silva, H. B., ... Agostinho, P. (2018). Blockade of adenosine A2A receptors recovers early deficits of memory and plasticity in the triple transgenic mouse model of Alzheimer's disease. *Neurobiology of Disease, 117,* 72–81.

Silva, C. G., Porciuncula, L. O., Canas, P. M., Oliveira, C. R., & Cunha, R. A. (2007). Blockade of adenosine A(2A) receptors prevents staurosporine-induced apoptosis of rat hippocampal neurons. *Neurobiology of Disease, 27*(2), 182–189.

Simo, M., Rifa-Ros, X., Rodriguez-Fornells, A., & Bruna, J. (2013). Chemobrain: A systematic review of structural and functional neuroimaging studies. *Neuroscience and Biobehavioral Reviews, 37*(8), 1311–1321.

Singh, A. K., Mahalingam, R., Squillace, S., Jacobson, K. A., Tosh, D. K., Dharmaraj, S., ... Heijnen, C. J. (2022). Targeting the A(3) adenosine receptor to prevent and reverse chemotherapy-induced neurotoxicities in mice. *Acta Neuropathologica Communications, 10*(1), 11.

Sitkovsky, M. V., Kjaergaard, J., Lukashev, D., & Ohta, A. (2008). Hypoxia-adenosinergic immunosuppression: tumor protection by T regulatory cells and cancerous tissue hypoxia. *Clinical Cancer Research: An Official Journal of the American Association for Cancer Research, 14*(19), 5947–5952.

Speidell, A. P., Demby, T., Lee, Y., Rodriguez, O., Albanese, C., Mandelblatt, J., & Rebeck, G. W. (2019). Development of a human APOE knock-in mouse model for study of cognitive function after cancer chemotherapy. *Neurotoxicity Research, 35*(2), 291–303.

Sperlagh, B., & Vizi, E. S. (2011). The role of extracellular adenosine in chemical neurotransmission in the hippocampus and Basal Ganglia: Pharmacological and clinical aspects. *Current Topics in Medicinal Chemistry, 11*(8), 1034–1046.

Spignoli, G., Pedata, F., & Pepeu, G. (1984). A1 and A2 adenosine receptors modulate acetylcholine release from brain slices. *European Journal of Pharmacology, 97*(3–4), 341–342.

Squassina, A., Manchia, M., Pisanu, C., Ardau, R., Arzedi, C., Bocchetta, A., ... Carpiniello, B. (2020). Telomere attrition and inflammatory load in severe psychiatric disorders and in response to psychotropic medications. *Neuropsychopharmacology: Official Publication of the American College of Neuropsychopharmacology, 45*(13), 2229–2238.

Stagg, J., & Smyth, M. J. (2010). Extracellular adenosine triphosphate and adenosine in cancer. *Oncogene, 29*(39), 5346–5358.

Stefancin, P., Cahaney, C., Parker, R. I., Preston, T., Coulehan, K., Hogan, L., & Duong, T. Q. (2020). Neural correlates of working memory function in pediatric cancer survivors treated with chemotherapy: An fMRI study. *NMR in Biomedicine, 33*(6) e4296.

Stork, P. J., & Schmitt, J. M. (2002). Crosstalk between cAMP and MAP kinase signaling in the regulation of cell proliferation. *Trends in Cell Biology, 12*(6), 258–266.

Sureechatchaiyan, P., Hamacher, A., Brockmann, N., Stork, B., & Kassack, M. U. (2018). Adenosine enhances cisplatin sensitivity in human ovarian cancer cells. *Purinergic Signalling, 14*(4), 395–408.

Takahashi, M., Li, Y., Dillon, T. J., & Stork, P. J. (2017). Phosphorylation of Rap1 by cAMP-dependent protein kinase (PKA) creates a binding site for KSR to sustain ERK activation by cAMP. *The Journal of Biological Chemistry, 292*(4), 1449–1461.

Tang, J., Oliveros, A., & Jang, M. H. (2019). Dysfunctional mitochondrial bioenergetics and synaptic degeneration in Alzheimer disease. *International Neurourology Journal, 23*(Suppl 1), S5–S10.

Temido-Ferreira, M., Coelho, J. E., Pousinha, P. A., & Lopes, L. V. (2019). Novel players in the aging synapse: Impact on cognition. *Journal of Caffeine and Adenosine Research, 9*(3), 104–127.

Temido-Ferreira, M., Ferreira, D. G., Batalha, V. L., Marques-Morgado, I., Coelho, J. E., Pereira, P., ... Lopes, L. V. (2020). Age-related shift in LTD is dependent on neuronal adenosine A2A receptors interplay with mGluR5 and NMDA receptors. *Molecular Psychiatry, 25*(8), 1876–1900.

Torre, M., Dey, A., Woods, J. K., & Feany, M. B. (2021). Elevated oxidative stress and DNA damage in cortical neurons of chemotherapy patients. *Journal of Neuropathology and Experimental Neurology, 80*(7), 705–712.

Trzeciakiewicz, H., Ajit, D., Tseng, J. H., Chen, Y., Ajit, A., Tabassum, Z., ... Cohen, T. J. (2020). An HDAC6-dependent surveillance mechanism suppresses tau-mediated neurodegeneration and cognitive decline. *Nature Communications, 11*(1), 5522.

Tseng, J. H., Xie, L., Song, S., Xie, Y., Allen, L., Ajit, D., ... Cohen, T. J. (2017). The deacetylase HDAC6 mediates endogenous neuritic tau pathology. *Cell Reports, 20*(9), 2169–2183.

Uziel, O., Lahav, M., Shargian, L., Beery, E., Pasvolsky, O., Rozovski, U., ... Yeshurun, M. (2020). Premature ageing following allogeneic hematopoietic stem cell transplantation. *Bone Marrow Transplantation, 55*(7), 1438–1446.

Van Calker, D., Muller, M., & Hamprecht, B. (1979). Adenosine regulates via two different types of receptors, the accumulation of cyclic AMP in cultured brain cells. *Journal of Neurochemistry, 33*(5), 999–1005.

Van Der Plas, E., Schachar, R. J., Hitzler, J., Crosbie, J., Guger, S. L., Spiegler, B. J., ... Nieman, B. J. (2017). Brain structure, working memory and response inhibition in childhood leukemia survivors. *Brain and Behavior, 7*(2), e00621.

Viana Da Silva, S., Haberl, M. G., Zhang, P., Bethge, P., Lemos, C., Goncalves, N., ... Mulle, C. (2016). Early synaptic deficits in the APP/PS1 mouse model of Alzheimer's disease involve neuronal adenosine A2A receptors. *Nature Communications, 7*, 11915.

Vogl, D. T., Raje, N., Jagannath, S., Richardson, P., Hari, P., Orlowski, R., ... Lonial, S. (2017). Ricolinostat, the first selective histone deacetylase 6 inhibitor, in combination with bortezomib and dexamethasone for relapsed or refractory multiple myeloma. *Clinical Cancer Research: An Official Journal of the American Association for Cancer Research, 23*(13), 3307–3315.

Wang, H., Muthu Karuppan, M. K., Devadoss, D., Nair, M., Chand, H. S., & Lakshmana, M. K. (2021). TFEB protein expression is reduced in aged brains and its overexpression mitigates senescence-associated biomarkers and memory deficits in mice. *Neurobiology of Aging, 106*, 26–36.

Wang, S., Prizment, A., Thyagarajan, B., & Blaes, A. (2021). Cancer treatment-induced accelerated aging in cancer survivors: Biology and assessment. *Cancers (Basel), 13*(3), 427.

Wardill, H. R., Mander, K. A., Van Sebille, Y. Z., Gibson, R. J., Logan, R. M., Bowen, J. M., & Sonis, S. T. (2016). Cytokine-mediated blood brain barrier disruption as a conduit for cancer/chemotherapy-associated neurotoxicity and cognitive dysfunction. *International Journal of Cancer. Journal International du Cancer, 139*(12), 2635–2645.

Wefel, J. S., Kesler, S. R., Noll, K. R., & Schagen, S. B. (2015). Clinical characteristics, pathophysiology, and management of noncentral nervous system cancer-related cognitive impairment in adults. *CA: A Cancer Journal for Clinicians, 65*(2), 123–138.

Wefel, J. S., Lenzi, R., Theriault, R., Buzdar, A. U., Cruickshank, S., & Meyers, C. A. (2004). 'Chemobrain' in breast carcinoma?: A prologue. *Cancer, 101*(3), 466–475.

Wefel, J. S., & Schagen, S. B. (2012). Chemotherapy-related cognitive dysfunction. *Current Neurology and Neuroscience Reports, 12*(3), 267–275.

Weiss, B. (2008). Chemobrain: A translational challenge for neurotoxicology. *Neurotoxicology, 29*(5), 891–898.

Welbat, J. U., Naewla, S., Pannangrong, W., Sirichoat, A., Aranarochana, A., & Wigmore, P. (2020). Neuroprotective effects of hesperidin against methotrexate-induced changes in neurogenesis and oxidative stress in the adult rat. *Biochemical Pharmacology, 178*, 114083.

Wen, J., Maxwell, R. R., Wolf, A. J., Spira, M., Gulinello, M. E., & Cole, P. D. (2018). Methotrexate causes persistent deficits in memory and executive function in a juvenile animal model. *Neuropharmacology, 139*, 76–84.

Wen, J., Patel, C., Diglio, F., Baker, K., Marshall, G., Li, S., & Cole, P. D. (2022). Cognitive impairment persists at least 1 year after juvenile rats are treated with methotrexate. *Neuropharmacology, 206*, 108939.

Williams, A. M., & Cole, P. D. (2021). Biomarkers of cognitive impairment in pediatric cancer survivors. *Journal of Clinical Oncology: Official Journal of the American Society of Clinical Oncology, 39*(16), 1766–1774.

Williams, A. M., Krull, K. R., Howell, C. R., Banerjee, P., Brinkman, T. M., Kaste, S. C., … Ness, K. K. (2021). Physiologic frailty and neurocognitive decline among young-adult childhood cancer survivors: A prospective study from the St Jude lifetime cohort. *Journal of Clinical Oncology: Official Journal of the American Society of Clinical Oncology, 39*(31), 3485–3495.

Willingham, S. B., Ho, P. Y., Hotson, A., Hill, C., Piccione, E. C., Hsieh, J., … Miller, R. A. (2018). A2AR antagonism with CPI-444 induces antitumor responses and augments efficacy to anti-PD-(L)1 and anti-CTLA-4 in preclinical models. *Cancer Immunology Research, 6*(10), 1136–1149.

Xie, Y., Zhi, K., & Meng, X. (2021). Effects and mechanisms of synaptotagmin-7 in the hippocampus on cognitive impairment in aging mice. *Molecular Neurobiology, 58*(11), 5756–5771.

Xu, D., Cao, F., Sun, S., Liu, T., & Feng, S. (2016). Inhibition of the Ras/Raf/ERK1/2 signaling pathway restores cultured spinal cord-injured neuronal migration, adhesion, and dendritic spine development. *Neurochemical Research, 41*(8), 2086–2096.

Yang, K., Cao, F., Sheikh, A. M., Malik, M., Wen, G., Wei, H., … Li, X. (2013). Up-regulation of Ras/Raf/ERK1/2 signaling impairs cultured neuronal cell migration, neurogenesis, synapse formation, and dendritic spine development. *Brain Structure & Function, 218*(3), 669–682.

Yin, J., Xu, K., Zhang, J., Kumar, A., & Yu, F. S. (2007). Wound-induced ATP release and EGF receptor activation in epithelial cells. *Journal of Cell Science, 120*(Pt 5), 815–825.

Yoo, K. H., Tang, J. J., Rashid, M. A., Cho, C. H., Corujo-Ramirez, A., Choi, J., … Jang, M. H. (2021). Nicotinamide mononucleotide prevents cisplatin-induced cognitive impairments. *Cancer Research, 81*(13), 3727–3737.

Young, A., Mittal, D., Stannard, K., Yong, M., Teng, M. W., Allard, B., ... Smyth, M. J. (2014). Co-blockade of immune checkpoints and adenosine A2A receptor suppresses metastasis. *Oncoimmunology, 3*(10) e958952.

Young, A., Ngiow, S. F., Gao, Y., Patch, A. M., Barkauskas, D. S., Messaoudene, M., ... Smyth, M. J. (2018). A2AR adenosine signaling suppresses natural killer cell maturation in the tumor microenvironment. *Cancer Research, 78*(4), 1003–1016.

Zheng, J., Xu, M., Walker, V., Yuan, J., Korologou-Linden, R., Robinson, J., ... Bi, Y. (2022). Evaluating the efficacy and mechanism of metformin targets on reducing Alzheimer's disease risk in the general population: A Mendelian randomisation study. *Diabetologia, 65*(10), 1664–1675.

Zhou, W., Kavelaars, A., & Heijnen, C. J. (2016). Metformin prevents cisplatin-induced cognitive impairment and brain damage in mice. *PLoS One, 11*(3), e0151890.

9780443186677